NASA/TP—2005–211519

Space Shuttle Operations and Infrastructure
A Systems Analysis of Design Root Causes and Effects

Carey M. McCleskey
Kennedy Space Center, Florida

National Aeronautics and
Space Administration

Kennedy Space Center

April 2005

Acknowledgments

The author would like to acknowledge the assistance of those individuals and organizations who aided in the research and preparation of this material.

First, NASA's Space Launch Initiative (SLI) program management for posing the hard questions to the operations community. At the Kennedy Space Center, this included the support of James Heald, Hugo Delgado, Robert Johnson, and Roger Hall. At the Marshall Space Flight Center, this included the support of Dennis Smith, Garry Lyles, Dr. Dale Thomas, and Chuck Smith.

The many individuals and organizations contributing to the research and execution of the study are acknowledged in Appendix A. The author would like, however, to make special acknowledgement to several key organizations and individuals. Frank Izquierdo and the KSC Shuttle systems engineers represent the backbone of NASA KSC's resident knowledge of Shuttle launch operations and infrastructure, and without their knowledge and experience in processing the Shuttle, this work would not have been accomplished. In addition, Kevin Ingoldsby is acknowledged for numerous hours, weeks, and months laboring with the author to capture the information and then designing and organizing a massive electronic database that is sure to be a treasured source of knowledge for many years to come. Mr. Ingoldsby was supported by the Lockheed Martin Michoud Space Systems organization. The author would also like to acknowledge the valuable inputs of his long-time associates, Russel Rhodes and Edgar Zapata. Their shared passion for "design for support" both inspired and encouraged the author to complete and disseminate the work presented here.

Finally, the engineering support of the professionals at Arctic Slope Research Corporation (ASRC) at KSC is acknowledged. In particular, David Johnson for his software expertise in aiding Mr. Ingoldsby in developing and then maintaining the electronic database for the author, and Deborah Doxey for her innovative and responsive support in creating, editing, and publishing this document.

Published by Books Express Publishing

ISBN 978-1-78266-056-9

Books Express publications are available from all good retail and online booksellers. For publishing proposals and direct ordering please contact us at: info@books-express.com

Foreword

This NASA Technical Publication explores and documents the nature of Space Shuttle operations and its supporting infrastructure in order to address fundamental questions often asked of the Space Shuttle Program—*why does it take so long to turn the Space Shuttle around for flight and why does it cost so much*? To accomplish this, the report provides an overview of the cause-and-effect relationships between generic flight and ground system design characteristics and resulting operations by using actual cumulative maintenance task times as a relative measure of direct work content. In addition, the paper provides an overview of how the Space Shuttle Program's operational infrastructure extends and accumulates from these design characteristics. Finally, learning from the experience of operating the Space Shuttle, the report derives a set of engineering and technology needs from which future space architects and technologists can revolutionize space travel from the inside out by developing and maturing more operable and supportable systems.

This page intentionally left blank.

Contents

Figures

Tables

1 BACKGROUND AND INTRODUCTION

1.1 Purpose

The primary purpose of this publication is to present the results of a function-by-function review of Space Shuttle ground processing work. The motivation for the analysis was to discover design root causes that have the potential of being acted upon through technical means to greatly improve the operational performance of future spaceflight systems. In addition, the analysis explored the extent of the ground infrastructure needed to support the processing work. Another objective was to identify, at a top level, the total scope of the national infrastructure required to execute the Space Shuttle program. This was done to help put processing costs and performance in perspective with respect to total space transportation affordability. The nation's partially reusable Space Shuttle provides us a unique empirical baseline for a detailed evaluation of its overall operational performance and, more specifically, its inherent operability and supportability characteristics. Therefore, understanding Space Shuttle accomplishments and operational shortfalls—from its operational payload delivery performance to the details of its servicing and repair—points the way for future flight and ground system design improvements.

1.2 Scope

Space Shuttle operations and infrastructure covers a wide variety of functions performed across numerous locations in the United States, and indeed in various places across the globe. The scope of this analysis focuses on the ground processing of the Space Shuttle Vehicle (SSV) and its supporting architectural elements. Specifically, the analysis was formulated to begin a structured process of identifying how particular design characteristics drive the intensity of ground processing work and the level of infrastructure required to routinely produce human spaceflights. While the general functions of recurring manufacturing, remanufacturing, mission planning, flight operations, and range operations were identified at a high level, these were not explored to the same depth as the ground processing operations and infrastructure. Nevertheless, the analysis provided groundbreaking insight into specific design factors inherent in overall operability and supportability of space transportation systems by examining how the Shuttle system is routinely assembled, maintained, repaired, serviced, fueled, and launched.

1.3 Space Shuttle Vehicle

The National Aeronautics and Space Administration (NASA) is currently considering future options for its space transportation needs in supporting our nation's vision for space exploration. At the same time, all sectors of the space economy (civil, military, and commercial) are in need of improvements in the responsiveness, safety, and operational affordability of space transportation systems. It is useful, therefore, to turn to our current experience in operating the Space Shuttle system. The Space Shuttle system is designed to launch a maximum of 28.8 metric tons (mt), or 63,500 pounds (lb), of payload and five to eight crew members. The Shuttle can launch into low Earth orbits that range from 203 to 602 kilometers (km), or 110 to 325 nautical miles. The Orbiter vehicles can be launched into orbital inclinations ranging from that of the International

Space Station (ISS) at 51.7° north to 28.5° due east from NASA's John F. Kennedy Space Center (KSC) off the east coast of Florida.[1]

1.4 Original Space Shuttle Operations Concepts Versus Actual Architecture

The original Space Shuttle operations concept was planned to achieve an annual flight rate of 40 launches from KSC, and 20 launches from Vandenberg Air Force Base (VAFB) off the coast of California. Turnaround concepts envisioned a 2-week-long process to prepare the SSV for launch (see Figure 1).[2]

[1] "Space Shuttle Program Description and Requirements Baseline," NSTS 07700.
[2] NASA SP-4XX Space Shuttle, ca. 1976.

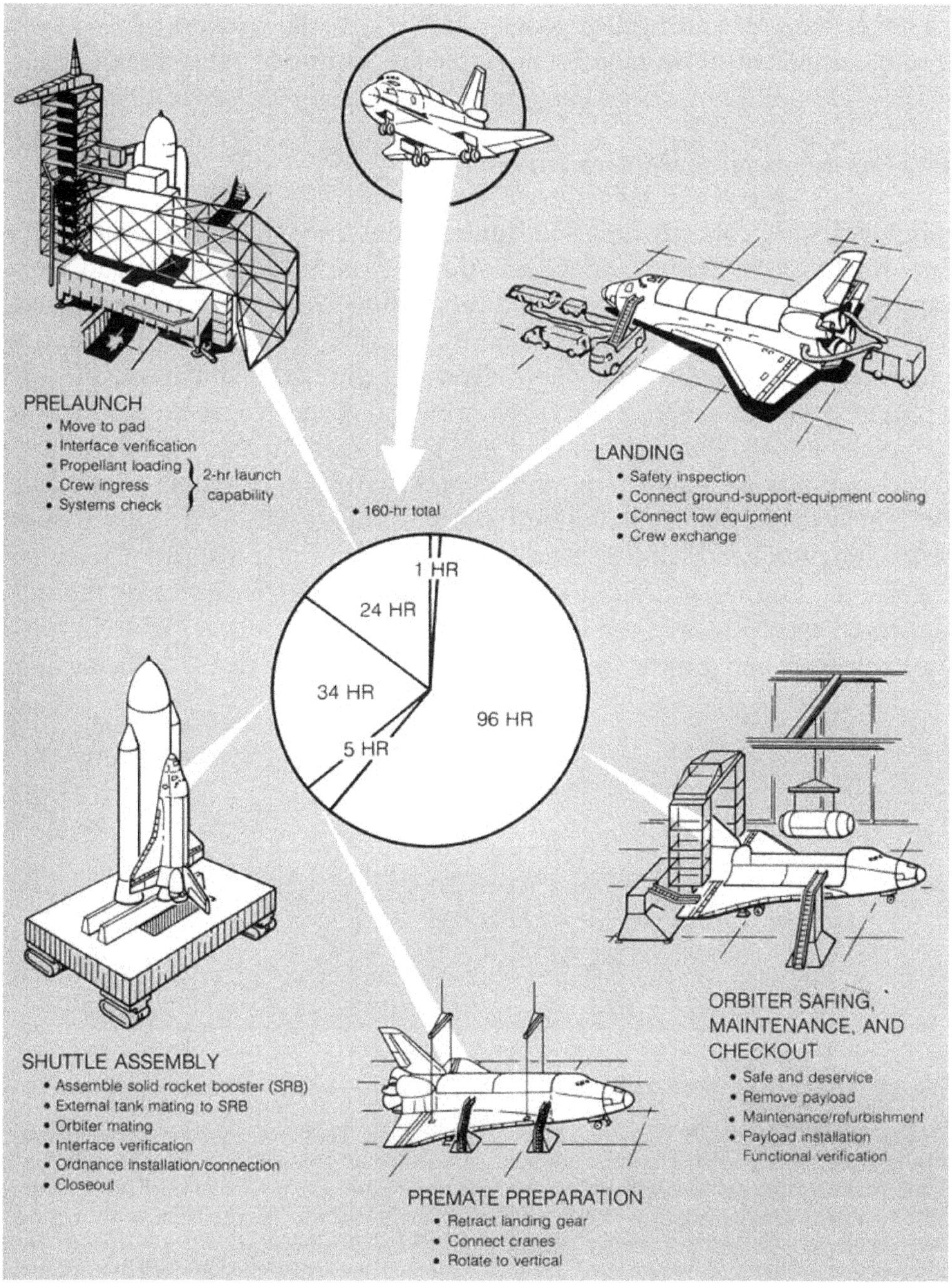

Figure 1. Original Concept of Shuttle Operations (ca. 1976)

However, on a sustained basis, the resultant system has demonstrated only seven to eight flights per year (achieved during the mid-1990s). A simplified view of the actual Space Shuttle architecture (as defined in the Space Shuttle Program Plan) is depicted in Figure 2. Portrayed in the figure are top-level *operations* for the hardware processing of Space Shuttle elements through dedicated Space Shuttle facilities.

In addition, a larger-than-expected set of geographically distributed workforces emerged, employing tens of thousands of full-time personnel that are required to sustain the Space Shuttle's unique capabilities. Figure 3 shows the extent of this Space Shuttle ground *infrastructure*.[3]

1.5 Shuttle Operations Accomplishments

Before examining specific concept versus design shortfalls, and the cause-and-effect relationships, it is important to recognize the Space Shuttle Program's accomplishments and acknowledge that it has been an important investment for the United States as it moves out into the space frontier. It was the first system to routinely reuse expensive spaceflight hardware. It was also the first system to retrieve hardware from space. Despite the previously mentioned shortfalls, the Shuttle also provided a great leap in capability for human space access by routinely sending crews of six to eight personnel to Earth orbit many times year after year for decades. To demonstrate this historically, Figure 4 shows the annual number of crew members delivered by U.S. human spaceflight systems. While excluding suborbital flights, the figure clearly shows the significant contribution provided by the Space Shuttle system. Figure 5 shows the annual payload carried by the Shuttle. Further utility of the Space Shuttle system is shown by the amount of deployed and retrieved cargo (such as the German Shuttle Pallet Satellite [SPAS]), as well as the in-space crew exchanges and returns by the reusable Orbiters (see Table 1, Table 2, and Figure 6).

[3] "Space Shuttle Program Plan," Updated November 13, 2002.

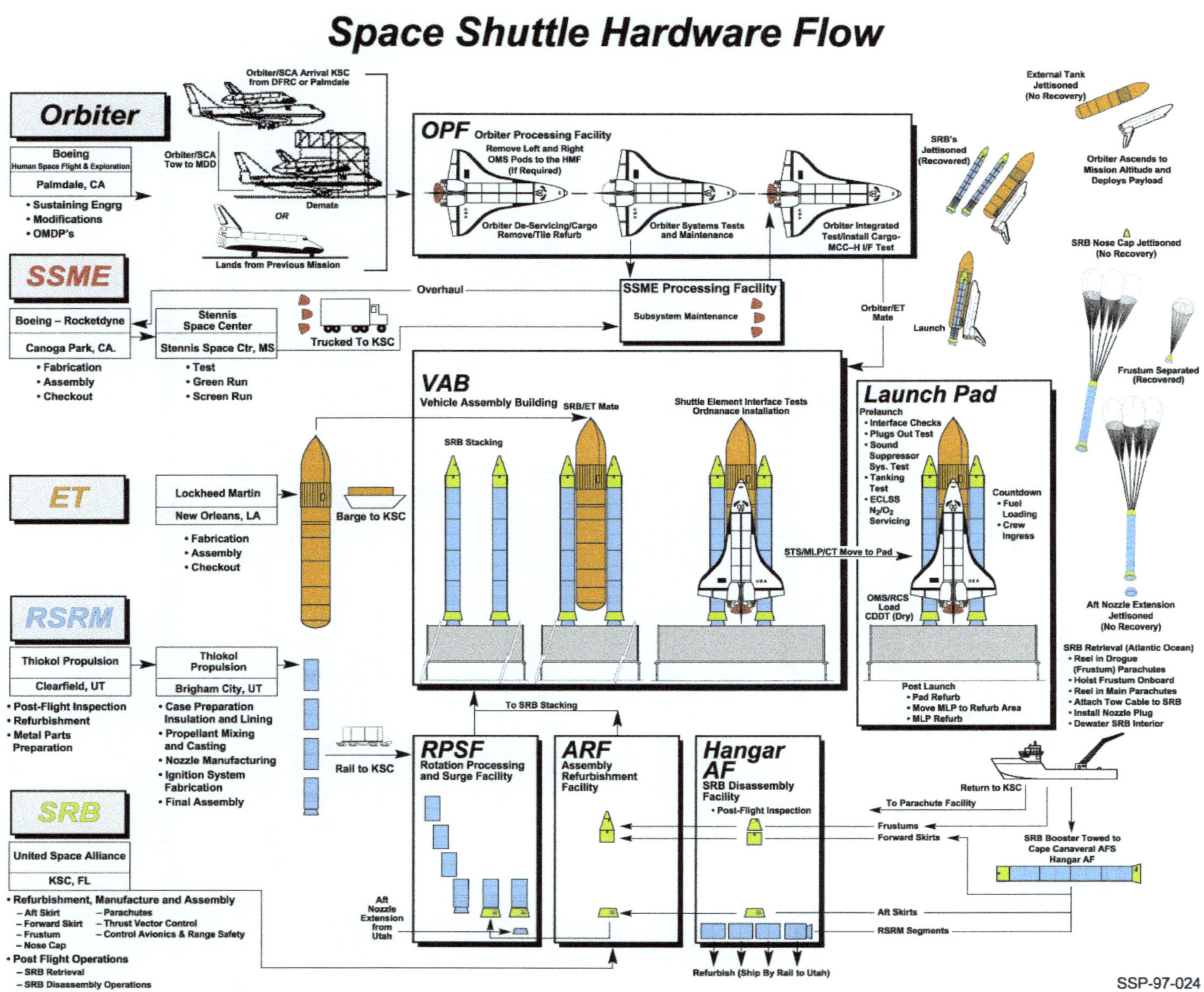

Figure 2. Space Shuttle Ground Processing Operations (ca. 1997)

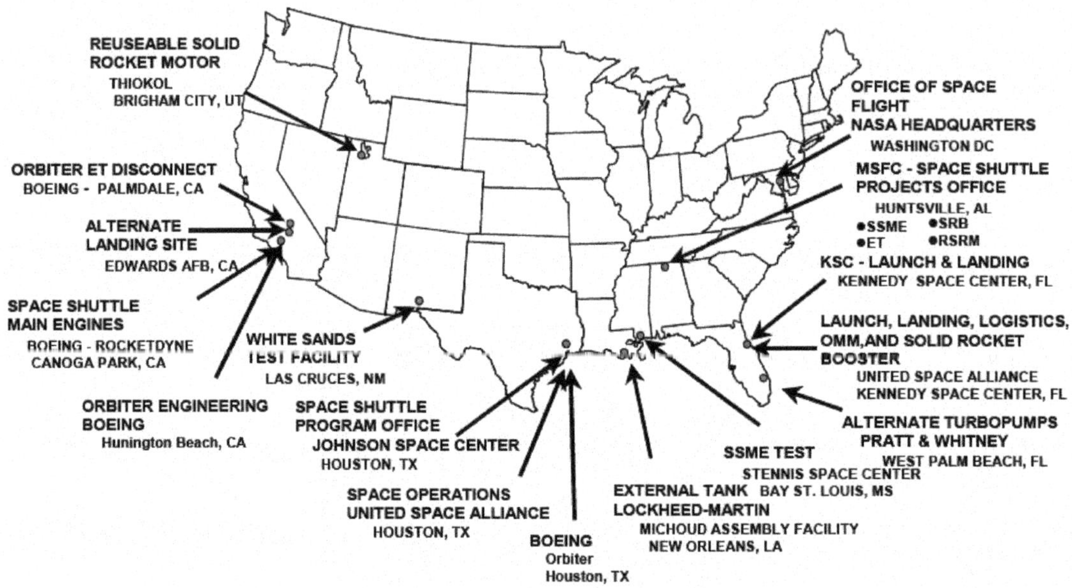

Figure 3. Space Shuttle Ground Infrastructure

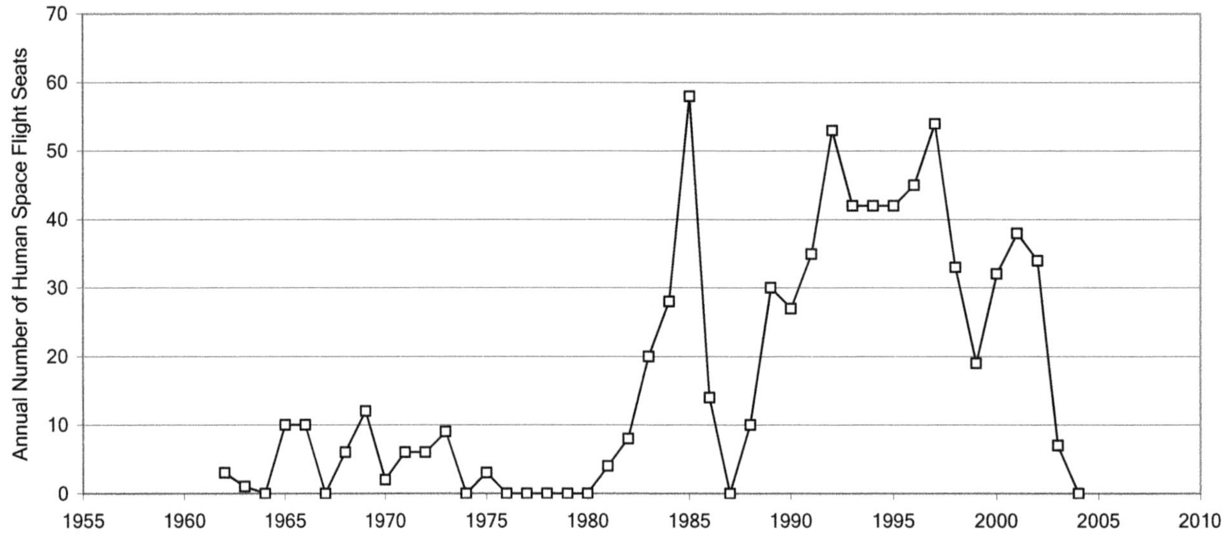

Figure 4. Historical Trend of U.S. Human Space Access

Figure 5. Historical Trend of Space Shuttle Cargo Throughput (1981 to 2004)

Table 1. Annual Summary of Space-Shuttle-Deployed and -Retrieved Cargo

Year	Total Annual Payload Liftoff Mass (1)		Deployed to Space		Deployed and Retrieved		Retrieved Only	
	Metric Tons (mt)	Pounds (lbs.)	Metric Tons (mt)	Pounds (lbs.)	Metric Tons (mt)	Pounds (lbs.)	Metric Tons (mt)	Pounds (lbs.)
1981	13.43	29,601	0.00	0	0.00	0	0.00	0
1982	25.03	55,184	6.62	14,585	0.00	0	0.00	0
1983	62.36	137,476	27.19	59,940	1.45	3,192	0.00	0
1984	72.15	159,060	42.76	94,268	0.00	0	1.08	2,381
1985 (2)	110.20	242,956	46.91	103,417	1.01	2,217	0.00	0
1986	35.04	77,258	5.60	12,351	0.00	0	0.00	0
1987	0.00	0	0.00	0	0.00	0	0.00	0
1988 (2)	20.23	44,601	17.02	37,514	0.00	0	0.00	0
1989 (2)	70.47	155,361	59.60	131,397	0.00	0	0.00	0
1990 (2)	45.11	99,450	27.99	61,699	0.00	0	0.00	0
1991	88.37	194,820	56.62	124,820	1.84	4,046	0.00	0
1992	93.62	206,387	27.04	59,613	0.67	1,486	0.00	0
1993	87.39	192,655	30.31	66,826	4.61	10,161	5.25	11,572
1994	72.58	160,009	0.08	171	4.53	9,996	0.00	0
1995	75.40	166,224	21.69	47,812	4.52	9,957	0.53	1,166
1996	64.82	142,898	3.89	8,582	7.71	17,004	6.04	13,321
1997	79.57	175,429	9.49	20,920	4.86	10,724	6.77	14,915
1998	61.01	134,499	15.39	33,931	1.35	2,973	3.09	6,807
1999	38.88	85,704	23.92	52,731	0.00	0	2.52	5,564
2000	64.99	143,274	30.57	67,404	0.00	0	1.30	2,859
2001	81.02	178,619	38.17	84,158	0.00	0	5.51	12,150
2002	58.25	128,410	--	--	--	--	--	--
2003	11.03	24,325	--	--	--	--	--	--
2004	0.00	0	0	0	0.00	0	0.00	0
Total	**1330.94**	**2,934,200**	**490.86**	**1,082,139**	**32.55**	**71,756**	**32.09**	**70,735**

Note 1: Weights listed are those chargeable to payload; taken from "Space Shuttle Flight Weight Summary," in Space Shuttle Missions Summary: Book 2, Next 100 Flights; NASA JSC/DA8, Rev. S, May 2002.

Note 2: Does not include Dept of Defense mission payloads flown on the Space Shuttle system from 1985 through 1990.

Table 2. Summary of Space Shuttle Crew Exchanges and Returns

Date	STS No.	Orbiter Tail No.	Number of Crew Exchanged Out	Number of Crew Exchanged In (Returned)	Exchange Point (Mir or ISS)
1995	STS-71	OV-104	2	3	MIR
1996	STS-76	OV-104	1	0	MIR
1996	STS-79	OV-104	1	1	MIR
1997	STS-81	OV-104	1	1	MIR
1997	STS-84	OV-104	1	1	MIR
1997	STS-86	OV-104	1	1	MIR
1998	STS-89	OV-105	1	1	MIR
1998	STS-91	OV-103	0	1	MIR
2001	STS-102	OV-103	3	3	ISS - Exp 1/2
2001	STS-105	OV-103	3	3	ISS - Exp 2/3
2001	STS-108	OV-105	3	3	ISS - Exp 3/4
2002	STS-111	OV-105	3	3	ISS - Exp 4/5
2002	STS-113	OV-105	3	3	ISS - Exp 5/6
Total STS Program			**23**	**24**	

Source: NASA web site, "Space Station Crew," http://www.spaceflight.nasa.gov/station/crew.

Figure 6. Station Crewman (in Logistics Module) Returns by Shuttle on STS–102

1.6 Identifying Operations and Infrastructure Cost and Performance Shortfalls

While producing major advances in the field of human astronautics, some shortfalls in expected operational performance occurred. For the Space Shuttle system, the most dramatic shortfall in performance can be seen by comparing the payload throughput capability envisioned for the Space Shuttle prior to its operation versus its current design throughput capability, and, its actual utilization (or demonstrated throughput). What changed during the course of designing the Space Shuttle that determined the outcome of today's Shuttle operations and infrastructure?

The system was originally conceived to carry 29.5 mt (65,000 lb) of payload per launch at a rate of 40 flights per year from KSC and was, therefore, envisioned to launch about 1,182 mt (2,640,000 lb) per year to space.[4] As shown in Figure 7, however, the Space Shuttle has demonstrated less than 20 percent of this capability, or an 80-percent shortfall in payload delivery performance. Several factors are shown to have contributed to this shortfall—the most significant of these being less than expected flight rate capability—with shortfalls in payload utilization and vehicle performance also contributing.

It should also be noted that most other space launch systems have, likewise, experienced significant shortfalls in payload throughput performance when comparing the original expectation against actual operational performance. The original Saturn launch vehicle concepts, for example, were initially justified in military studies on a requirement for 100 launches per year![5] In fact, this was one of the original motivating factors inspiring the Launch Complex 39 "mobile launch conept." Understanding the many factors, including the design characteristics that influence a system's ultimate operational effectiveness, will be even more important in the future as commitments for more ambitious and enduring space enterprises are accepted. Remarkably, the quantification of such fundamental operations as servicing, inspection, and assembly, as well as

---

[4] "Space Shuttle,", NASA SP-407, Lyndon B. Johnson Space Center, NASA, Washington, D.C., 1976, pp. 4–5, 76.

[5] H.H. Koelle, *Saturn System Study,* Redstone Arsenal, AL, November 13, 1959, pp. 4–5, 21.

unplanned troubleshooting and repair, remains largely unanalyzed and unpublished—for any space transportation system, not just the Shuttle. Yet, spaceflight architects will require this important benchmark information if the inherent operability and supportability of a proposed concept or design are not to be left to chance. This work was initiated to provide important insight into key technical factors that can positively influence the operational outcome of future space transportation architectures.

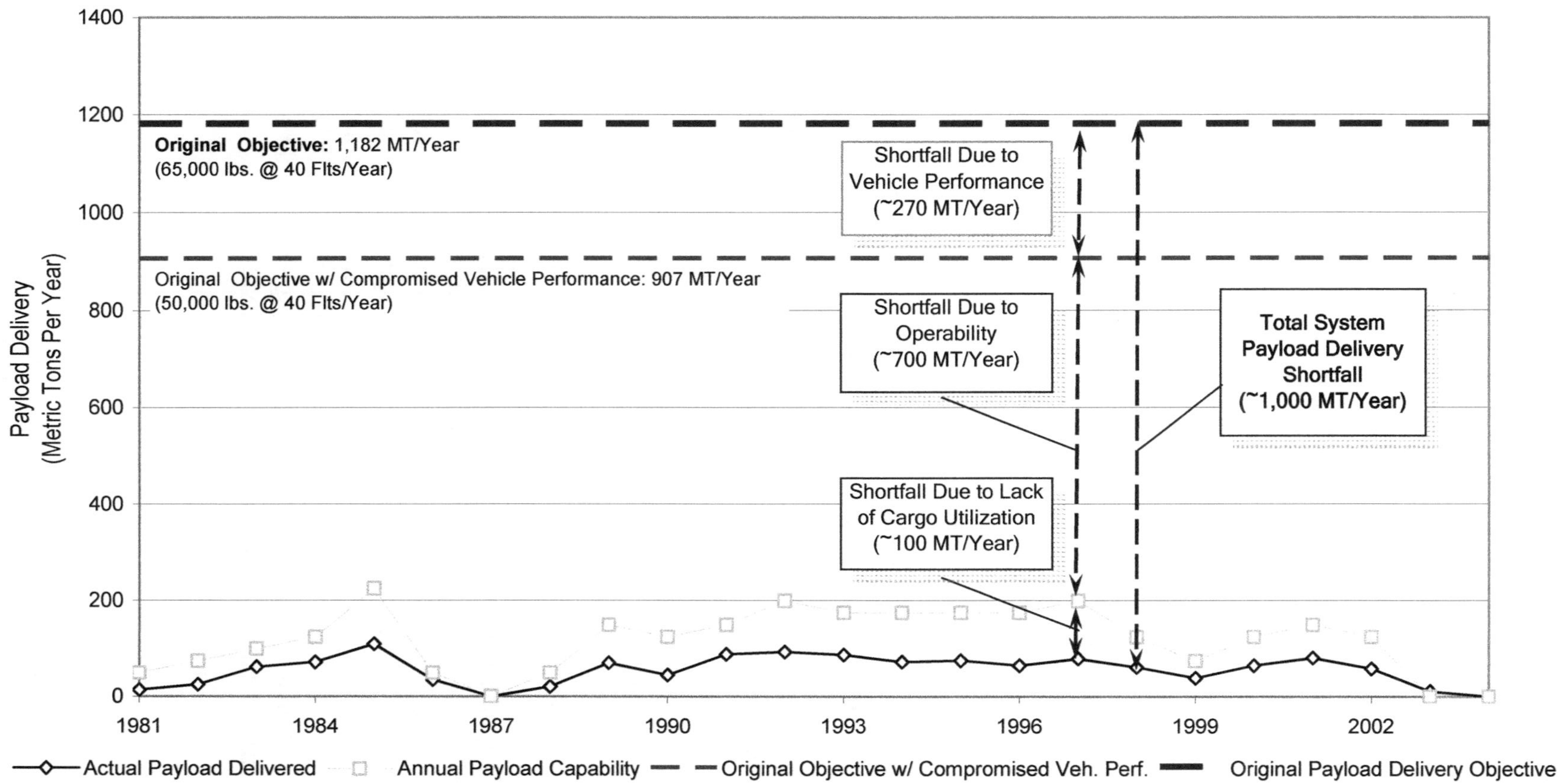

Figure 7. Summary of Space Shuttle Payload Delivery Shortfalls

2 ANALYSIS FORMULATION

The primary purpose of the analysis presented in this publication was to conduct a function-by-function review of Space Shuttle ground turnaround processing to discover design root causes that have the potential of being acted upon through technical means to greatly improve the operational performance of future spaceflight systems.

In order to relate Space Shuttle design characteristics to generic spaceflight operations functions, NASA KSC personnel conducted an in-house Space Shuttle Design Root Cause Analysis (RCA) from the spring of 2002 through calendar year 2003. The analysis was conducted to answer two fundamental questions posed by the management of NASA's Space Launch Initiative (SLI):[6]

- Why does it take so long to process the Shuttle?
- Why does it cost so much?

The approach to answering these questions was to confine the response to that which the technical community could address, and avoid delving into managerial and organizational issues that have been thoroughly explored over the life of the Space Shuttle program.

The first question addresses the direct flight hardware processing operations and requires detailing sources of vehicle maintenance, repair, and assembly activities. This was an immense task considering the thousands of scheduled operations and maintenance tasks required to produce a typical Space Shuttle flight. The second question was equally daunting and required thoroughly understanding the functions of the national infrastructure required to support Space Shuttle operations.

In Figure 8, an iceberg represents the overall multibillion-dollar annual cost burden (i.e., there is a small, highly visible set of direct costs, but a much larger hidden set of fixed costs). These less visible functions and their relatively large fixed cost contribution are needed to support and enable the smaller but more visible launch processing functions and their direct costs.

The analysis approach for answering "Why does it take so long?" focused on creation of a structured data evaluation process prior to any causal analysis. NASA systems engineers responsible for and well-experienced with Space Shuttle operations performed the data review and evaluation. The resulting source data provided a uniform, quantified platform for subsequent evaluation, discussion, identification of work drivers, and root cause consensus that was traceable to actual processing tasks. The results—both the source data and the cause-and-effect relationships—were ultimately captured in user-friendly, highly portable platforms (such as commercial desktop electronic database files and electronic spreadsheets).

[6] C.M. McCleskey et al., "STS Root Cause Analysis to Derive SLI Operations Requirements," NASA SLI Board Presentation, CR KSC-013, April 2002.

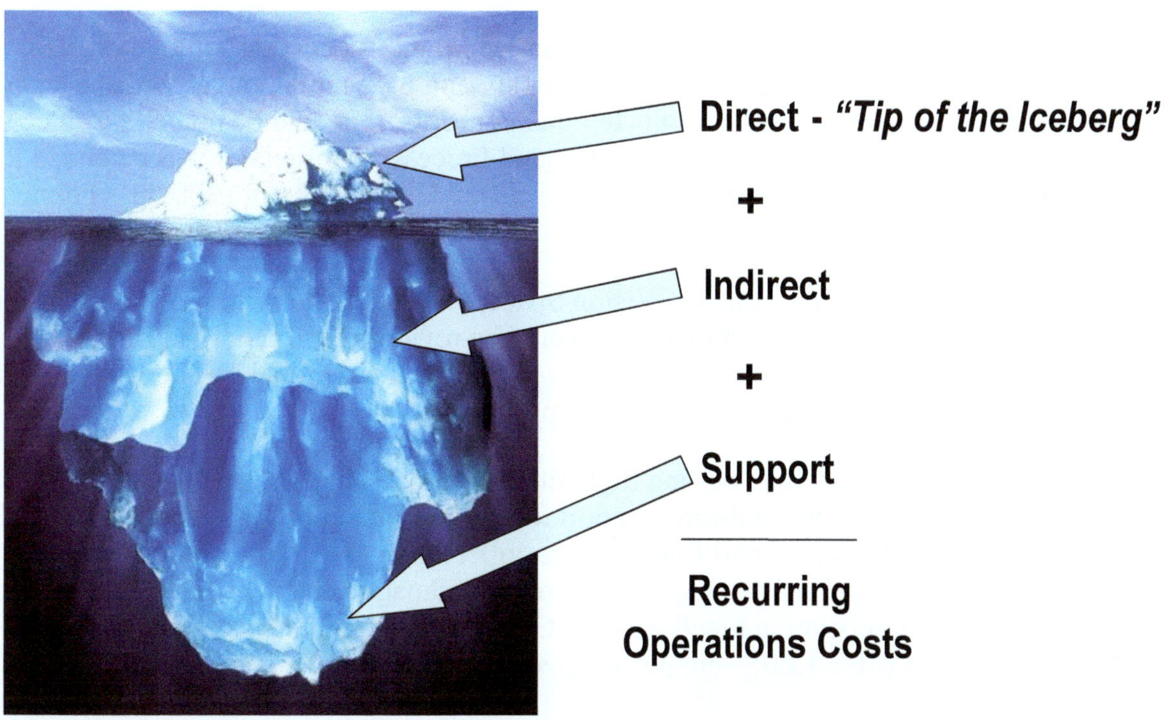

Figure 8. Levels of Recurring Space Transportation Costs

3 OPERATIONAL CONSTRUCTS CONSIDERED FOR ANALYSIS

To answer the questions posed in the formulation of the analysis, it was necessary to define a top-level, generic model of operations that might be applied to any space transportation architecture. Two were considered at the outset of the analysis, and a third emerged during the analysis.

The first operational construct considered was the Vision Spaceport functional module definition created in 1997 by the Spaceport Synergy Team—a KSC group of NASA and industry technologists supporting NASA Headquarters' Highly Reusable Space Transportation, or HRST, study.[7] This model defines 12 concept-generic operations and infrastructure functions. The advantage of this model was that the functions and subfunctions could easily be mapped to both the Space Shuttle and other advanced space transportation concepts being considered. A disadvantage was the lack of relationships defined between the functions for discrete-event task modeling.

Another model considered was the Space Shuttle Ground Processing Simulation. This detailed Shuttle processing flow model was ideal for analyzing the Shuttle, but at the time had disadvantages in translating the analysis results to other concepts. It should be noted that since the completion of this analysis work, the Shuttle Ground Processing Simulator was extended to include a more concept-generic capability and is known as the Generic Model for Future Launch Operations, or GEM-FLO.[8]

The analysis was initiated using the Vision Spaceport module definitions (see Figure 9).[9] However, the generic operations subfunctions would be reviewed and modified in detail by the analysis team as the work progressed. The result was a refined functional breakdown structure, or FBS, and is documented to three levels in Appendix A.

The direct operations include on-line flight element processing tasks (FBS 1.0 through 5.0). Indirect functions required for real-time support to the direct functions are identified in FBS items 6.0 through 11.0. Major support functions are collected in FBS 12.0, which include sustaining engineering, safety and risk management, program management, scheduling, training, mission planning, and administrative functions. The Connecting Community Infrastructure and Public Support Services (FBS 13.0) were not analyzed in this effort.

A third model emerged well into the effort, when NASA and contracted RCA project members were brought in for consultation by the Advanced Spaceport Technology Working Group (ASTWG) to help baseline a new generic spaceport model for that group. The objective of the new ASTWG model was to gain broader national consensus from spaceport stakeholders across government, industry, and academia.[10] The Vision Spaceport model of operations and infrastructure was used as the starting point for the ASTWG effort but was refined and modified to include functional and quasitemporal relationships among the top-level elements (see Figure 10 and

---

[7] Spaceport Synergy Team, "Catalog of Spaceport Elements with Functional Definition," NASA *Highly Reusable Space Transportation (HRST) Study,* October 1997.

[8] G.R. Cates et al., "Modeling the Space Shuttle," in E. Yucesan et al., (eds.), *Proc. 2002 Winter Simulation Conference.*

[9] C.M. McCleskey, "Strategic Space Launch Concept and Technology Roadmaps to Develop Visionary Spaceports," IAF-99-V-5-06, *50th International Astronautical Congress,* Amsterdam, The Netherlands, October 1999.

[10] C. Guidi, et al, *Advanced Spaceport Technology Working Group: Baseline Report*, November 2003.

Figure 11). At the conclusion of the RCA project, the subfunctions that were derived from the Vision Spaceport model were also remapped into the top-level ASTWG model. A three-level ASTWG functional breakdown structure is also provided in Appendix A.

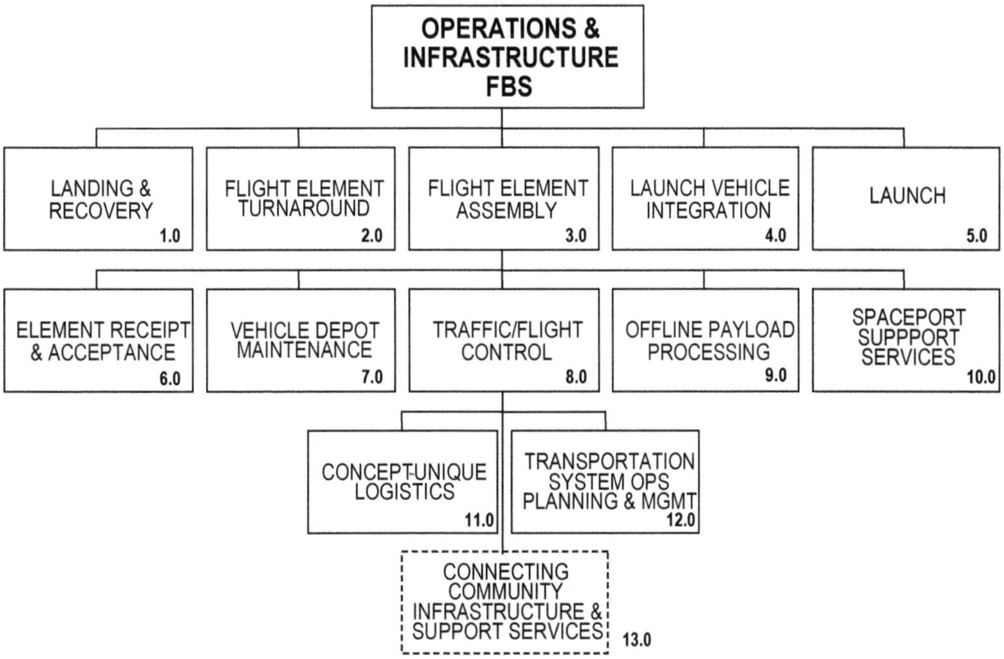

Figure 9. Generic Operations Functional Breakdown Structure (FBS)

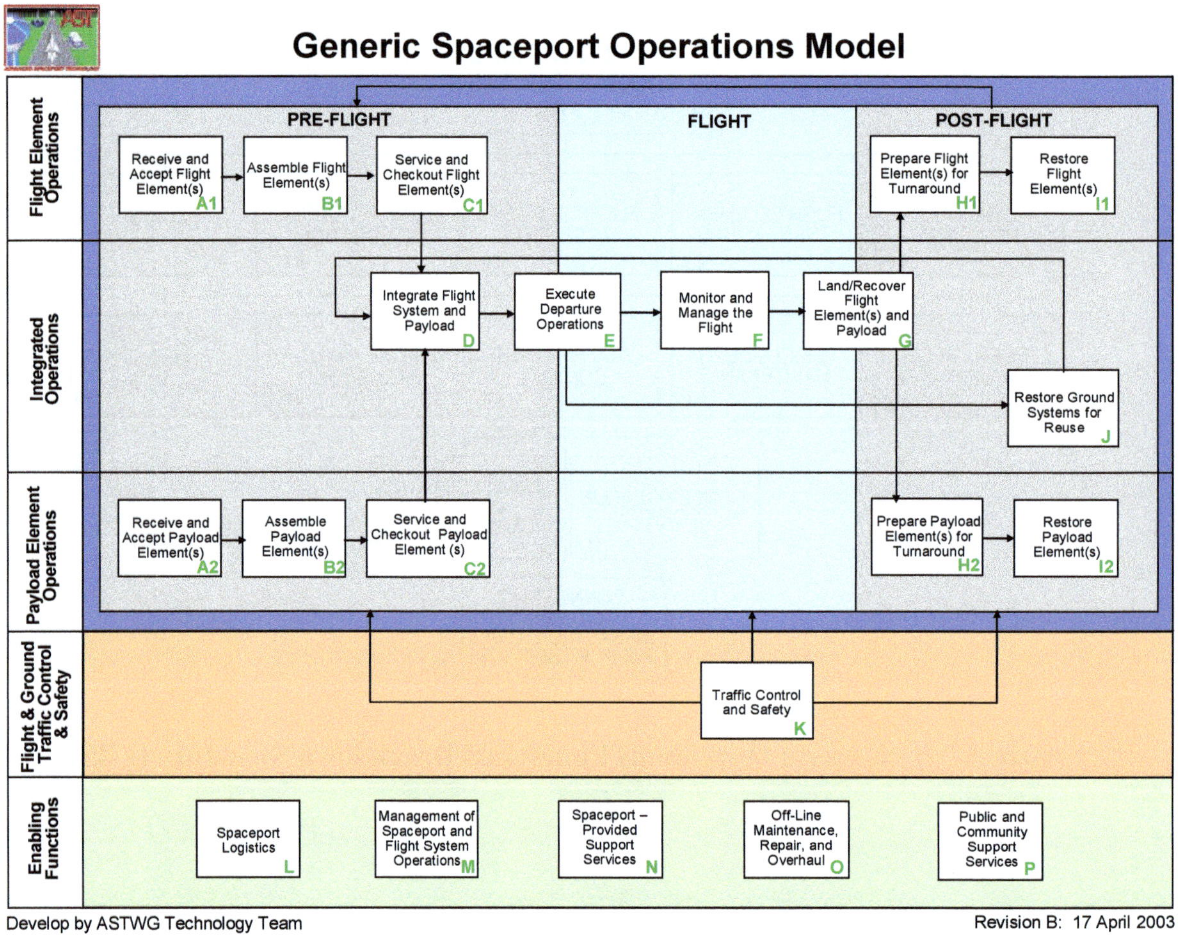

Figure 10. ASTWG Model of Generic Spaceport Functions

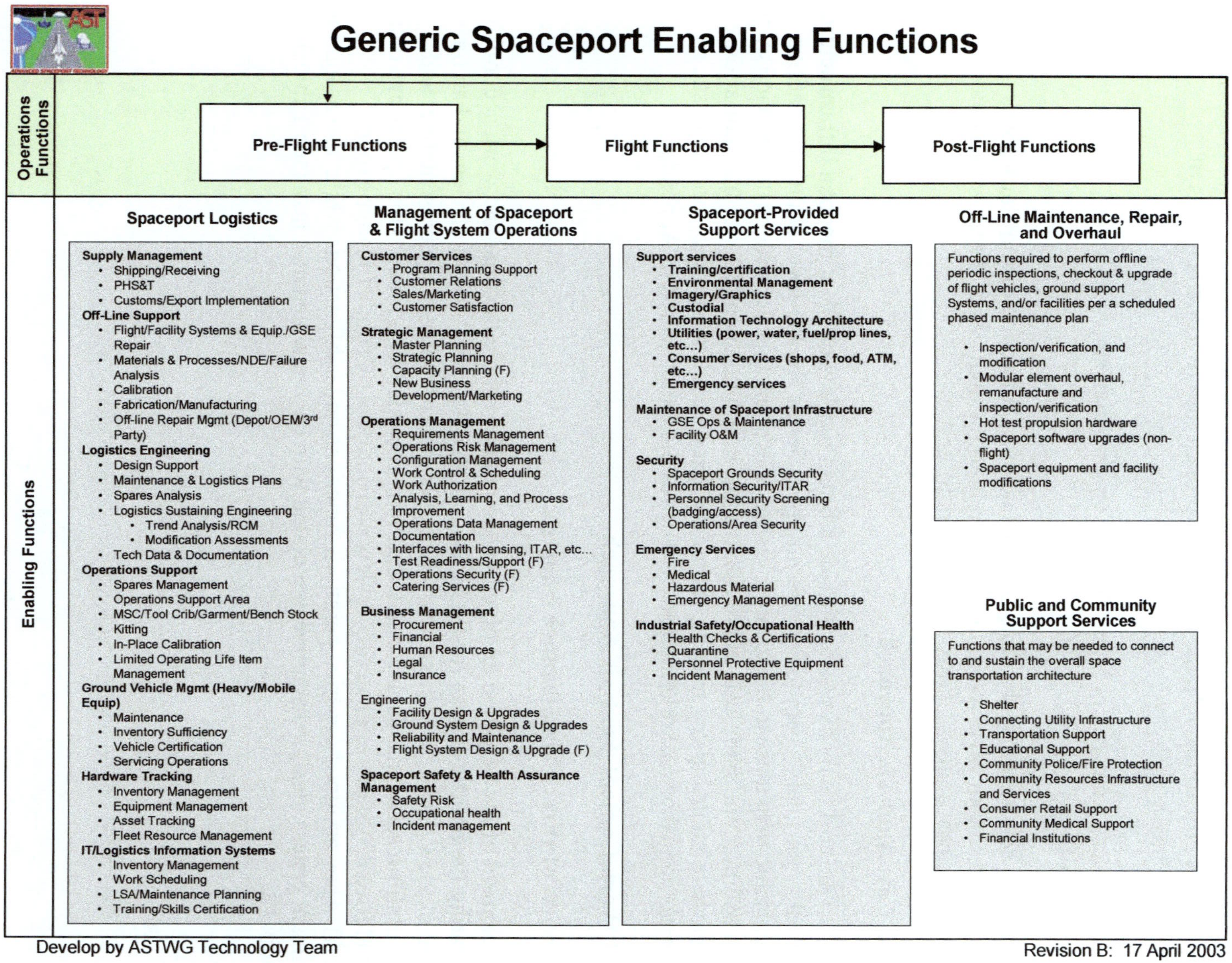

Figure 11. ASTWG Spaceport Model—Support Functions

4 ANALYSIS APPROACH

With the fundamental questions formulated (i.e., *Why does it take so long to process the Shuttle?* and *Why does it cost so much?*), the effort was broken into two distinct areas of investigation:

- Responsiveness Analysis, or Direct Work Content Analysis
- Cost Analysis

Both analyses were conducted, at the conclusion of which a collection of root causes was also analyzed. Most of the labor involved in the effort went to the analysis of direct work content. The approach to the cost analysis, however, is taken up first.

4.1 Overall Cost Analysis Approach

The approach to the cost analysis was to provide a broad top-level look at the total infrastructure costs. Specifically, the analysis approach was simply to define the top-level Shuttle operations budget elements and map these to the generic Vision Spaceport functions identified previously in Figure 9.[11] In addition, the study was to provide insight into fundamental space transportation cost-performance relationships by making visible the fixed versus flight-rate-variable costs, as well as materials and labor cost relationships.[12] The results are provided in Appendix B.

4.2 Direct Work Content Analysis Approach

4.2.1 Overview of Direct Maintenance Analysis

The second element of investigation involved understanding where the time and work accumulates in preparing the Shuttle system for flight, from both a designer's point of view and an operator's point of view. Here the approach was to gather a comprehensive, representative, and functionally uniform set of maintenance task data with recorded task times. This operations database then underwent systems engineering review and analysis by subject matter experts (Appendix A). They assigned generic operations functions and subfunctions while also assigning the predominant generic design discipline associated with each maintenance task (see Table 3 and Appendix E).

---

[11] NASA MSFC Engineering Cost Group, Access to Space Study, "STS Baseline Ops Cost Based on FY94 OMB STS Cost Per Flight (with Updates from Project Offices)." Applied Research, Inc., Huntsville, AL, April 13, 1993.
[12] "Shuttle Operations Zero-Base Cost Study," NASA presentation to Dr. Lenoir, NASA Headquarters, Office of Space Flight, Washington, D.C., July 2, 1991.

Table 3. Generic Design Discipline Assignments

Generic Design Discipline	*Shuttle Operations System Codes*
Command, Control & Health Management	DPS, INS, SOF
Cockpit & Crew Cabin	CCE, FCS
Communications	COM
Environmental Control & Life Support	ECL
Ground Systems & Facilities	GSE
Guidance, Navigation & Flight Controls	GNC
Payload Accommodations	PLO (CM, CE, TP)
Power Management	APU, HYD, FCP, EPD, OEL, OTC (Vehicle power-up switch lists)
Propulsion	MPS, OMS-RCS, SME
Safety Management & Control	PVD, MPS, SME (safety purges)
Structures, Mechanisms & Vehicle Handling	MEQ, PYR, STR, OHE, OPT, GSE, OSO, VPL, QC
Thermal Protection	TPS
Thermal Management	TPS, TCS, ECL (Freon and water cooling loops)

In this manner, both the designer and the operations technologist would have the ability to clearly identify concentrations of direct operations work from each of their perspectives—all from a common, and relatively uniform, data source.

Previous studies by NASA to quantify and characterize actual Space Shuttle ground processing provided some insight during the 1990s.[13,14] work tended to mix both direct and indirect work, with some technical disciplines (most often main engine and thermal tile) accumulating off-line work, while the off-line work of many other disciplines went unaccounted. In this effort, the direct work (i.e., the on-vehicle work), was accumulated homogeneously across technical disciplines, while any off-line work identified was carefully segregated and analyzed separately in the infrastructure support and cost analyses. The use of the FBS, with documented definitions and descriptions of the functions, helped the analysis team to maintain this distinction between direct and indirect work as issues arose about the classification of certain data sets.

The direct work was measured by the accumulation of actual planned and unplanned tasks and the time taken to perform the tasks. While this data does not directly include the laborhours (that is, does not take into account the crew size of a task), the quantified accumulation of this work nevertheless represents a figure-of-merit of relative work content and thus provides new insight into the inherent operability of the overall system, particularly for flight element processing. The results of this portion of the effort are documented in 5.2

4.2.2 The Eight-Flow STS Processing Data Set

The data set used for the direct maintenance analysis encompassed eight processing flows in the 1997 time frame. The data used for the analysis was extracted from as-run task scheduling data that documented the system function, a direct maintenance task title, and the duration of the task. A summary of these processing flows is provided in Table 4.

[13] W.D. Morris et al., "Analysis of Shuttle Orbiter Reliability and Maintainability Data for Conceptual Studies;" AIAA 96-4245, *1996 AIAA Space Programs and Technologies Conference,* Huntsville, AL, September 24-6, 1996.
[14] R.L. Christenson and D.R. Komar, "Reusable Rocket Engine Operability Modeling and Analysis," NASA TM/TP-1998-208530, Marshall Space Flight Center, Huntsville, AL, July, 1998.

Specifically, the data set covered the launch preparations for the following missions in chronological order of launch: STS–81, STS–82, STS–83, STS–84, STS–94, STS–85, STS–86, and STS–87. The data set was chosen to represent a period when the program experienced its maximum sustained annual throughput following the *Challenger* accident. It also contains a processing flow (STS–82) performed following a periodic inspection, or Orbiter Maintenance Depot Period (OMDP) at its manufacturing plant in Palmdale, California. These processing flows have tended to be much higher in work content because of the level of intrusive maintenance and the amount of extra work needed to fully restore the vehicle to flight-worthy condition. It also contained the shortest post-*Challenger* Orbiter processing flow, STS–94, which was a reflight of the previous mission cut short by a fuel cell failure.

4.2.3 Root Cause Analysis Data Field Descriptions

An example of the source data for the analysis is shown in Appendix A. The appendix presents the source data used in analyzing the STS–81 flow. The following subsections describe each of the fields in Appendix A.

Table 4. Summary of Eight Processing Flows Used in Analysis

Flow	Launch Date	Orbiter / Flight No.	OPF (Work Days)	Pad (Work Days)	Total Cycle Time (Work/Cal Days)	Primary Payload	Notes
STS81	01/12/1997	OV-104/18	62	24	91/108	SpaceHab 6	
STS82	02/11/1997	OV-103/22	147	26	178/223	HST SM-02	Post-OMDP
STS83	04/04/1997	OV-102/22	73	24	103/118	Spacelab MSL-1/LM	Min Duration Flight (MDF)/Fuel Cell #2 Failure
STS84	05/15/1997	OV-104/19	77	21	102/113	SpaceHab 7	
STS94	07/01/1997	OV-102/23	53	21	81/84	Spacelab/MSL-1/LM	STS 83 Reflight
STS85	08/07/1997	OV-103/23	102	23	130/168	CRISTA-SPAS-02	
STS86	09/25/1997	OV-104/20	60	29	94/124	SpaceHab 8	
STS87	11/19/1997	OV-102/24	93	22	120/125	Spacelab/Spartan/USMP	

4.2.3.1 Data Set

MLP81 — The STS–81 direct work involved in operating and turning around the Mobile Launcher Platform (MLP).

OPF81 — The STS–81 direct work involved in operating and turning around the Space Shuttle Orbiter in its Orbiter Processing Facility (OPF).

PAD81 — The STS–81 direct work involved in operating and turning around Launch Pad 39.

PSF81 — The STS–81 direct work involved with the receiving, acceptance, and assembly activity associated with the Solid Rocket Booster (SRB) segments in the Rotation, Processing, and Surge Facility (RPSF).

SSV81 — The STS–81 direct work involved in preparing the fully assembled Space Shuttle Vehicle on its MLP—whether at the Vehicle Assembly Building (VAB) or at the Launch Pad.

STK81 — The STS–81 direct work involved in stacking the SRB elements on the MLP.

TKC81 — The STS–81 direct work involved in mating the External Tank (ET) and the Orbiter elements with the SRB elements on the MLP.

TKP81 — The STS–81 direct work involved in preparing the ET in its checkout cell prior to stacking the SRB elements.

4.2.3.2 Start Time

This field defines the day (DD), month (MMM), year (YYYY), and 24-hour (H) start time of the work item.

4.2.3.3 Procedure Name

This field identifies the planned or unplanned work procedure designation. A shorthand designation is used for many one-time-use procedures. These may include such items as problem reports and special test procedures.

4.2.3.4 Description

Descriptor field associated with the work item.

4.2.3.5 Duration

This field defines the duration of the work item in hours.

4.2.3.6 System

This field defines the KSC organizational system responsible for the direct work item. The three-character codes are defined in Appendix E.

4.2.3.7 Design Discipline

This field defines the generic design discipline assigned by the analysis team to the work item. Definitions of these disciplines can be found in Table 3.

4.2.3.8 Generic Operations FBS Number

This field defines the generic operations subfunction assigned by the analysis team to the work item. Definitions of these functions are found in Appendix A.

4.3 The STS Design Root Cause Knowledge Base

The eight-flow STS processing data set described in 4.2 was eventually compiled into a Microsoft Access XP database. Once in the database, a structured process of assigning the operations functions and design disciplines was performed, followed by an itemization of design causes for the existence of the work task or major influences to the length of time of the task. In addition, technical need statements associated with the identified cause statements were cataloged. The figures in Appendix A document the STS Access Knowledge Base.

5 OVERVIEW OF RESULTS

5.1 Overall Results of Cost Analysis

The question *Why does it cost so much to turn the Shuttle around for flight?* can be answered several ways. Knowing the annual costs yields a certain level of insight. However, knowing these costs as a function of flight rate capability can produce quantifiable insight into overall system affordability. As we shall see later, knowing the flight rate capability can also answer why the average cost per flight is so high (approximately $500M) or what the marginal or incremental costs per flight tend to be (approximately $100M).

5.1.1 Cost-Performance Relationships

The concept of a cost-performance curve that relates the Shuttle annual operations cost to flight rate performance is shown in Figure 12 and Figure 13.[15] Of interest for affordable access to space is the fact that the slope of the curve can easily yield a cost-per-pound figure of merit since the units of the ordinate are in dollars per year, while those of the abscissa can be converted to payload pounds per year, if the annual payload throughput is known. This curve, then, represents the economic value of a space transportation system by clearly showing the affordability of the system relative to its productiveness. This relationship can be illustrated by plotting the flight rate, payload throughput, and annual budgets for the Shuttle system in the operating years of 1992 through 1998 (see Figure 14 and Figure 15).[16]

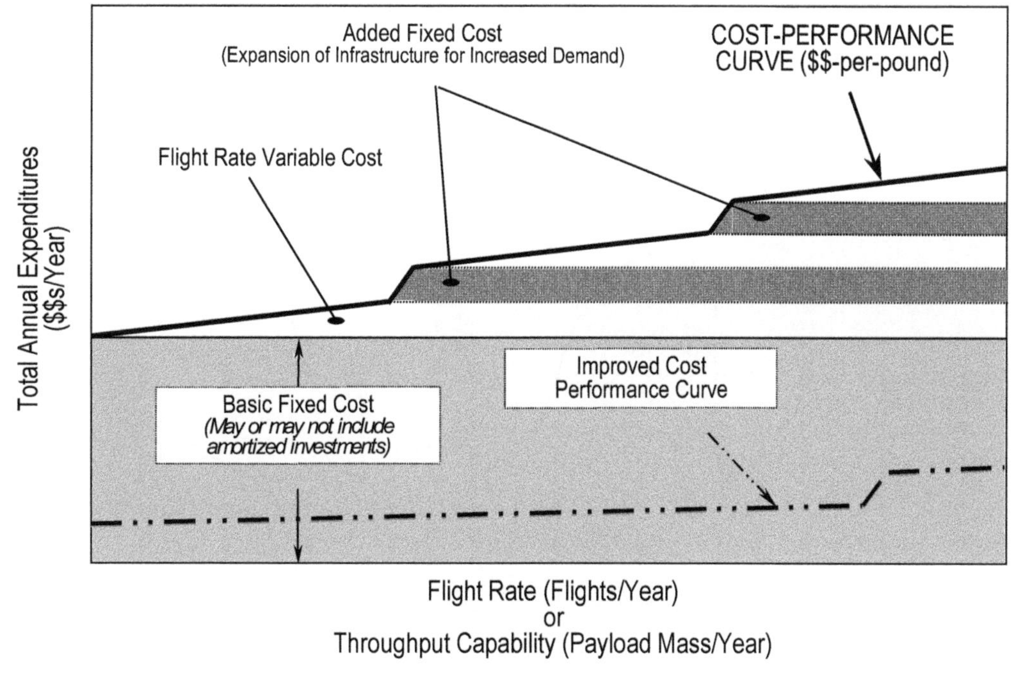

Figure 12. Conceptual Space Transportation "Cost-Performance" Curve

[15] "Shuttle Operations Zero Base Cost Study," NASA presentation to Dr. Lenoir, July 2, 1991.
[16] D. Morgan and A. Jacobs, "NASA's Space Shuttle Program: Space Shuttle Appropriations FY1992-FY2002," Congressional Research Service, Library of Congress, RS21411, February 13, 2003.

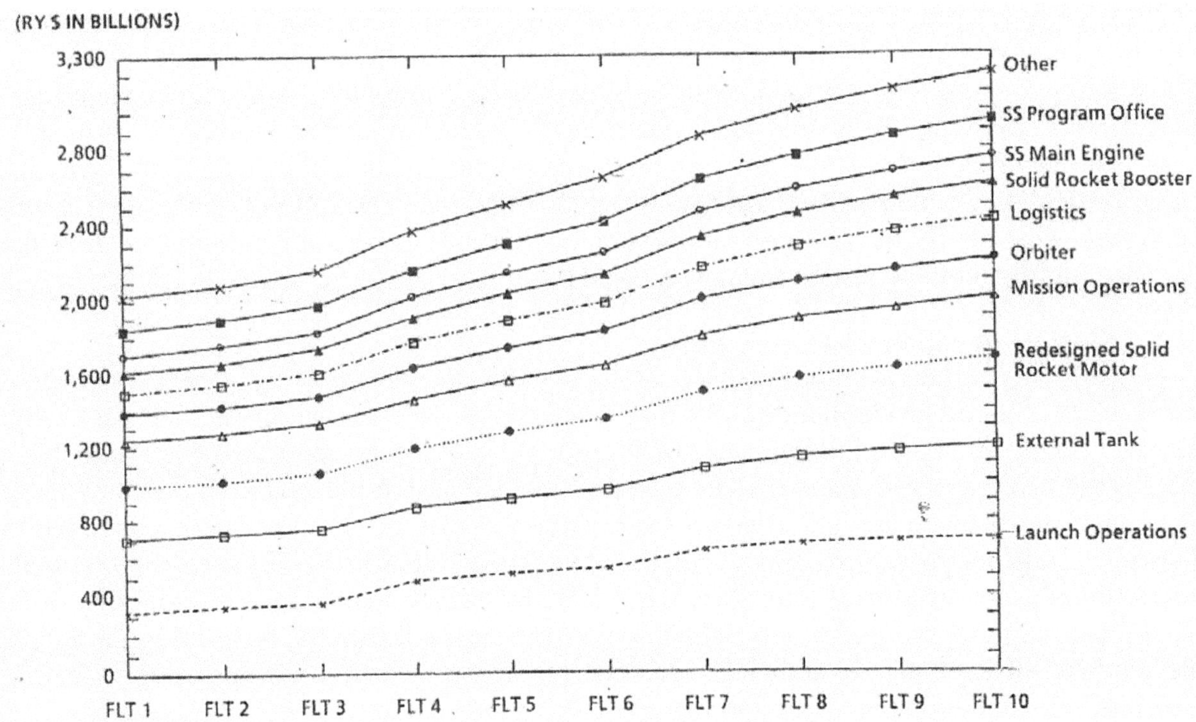

Figure 13. Example Cost–Performance Estimate (Space Shuttle Program ca. 1992)

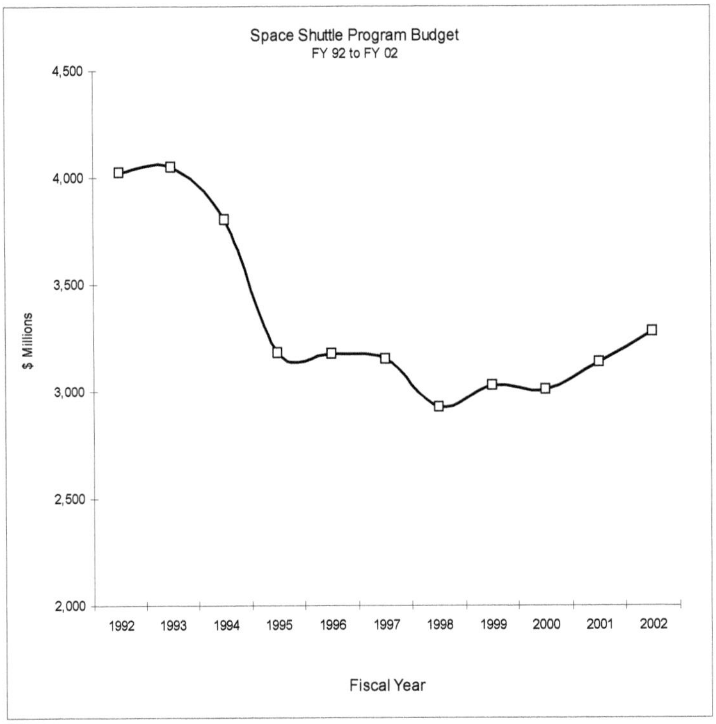

Figure 14. Example Operating Budget Trend (Space Shuttle 1992 Through 2002)

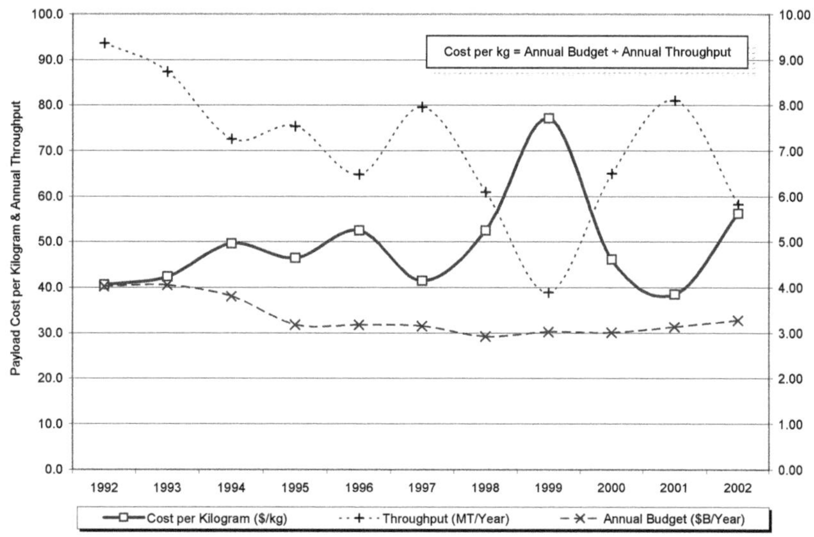

Figure 15. Cost-Per-Pound Trend for Space Shuttle, 1992 to 2002

What these figures reveal is the high degree of sensitivity to flight rate for overall mission costs and payload delivery affordability.

5.1.2 Shuttle Budget Breakdown by Operations and Infrastructure Functions

The cost analysis allocated various Shuttle budget elements to operations and infrastructure functions (see 4.1). The results are shown in Table 5. The table used the most detailed STS budget data available at the time of the analysis and was derived from NASA Access to Space Study data performed against the Fiscal Year 1994 Space Shuttle Budget.[17] Each budget line item was then mapped to the top-level generic functions described in Figure 9. The table is sorted from top to bottom in order of increasing cost contribution. While the data is somewhat outdated, the relative distribution of functions is considered reflective of program cost concentrations.

The "iceberg" nature of the annual recurring program costs are reflected in Table 5. The more visible and well-known direct vehicle operations costs, for example, tend to be at the top, while the indirect and support functions are at the bottom with the highest cost contribution. Note that the Transportation System Operations Planning and Management function (FBS 12.0) is close to half the program cost. Referring to Appendix A, it can be seen that these functions include sustaining engineering and safety and mission assurance functions. As will be shown in 5.2, because of the high degree of unplanned work and intrusive assembly/disassembly operations, engineering workforces across multiple program elements across the nation are required to regain engineering and management confidence and certify the system is ready for flight (i.e., is *flight-worthy*). To visualize the interrelationships between the direct and the indirect/support functions, Figure 16 shows the linkages at common flight readiness milestones.[18] While not representing all the sublevel reviews, the figure demonstrates how flight readiness milestones intersect complex

[17] NASA MSFC Engineering Cost Group, Access to Space Study, "STS Baseline Ops Cost Based on FY94 OMB STS Cost Per Flight (with Updates from Project Offices)," Applied Research, Inc., Huntsville, AL, April 13, 1993.
[18] Space Shuttle Program's "Requirements and Procedures for Certification of Flight Readiness," NSTS 08117 Rev. M.

and labor-intensive flight production activity, thus providing engineering confidence in the vehicle's flight readiness.

Table 5. Space Shuttle Budget Allocations to Generic Operations and Infrastructure Functions

(Based on Access to Space Study STS Budget Breakdown, ca. 1994 and Spaceport Synergy Team analysis, ca. 1998)

Generic Operations Function	FBS No.	Program Integration, Management Support, Institution & Network Support	Launch and Landing	Solid Rocket Motor (SRM) Project	External Tank (ET) Project	Orbiter Project and Logistics	Mission Operations	Solid Rocket Booster (SRB) Project	Space Shuttle Main Engine (SSME) Project	Crew Operations & Training	Launch Site Payload Support	Propel-lants (Cryos)	Total $M FY94	Total (%)
Element Receipt & Acceptance	6.0	(A)	0.4	(A)	1.0	(A)	(A)	(A)	(A)	(A)	(A)	(A)	1.4	0.0%
Landing & Recovery	1.0	(A)	19.6	(A)	(A)	(A)	(A)	(A)	(A)	(A)	(A)	(A)	19.6	0.6%
Assembly & Integration	3.0/4.0	(A)	27.1	(A)	(A)	(A)	(A)	(A)	(A)	(A)	(A)	(A)	27.1	0.8%
Launch	5.0	(A)	51.5	(A)	(A)	(A)	(A)	(A)	(A)	(A)	(A)	(A)	51.5	1.5%
Offline Payload/Crew Processing	9.0	(A)	(A)	(A)	(A)	40.2	(A)	(A)	(A)	(A)	35.7	(A)	75.9	2.3%
Flight Element Turnaround	2.0	(A)	112.3	(A)	(A)	(A)	(A)	(A)	(A)	(A)	(A)	(A)	112.3	3.3%
Traffic/Flight Control	8.0	72.3	49.4	(A)	(A)	(A)	77.7	(A)	(A)	(A)	(A)	(A)	199.4	5.9%
Vehicle Depot Maintenance	7.0	(A)	1.5	(A)	(A)	108.3	0.0	53.7	74.0	(A)	(A)	(A)	237.5	7.1%
Spaceport Support Services	10.0	0.2	147.9	(A)	76.8	(A)	93.7	(A)	(A)	(A)	(A)	(A)	318.6	9.5%
Concept-Unique Logistics	11.0	0.9	0.7	337.0	263.6	177.8	(A)	46.2	(A)	(A)	(A)	16.5	842.7	25.1%
Transp. Sys. Ops Planning & Mgmt	12.0	848.7	209.1	67.2	31.0	25.0	121.2	52.1	51.3	71.8	(A)	(A)	1477.4	43.9%
Total ($M FY94)		922.1	619.5	404.2	372.4	351.3	292.6	152.0	125.3	71.8	35.7	16.5	3363.4	100.0%
Percent		27.4%	18.4%	12.0%	11.1%	10.4%	8.7%	4.5%	3.7%	2.1%	1.1%	0.5%	100.0%	

(A) Costs for these items were found to be either negligible or not separately identified in the budget breakdown. Some budgeted items may be accounted for in other items.

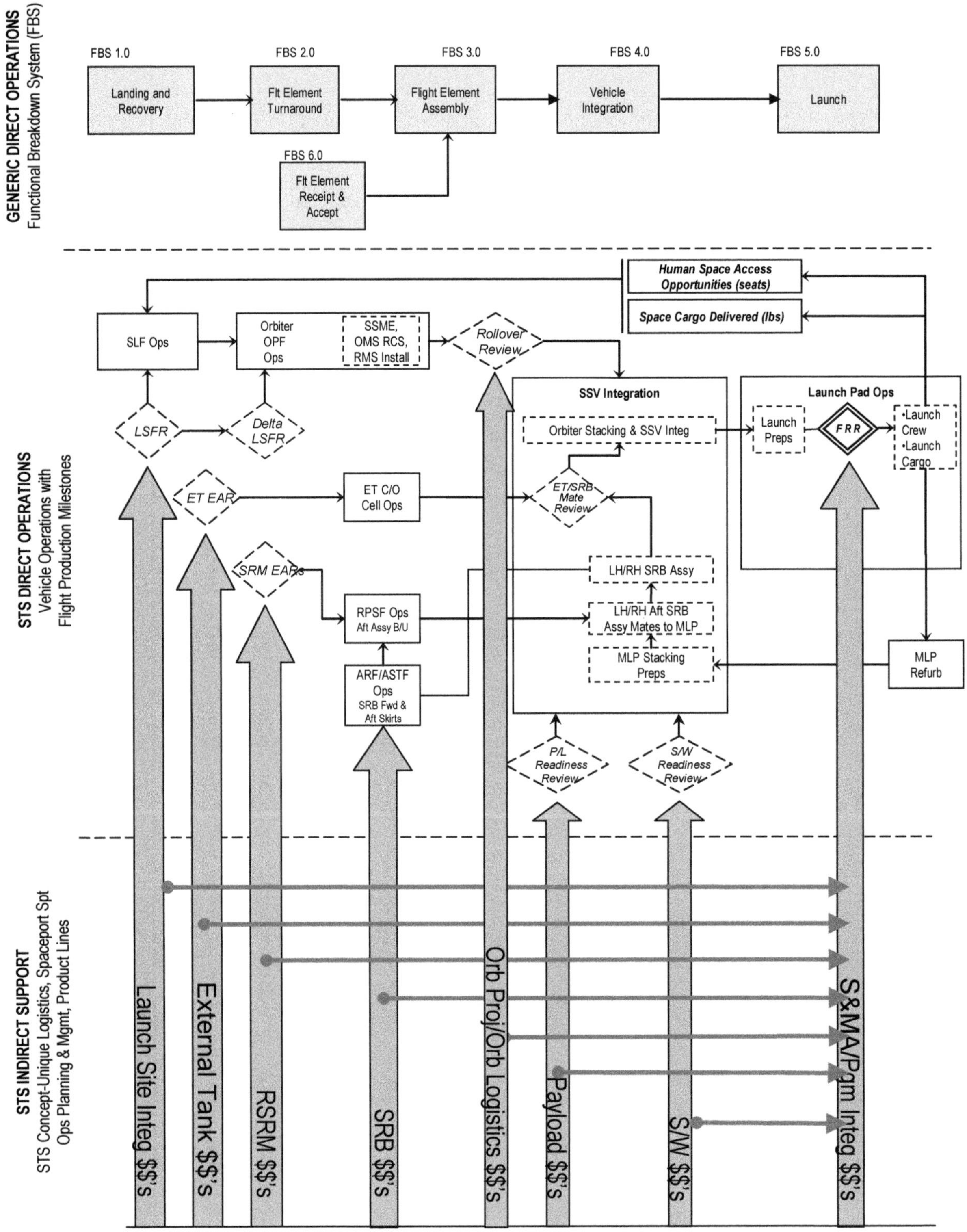

Figure 16. Connectivity Between Direct Vehicle Processing and Infrastructure Costs

5.1.3 Flight Preparation and Planning Work

The process of creating a Shuttle flight does not begin and end with the processing of the flight elements at the launch site. There are many important functions creating indirect work required to produce a flight. These sustaining engineering and planning functions require infrastructures and workforces that are intimately interrelated to the direct work that processes the vehicle and its payload.

For example, a typical 400-day "planning template" includes customized flight planning functions, mission-unique flight software development, and payload-planning activities—as well as the 126 days typically required for flight element and integrated Space Shuttle Vehicle hardware processing.[19]

Referring back to the iceberg chart in Figure 8, these functions are "below the waterline." The flight preparation and planning functions, however, must be accounted for in the overall calculation of responsiveness and affordability (i.e., *Why does it take so long and cost so much?*). In addition to cycle time performance and cost, these functions are critical to providing engineering confidence that a flight-ready product and a flight-ready crew can carry out the intended missions.

While this analysis did not attempt to identify all of these critical infrastructure functions at an equal level of detail, it did, however, identify their overall effect on a typical program operating budget. The major Flight Preparation and Process Planning (FPPP) functions, as identified in the Space Shuttle Program's *Requirements and Procedures for Certification of Flight Readiness,* NSTS 08117 Rev. M, are as follows:

- ***Vehicle Processing***
- ***Payload Processing***
- ***Configuration Management & Requirements Definition***
- ***Flight Certification***
- ***Facility, Equipment, and GSE Certification***
- ***Personnel Certification***

- *Special Testing & Analysis*
- *Material Review (MR)*
- *Hazard Analysis*
- *Failure Modes and Effects Analysis (FMEA)/Critical Items List (CIL)*
- *Flight Crew Training & Medical Certification*
- *Validation that External Inputs are Appropriate for a Designated Flight*

[19] "Flight Production Generic Templates,"JSC 25187, Appendix A., NASA Johnson Space Center, Houston, TX.

5.2 Overall Results of Direct Work Content Concentrations

5.2.1 Overall Flow

The direct, on-vehicle processing that was analyzed is presented in block diagram form in Figure 17. Each major process element will be reviewed in terms of accumulated work across eight processing flows that led to launches in 1997.

The serial time required to produce a Shuttle launch from the landing of the Orbiter to its next launch is trended in Figure 18. The time period is from 1988 with the return to flight of STS–26R, following the *Challenger* accident, through the processing of *Columbia* for mission STS–107 in 2003 prior to its accident. The time period of the analysis is shown for the eight processing flows highlighted leading to launches during 1997. The number of calendar days required for each processing flow shows a large range, with most turnaround flows averaging less than 200 days. In general, the processing flows exceeding 300 days were the result of bringing an Orbiter back into service following an extended maintenance, inspection, and modification period at the Orbiter's manufacturing plant in Palmdale, California. (The times shown do not reflect the time spent in California, but rather the time spent at KSC bringing the Orbiter back to a flight-certifiable condition and preparing it for its next flight).[20]

The highest-level summary of the eight-flow STS processing data analyzed is provided in Table 6. This table totals the task times of all the direct work for all eight processing flows examined. Table 6 accumulates these task times (designated as "task-hours" in this analysis) according to the top-level direct-operations functions and by the generic design discipline, regardless of critical path.

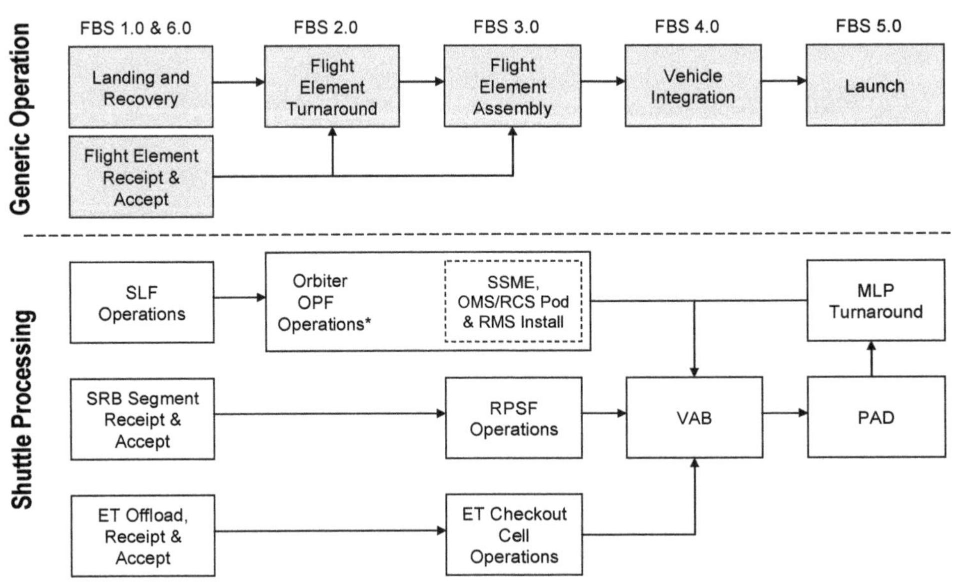

Figure 17. Relationship Between Analysis Classifications and Shuttle Direct Vehicle Processing

[20] C. Cates, "Space Shuttle Processing and Metrics Overview," NASA John F. Kennedy Space Center, July 26, 2003.

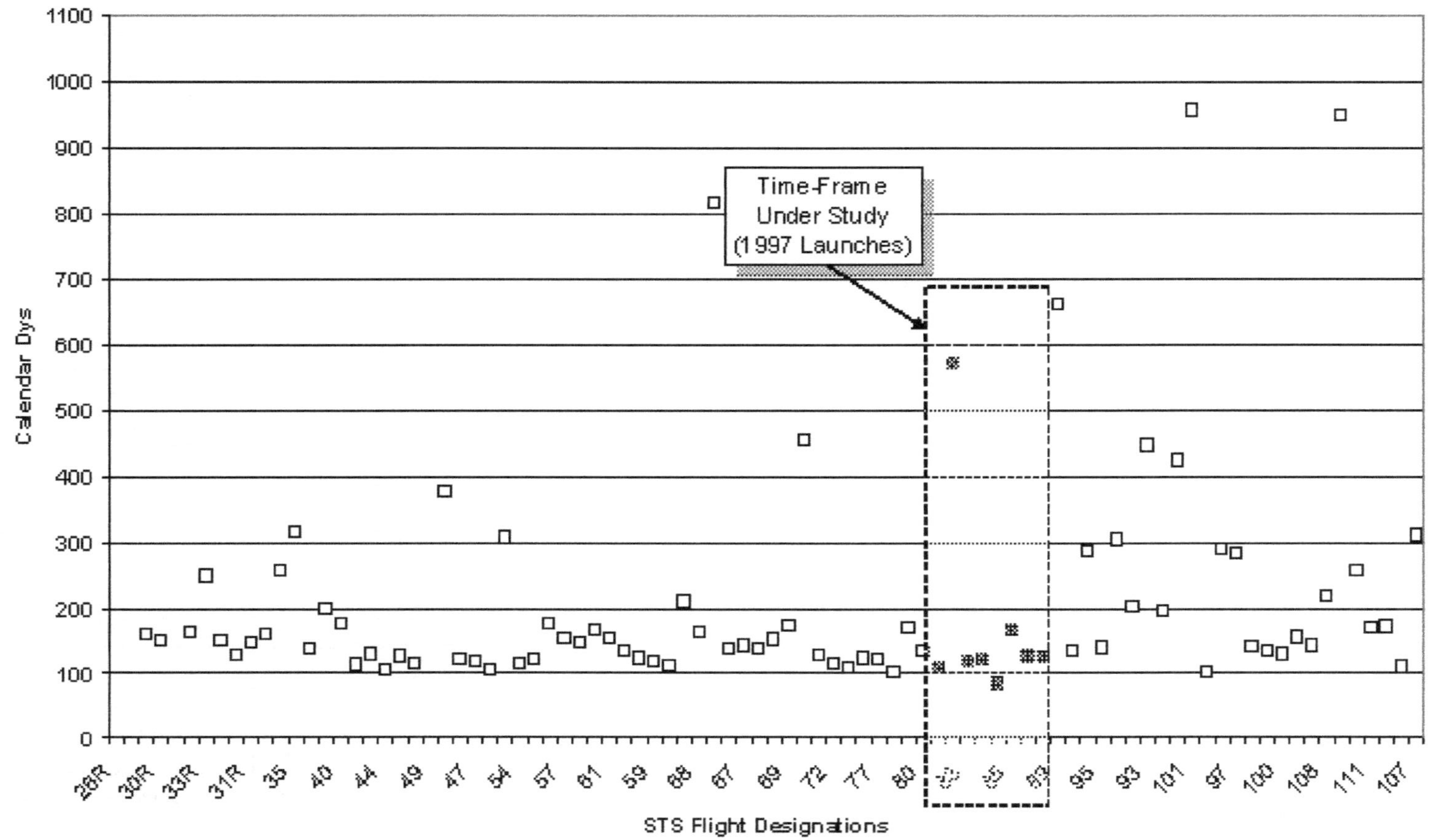

Figure 18. Shuttle Processing Intervals—From Orbiter Landing to Next Launch

Table 6. Overall Results of Shuttle Direct Work Content

Concentration of Space Shuttle Direct Work Content for 1997 Launches
(As measured by cumulative task-hours for eight processing flows)

GENERIC OPERATIONS FUNCTION	ID	Structures , Mechanisms and Vehicle Handling	Liquid Propulsion	Thermal Management	Power Management	Safety Management & Control	Ground Interfacing Systems and Facilities	Payload Accommodations	Environmental Control and Life Support	Command, Control & Health Management	Communications	Guid, Nav & Ctl	Cockpit & Crew Cabin	Total (Task-hours)	Percentage
		GENERIC DESIGN DISCIPLINE													
		1	2	3	4	5	6	7	8	9	10	11	12		
Flight Element Turnaround	A	57,843	33,606	42,514	30,025	17,318	5,825	11,894	13,632	3,703	3,159	1,888	1,729	**223,136**	**55.98%**
Launch (C)	B	8,870	12,965	44	4,638	9,429	19,388	2,841	627	5,289	(A)	289	410	**64,790**	**16.26%**
Flight Element Assembly (D)	C	39,718	9,085	2,691	427	317	613	173	221	3,343	292	301	(A)	**57,181**	**14.35%**
Vehicle Integration	D	22,523	5,370	1,244	4,927	6,051	3,035	1,348	57	1,341	(A)	(A)	40	**45,936**	**11.53%**
Flight Element Shipping, Receiving & Acceptance	E	5,303	1,558	100	54	(A)	72	(A)	(A)	32	(A)	(A)	(A)	**7,119**	**1.79%**
Landing & Recovery	F	(B)	(B)	(B)	(B)	(B)	(B)	(B)	(B)	(B)	(B)	(B)	(B)	**402**	**0.10%**
Total (Task-hours)		**134,257**	**62,584**	**46,593**	**40,071**	**33,115**	**28,933**	**16,256**	**14,537**	**13,708**	**3,451**	**2,478**	**2,179**	**398,564**	**100.00%**
Percent		**33.69%**	**15.70%**	**11.69%**	**10.05%**	**8.31%**	**7.26%**	**4.08%**	**3.65%**	**3.44%**	**0.87%**	**0.62%**	**0.55%**		

(A) System disciplines not showing dedicated concentrations of direct work. This should not be interpreted that no direct work occurs. Some of the tasks involving the design discipline may have been identified, for the purposes of this analysis, as being dominated by another discipline. Proper interpretation is that these items have no identifiable direct work concentrations relative to the total annual direct work content.

(B) Direct Shuttle Landing Facility (SLF) Landing and Recovery tasks not accumulated by design discipline in this table. Also, does not reflect direct work for SRB recovery operations and contingency landing site coverage.

(C) The Launch tasks accumulated in the table do not include direct launch tasks performed by the Eastern Range for Command and Control, Safety Management and Communications functions.

(D) Does not include direct work for ET manufacturing and SRM/SRB Remanufacturing.

Figure 19. Orbiter *Atlantis* Transported Out of Orbiter Processing Facility (OPF)

5.2.2 Flight Element Turnaround Direct Work Content

Orbiter turnaround (Figure 19) was found to represent roughly half (approximately 56 percent) of the direct, on-line work involved in producing a Shuttle launch (see Table 6). In order to produce a year's worth of flights in 1997, the scheduling of over 13,700 tasks was required, totaling over 223,000 task-hours, necessitating an annual investment of millions of labor hours. Note that this excludes off-line repair and refurbishment (e.g., tile shop work, main engine shop work, Orbital Maneuvering System/Reaction Control System [OMS/RCS] pod maintenance, component repairs at the NASA Shuttle Logistics Depot (NSLD), and SRB refurbishment). Again, these were excluded so not to inadvertently include off-line maintenance functions in the accounting of on-line vehicle preparations. An eight-flow statistical breakdown of Table 6 is provided in Table 7.

Table 7. Space Shuttle Eight-Flow Direct Work Concentration

Operations Functions Associated with Design Disciplines (Table 6 Cells in Descending Order)			Eight (8) Flow Analysis Statistical Summary (Values in Cumulative Task-Hours)												
Table 6 Cell ID	Operational Functions for Generic Design Disciplines	Eight Flow (1997) Total	STS-81	*(OMDP) STS-82*	STS-83	STS-84	STS-85	STS-86	STS-87	STS-94	Eight (8) Flow Min	Eight (8) Flow Max	**Normal Flow Avg**	Normal Flow Std Dev	Normal Flow 90% Conf Interval
A1	Structures, Mechanisms and Handling Systems Turnaround	57,843	5,850	*12,532*	6,211	6,340	8,781	5,057	8,560	4,512	4,512	12,532	**6,473**	1,633	1,015
A3	Thermal Management Systems Turnaround	42,514	3,674	*8,916*	2,498	4,590	8,138	5,070	5,544	3,409	2,498	8,916	**4,703**	1,837	1,142
C1	Structures, Mechanisms System Assembly and Flight Element Handling	39,718	4,828	*4,857*	5,063	5,195	5,321	4,893	5,098	4,463	4,463	5,321	**4,980**	283	176
A2	Liquid Propulsion Systems Turnaround	33,606	3,408	*4,389*	6,350	3,654	6,550	2,871	4,446	2,158	2,158	6,550	**4,205**	1,687	1,049
A4	Power Management Systems Turnaround	30,025	3,721	*9,533*	3,173	2,396	3,083	1,978	3,992	2,392	1,978	9,533	**2,962**	743	462
D1	Structures, Mechanisms System Assembly and Vehicle Integration Handling	22,523	2,791	*2,988*	3,015	3,715	2,751	2,339	2,502	2,478	2,339	3,715	**2,799**	464	288
B6	Ground Interfacing Systems and Facilities for Launch Operations	10,000	2,702	*4,662*	1,218	671	2,011	2,010	4,130	2,015	671	4,553	**2,121**	1,114	692
A5	Safety Management & Control Systems Turnaround	17,318	1,732	*905*	2,930	2,188	3,628	1,863	3,037	1,467	905	3,628	**2,406**	800	498
A8	Environmental Control and Life Support Systems Turnaround	13,632	808	*3,706*	1,681	735	1,919	1,546	1,941	1,296	735	3,706	**1,418**	494	307
B2	Liquid Propulsion Systems Launch Operations	12,965	1,232	*2,014*	1,324	974	1,079	941	3,640	1,279	941	3,640	**1,496**	957	595
A7	Payload Accommodations Turnaround	11,894	828	*1,922*	1,692	954	2,963	854	2,289	392	392	2,963	**1,425**	927	576
B5	Safety Management & Control Systems Launch Operations	9,429	1,736	*1,762*	337	270	422	394	2,044	805	270	2,044	**858**	731	454
C2	Liquid Propulsion System Assembly	9,085	908	*2,163*	1,276	1,142	958	693	765	1,180	693	2,163	**989**	219	136
B1	Structures, Mechanisms Launch Operations	8,870	910	*1,598*	840	423	739	1,016	1,591	1,697	423	1,697	**1,031**	459	285
D5	Safety Management & Control System Integration	6,051	239	*524*	1,017	636	799	248	1,268	1,565	239	1,565	**825**	499	310
A6	Ground Systems and Facilities Work for Turnaround	5,825	1,197	*69*	3,240	49	83	50	1,011	126	49	3,240	**822**	1,172	729
D2	Liquid Propulsion System Integration	5,370	587	*528*	461	406	349	388	524	2,389	349	2,389	**729**	737	458
E1	Structures, Mechanisms and Handling for Element Receiving and Acceptance	5,303	323	*1,592*	612	776	588	549	381	482	323	1,592	**530**	152	94
B9	Command, Control & Health Management Systems Launch Operations	5,289	435	*370*	353	416	298	594	338	894	298	894	**475**	208	129
D4	Power Management System Integration	4,927	665	*1,081*	172	1,255	1,092	1,056	752	979	172	1,255	**853**	362	225
B4	Power Management Systems Launch Operations	4,638	376	*574*	0	327	271	213	632	213	0	632	**290**	192	120
A9	Command, Control & Health Management Systems Turnaround	3,703	358	*593*	520	341	573	139	869	310	139	869	**444**	235	146
C9	Command, Control & Health Management Systems Assembly	3,343	434	*488*	340	356	545	326	424	430	326	545	**408**	76	47
A10	Communication Systems Turnaround	3,159	714	*931*	264	330	298	248	282	92	92	931	**318**	190	118
D6	Ground Interfacing Systems and Facilities For Vehicle Integration	3,035	988	*406*	169	459	279	199	469	66	66	988	**376**	308	192
B7	Payload Accommodations Launch Operations	2,841	578	*560*	108	412	459	274	217	233	108	578	**326**	163	101
C3	Thermal Management Systems Assembly	2,691	440	*514*	64	340	260	242	243	268	64	514	**265**	114	71
A11	Guidance, Navigation & Control Systems Turnaround	1,888	100	*560*	350	246	279	74	129	150	74	560	**190**	103	64
A12	Cockpit & Crew Cabin Systems Turnaround	1,729	232	*357*	118	136	540	196	98	52	52	540	**196**	163	101
E2	Receiving/Acceptance of Flight Element Liquid Propulsion Systems	1,558	430	*319*	116	162	122	121	134	154	116	430	**177**	113	70
D7	Payload Accommodations System Integration	1,348	104	*36*	347	88	129	84	366	206	36	366	**189**	122	76
D9	Command, Control & Health Management System Integration	1,341	386	*161*	230	242	130	185	132	167	130	386	**210**	89	55
D3	Thermal Management System Integration	1,244	130	*162*	884	225	154	160	81	160	81	884	**256**	280	174
B8	Environmental Control and Life Support Systems Launch Operations	627	88	*8*	0	55	192	212	44	28	0	212	**88**	82	51
C6	Ground Interfacing Systems and Facilities During Assembly	613	50	*41*	64	121	83	113	59	82	41	121	**82**	27	17
C4	Power Management Systems Asembly	427	36	*287*	384	8	16	16	16	24	8	384	**71**	138	86
B12	Cockpit & Crew Cabin Systems Launch Operations	410	55	*66*	75	59	27	51	47	30	27	75	**49**	17	10
C5	Safety Control System Assembly	317	53	*124*	24	40	0	0	0	36	0	124	**22**	22	14
C11	Guidance, Navigation & Control Systems Assembly	301	0	*301*	0	0	0	0	0	0	0	301	**0**	0	--
C10	Communications Systems Assembly	292	0	*292*	0	0	0	0	0	0	0	292	**0**	0	--
B11	Guid, Nav & Ctl Systems Launch Operations	289	40	*33*	44	34	32	36	42	28	28	44	**37**	6	4
C8	Environmental Control and Life Support Systems Assembly	221	0	*117*	0	0	104	0	0	0	0	117	**15**	39	24
C7	Assembly of Payload Accommodations	173	0	*17*	30	0	0	0	124	2	0	124	**22**	46	29
E3	Receiving/Acceptance of Thermal Management Systems	100	0	*100*	0	0	0	0	0	0	0	100	**0**	0	--
E6	Ground Interfacing Systems and Facilities for Receiving/Acceptance of Flight Element	72	8	*8*	8	8	16	8	8	8	8	16	**9**	3	2
D8	Environmental Control and Life Support System Integration	57	8	*8*	8	1	8	8	8	8	1	8	**7**	3	2
E4	Receiving/Acceptance of Flight Element Power Systems	54	0	*54*	0	0	0	0	0	0	0	54	**0**	0	--
B3	Thermal Management Systems Launch Operations	44	40	*0*	1,884	0	0	0	0	4	0	1,884	**275**	709	441
D12	Cockpit & Crew Cabin System Integration	40	8	*0*	0	0	0	32	0	0	0	32	**6**	12	7
E9	Receiving/Acceptance of Flight Element Cmd, Ctl & Health Mgmnt Systems	32	0	*32*	0	0	0	0	0	0	0	32	**0**	0	--
	Total (Less Landing and Recovery)	398,162	44,320	*73,151*	49,494	40,775	56,069	37,347	58,277	38,729	37,347	73,151	**46,430**	8,375	5,207
	Landing and Recovery	402	40	*75*	49	53	64	43	52	26	26	75	**47**	12	7
	Grand Total	**398,564**	**44,360**	***73,226***	**49,543**	**40,828**	**56,133**	**37,390**	**58,329**	**38,755**	37,390	73,226	**46,477**	8,384	5,212

5.2.2.1 Orbiter Processing Facility Background

For routine landings at KSC, the Orbiters are towed to one of three Orbiter Processing Facilities (OPFs) shortly after their arrival (see Figure 20).[21]

The OPF complex has three almost identical high bays, each of which is 60 meters (m) (197 feet [ft]) long, 46 m (150 ft) wide, and 29 m (95 ft) high, and encompasses a 2,694-m (29,000-sq-ft) area. A low bay connects high bays 1 and 2. It is 71 m (233 ft) long, 30 m (97 ft) wide, and nearly 8 m (25 ft) high. High bay 3, built last, also has an adjacent low bay. Annexes and portable buildings provide additional shop and office space.

Each high bay is equipped with a 30-ton (27-metric-ton) bridge crane with a hook height of approximately 20 m (66 ft). Platforms, a main access bridge, and two rolling bridges with trucks provide access to various parts of the Orbiter. The trucks have telescoping arms with rotating buckets to hold workers. The high bays have an emergency exhaust system in case of a hypergolic fuel spill. The low bay contains areas for electronic, mechanical, and electrical equipment; a communications room; offices; and supervisory control rooms. All bays have fire protection systems.

In addition to routine postflight servicing and checkout, many of the vehicle modifications needed for future flight requirements, or to enhance vehicle performance and correct deficiencies, are performed in the OPF.

Spacecraft or payloads processed through checkout in a horizontal position, usually the larger ones such as Spacelab, are installed in the Orbiter in this facility. Spacecraft handled in a vertical position normally are installed at the launch pad. After processing, the Orbiter is usually towed into the VAB transfer aisle.

5.2.2.2 Orbiter Turnaround Work Content Summary

A more detailed presentation of the results of the Orbiter turnaround processing review is summarized in Table 8. Similar to Table 6, it shows the concentration of work required to turn around the Orbiter sorted by design discipline. Subsequent subsections detail the analysis results in descending order of total work content.

[21] "Countdown! NASA Space Shuttles and Facilities," IS-2003-09-006-KSC, NASA Information Summaries, September 2003.

Figure 20. Orbiter Processing Facilities at Launch Complex 39

Table 8. Summary of Shuttle Orbiter Turnaround Work Content

Concentration of Space Shuttle Orbiter Turnaround Direct Work Content for 1997 Launches (As measured by cumulative task-hours for eight processing flows)	FBS	ID	Structures , Mechanisms, Veh Handling (1)	Thermal Management (2)	Propulsion (3)	Power Management (4)	Safety Management & Control (B) (5)	Environmental Ctl & Life Spt (6)	Payload Accommodations (7)	Ground Systems & Facilities (8)	Cmd, Ctl & Health Mngmt (9)	Communications (10)	Guid, Nav & Ctl (11)	Cockpit & Crew Cabin (12)	Grand Total	Percentage
			Generic Design Discipline													
Generic Operations Function																
Turnaround Unplanned Troubleshooting and Repair	2.08	A	17,659	23,448	4,652	13,791	353	833	6	916	2,004	804	1,165	327	**65,958**	**29.56%**
Vehicle Element Systems Servicing	2.04	B	11,284	12,579	8,429	8,880	283	2,656	424	4	829	112	155	931	**46,566**	**20.87%**
Processing Support Systems and Functions (B)	2.07	C	1,075	24	7,252	304	13,454	7,137	(A)	2	25	48	10	48	**29,379**	**13.17%**
Turnaround Inspection and Checkout	2.05	D	8,762	2,659	6,687	3,907	2,580	1,450	125	(A)	663	1,276	494	292	**28,895**	**12.95%**
Vehicle Payload Accomodations Turnaround	2.06	E	5,106	1,687	(A)	69	18	431	11,210	5	9	8	(A)	115	**18,658**	**8.36%**
Turnaround Facility and Equipment Periodic Maintenance	2.11	F	3,105	(A)	2,128	74	40	520	(A)	4,380	2	(A)	(A)	(A)	**10,249**	**4.59%**
Removal of Access, Umbilicals and Closeout	2.10	G	5,219	1,534	971	36	438	32	102	(A)	(A)	141	(A)	(A)	**8,473**	**3.80%**
Modifications and Special Tests	2.09	H	2,224	397	1,093	1,821	(A)	40	4	2	171	752	42	16	**6,562**	**2.94%**
Vehicle Positioning, Connection to Services, Gaining Access, and Protection	2.02	I	2,769	186	1,065	116	17	189	23	120	(A)	18	22	(A)	**4,525**	**2.03%**
Vehicle Element Safing	2.03	J	346	(A)	1,245	462	119	98	(A)	(A)	(A)	(A)	(A)	(A)	**2,270**	**1.02%**
Facility Preps for Vehicle Turnaround	2.01	K	294	(A)	84	565	16	246	(A)	396	(A)	(A)	(A)	(A)	**1,601**	**0.72%**
Total (Task-Hours)			**57,843**	**42,514**	**33,606**	**30,025**	**17,318**	**13,632**	**11,894**	**5,825**	**3,703**	**3,159**	**1,888**	**1,729**	**223,136**	**100.00%**
Percent			**25.92%**	**19.05%**	**15.06%**	**13.46%**	**7.76%**	**6.11%**	**5.33%**	**2.61%**	**1.66%**	**1.42%**	**0.85%**	**0.77%**	**100.00%**	

(A) System disciplines not showing dedicated concentrations of direct work. This should not be interpreted that no direct work occurs. Some of the tasks involving the design discipline may have been identified, for the purposes of this analysis, as being dominated by another discipline. Proper interpretation is that these items have no identifiable direct work concentrations relative to the total annual direct work content.

(B) A large contributor to the work accumulated in cell (C5) of the table is due to Orbiter/OPF purge circuit monitoring. While the task-hours accumulated are large, it should be noted that this work involves very little labor.

5.2.2.3 Orbiter Unplanned Troubleshooting and Repair

The highest concentration of Orbiter turnaround work was found in the Unplanned Troubleshooting and Repair category, with 66,018 task-hours accumulated for the eight flows examined. In addition to the 30-percent contribution of thermal protection tile work on the vehicle (see Figure 21), troubleshooting and replacement of system components, or line replaceable units (LRUs), also contributed quite heavily (22 percent). Half of this work originated with the power management components; while propulsion, avionics, and other systems contributed to the other half of the work required to troubleshoot and replace faulty hardware.

Five other categories were identified as major contributors to unplanned troubleshooting and repair work: structural repair and refurbishment; troubleshooting, retest and repair of electrical cables and connectors; troubleshooting and repair of various mechanisms and thermal/pressure seals; troubleshooting and repair of ground equipment; and troubleshooting and repair of various plumbing systems.

Figure 21. Thermal Protection Repair Activity Around Orbiter Elevon Aerosurfaces

5.2.2.4 Orbiter Servicing

The routine servicing of the Orbiter's fluid and mechanical systems between flights also accumulated a very high level of work content (21 percent), with 46,570 task-hours accumulated across the eight flows analyzed.

This work includes routine Orbiter cleaning and fluid filter servicing (28 percent of the servicing work). Dominating this work were the continual RCS hypergolic thruster desiccant inspections and routine forward, mid, and aft compartment cleaning operations made necessary by the large technician traffic driven by an equally large volume of intrusive maintenance required by the design.

This category also accumulates work involved in applying, monitoring, and controlling vehicle power and establishing data and health management services (16 percent). For the thermal protection systems (TPS), tile and blanket insulation waterproofing work was classified as a flight element servicing function (10.5 percent), as were routine TPS mold-line penetration and leading-edge/nose cap and aerosurface hinge-line servicing and rework (17 percent). Other contributors included servicing of flight crew systems, routine ordnance handling and installation, mechanical servicing of the landing gear mechanisms, and routine replacement of expendable and limited-life items.

Also, because of the numerous operational fluids designed into the Orbiter (see 7.2.3), the following operations were included in the flight element systems servicing function: draining of fluid systems, deservicing, servicing, fill, and pressurizations—many in the Orbiter's aft compartment (Figure 22). These operations accumulated over 5,600 task-hours of work in the OPF as identified in the analysis.

5.2.2.5 Orbiter Turnaround Inspection and Checkout

The Orbiter airframe and critical systems undergo routine inspection and checkout (also referred to in this analysis as *functional verification*) between flights (Figure 23). This contributes to the overall flight element turnaround work content (13 percent), with 29,055 task-hours involving nearly 2,800 separately scheduled tasks accumulated across the eight flows analyzed.

There were three major subfunctions that contributed 97 percent of the functional verification work. The first was inspection operations associated with verifying the structural integrity of the Orbiter's airframe and mechanisms. This includes inspections for cracks, dings, corrosion, binding, excessive wear, and other structural damage or degradation. Thesubfunction of inspections contributed over 42 percent of the functional verification work. The second major subfunction, active component and subsystem checks (exclusive of power-up checks and switch list verifications), contributed another 31 percent of this work. Verifying the functional integrity of vehicle propellant, fluid, and gas systems (primarily for leaks and contamination) was the third major subfunction and accounted for nearly 24 percent.

Figure 22. Many Servicing Actions Are Concentrated in the Orbiter Aft Compartment

Figure 23. Technicians Conducting Inspections and Checkout in the OPF

5.2.2.6 Vehicle Payload Accommodations Turnaround

One of the most complex operational functions involves teardown of the Orbiter's previous flight's payload accommodations and construction of the payload accommodations for the next flight (Figure 24). A great deal of mission-to-mission customization of the payload bay has been required to meet the Shuttle Program's manifest. An alternative approach might dedicate an Orbiter to a minimum set of standard and easy-to-attach payload accommodations. Such an "express" Orbiter could optimize payload throughput capability and thus create an "express lane" for certain heavy cargo volume customers. While not directly surfacing the critical path nature of these operations,[22] the analysis found that vehicle payload accommodations turnaround contributed over 8 percent of the overall flight element turnaround work content, with over 18,600 task-hours involving over 1,600 separately scheduled tasks accumulated across the eight flows analyzed.

No one subfunction dominated, but the labor included: reconfiguration of Orbiter-supplied electrical and information services; troubleshooting and repair of leaks, payload fluid/pneumatic system decontamination and cleaning; reconfiguration of unique payload thermal protection insulation; personnel access and lighting setups and teardowns; payload accommodations (Orbiter payload bay) cleaning and closeout; repositioning of Orbiter fixed mechanisms (such as antennas, hatches, radiators, and payload bay doors); reconfiguration of Orbiter-supplied fluid services; and unplanned troubleshooting and repair or payload accommodations.

It should be emphasized that the physical installation, removal, and integrated testing of the payloads with the Orbiter's payload bay are not accounted for in this function (see 5.2.5). This function is for the Orbiter's payload accommodations—not the payload operations themselves.

5.2.2.7 Other Orbiter Turnaround Work

A large numerical contribution to the total turnaround task-hours in the data set (about 29,000 task-hours) was attributed to "Processing Support Systems and Functions," for about 13 percent of the activity. The large number of task–hours in this category was Orbiter purge circuit monitoring. The Orbiter has several purge circuits to control temperature and humidity to protect vehicle equipment from corrosion, as well as for technician comfort. This was the one area identified by the analysis team where the approach of using the available task-hours, and not the labor-hours associated with each task, displayed a weakness. The amount of "work" involved in performing the Orbiter purge circuit monitoring is actually rather small compared to the average labor required per task-hour. It routinely appears on the schedules to show when the purge circuit is required to be up and functional, with infrequent technician monitoring required. Other important work that accumulates in this category is system operation required to support another system's maintenance and repair. For example, complex and labor-intensive hydraulics and flight control system operations are required for landing gear extensions and retractions, main engine positioning, and aerosurface positioning for propulsion and tile work. While Table 8 does not attempt to adjust for this, one could factor out the monitoring (cell ID C5 in Table 8), making the total of the "Processing Support Systems and Functions" closer to 16,000 task-hours. The

[22] C. Guidi, pp. 56–58.

analysis team was not able to identify other obvious examples, such as scheduled purge circuit monitoring, where the cumulative task-hours did not appear to reflect real-world work content.

Figure 24. Customized Orbiter Payload Accommodations are the Norm

In addition, over 10,000 task-hours were concentrated around OPF facility and equipment maintenance, including both maintenance and occasional modifications and process changes for the ground support systems. Examples of this type of work include weather protection operations for hydraulic service units, facility maintenance outages for oxygen analyzers, 400-cycle ac power converters, Firex alarms, lighting, air-handling units, hypergolic and ammonia servicing systems, hazardous waste removal, and so forth.

Next highest on the list of turnaround functions involves preparing the vehicle for rollout following the numerous maintenance actions that have destroyed the previous flight's certified configuration. This comprises the removal of access, ground umbilicals, and ground-provided services and closeouts and accounted for over 8,400 task-hours of work for a 3.8 percent overall contribution to turnaround work. Much of this work involves installation of flight panels, aerosurface flipper doors, hinge line cove seals, access doors, and avionics bay panels and the closeout of thermal/pressure seals for each of these. This also includes the final weight and center-of-gravity determination and transfer operations to the Vehicle Assembly Building for vehicle integration.

Modifications and special tests accounted for 3 percent of the turnaround work content, while vehicle positioning for roll-in to the OPF, connection to facility and equipment services, access installation, and protection accounted for another 2 percent.

The important turnaround function of vehicle element safing, while only accounting for a little over 1 percent of the task-hours, nevertheless clears the facility of personnel and locks out other work. In addition, these operations draw on expensive-to-maintain Self-Contained Atmospheric Protection Ensemble, or SCAPE suit, infrastructure

Finally, about 0.7 percent of the turnaround work content is devoted to facility preparations.

5.2.3 Overall Launch Operations Work Content

Having reviewed the flight element turnaround and the flight element assembly functions, launch operations functions (Figure 25) accumulated the third highest level of direct vehicle work (over 67,700 task-hours), contributing about 17 percent to the overall level of direct work. Not all of this work involves the preparation for launch but also includes the work required at the launch facility (the "pad") to restore the fixed facilities, mobile launcher, and other equipment to a certifiable condition for the next set of launch activities.

5.2.3.1 Launch Operations Work Summary

A breakdown of the results of the launch processing review is summarized in Table 9. Similar to previous tables, it shows the concentration of work required at the launch pad facility and for activities leading to launch, as well as postlaunch activities and prelaunch preparations by design discipline.

Figure 25. Shuttle Vehicle on its Mobile Launcher Platform (MLP) at Launch Pad 39

Table 9. Summary of Space Shuttle Launch Operations Work Content

Concentrtion of Space Shuttle Launch Work Content
for 1997 Launches
(As measured by cumulative task-hours for eight processing flows)

Generic Operations Functions	FBS	ID	1 Ground Systems & Facilities	2 Liquid Propulsion	3 Safety Management & Control	4 Structures , Mechanisms, Veh Handling	5 Cmd, Ctl & Health Mngmt	6 Power Management	7 Payload Accommodations	8 Environmental Ctl & Life Spt	9 Cockpit & Crew Cabin	10 Guid, Nav & Ctl	11 Thermal Management	12 Communications	Grand Total	Percentage
Facility/GSE,Refurbishment, Reservicing & Preps for Vehicle Arrival	5.01	A	7,921	3,186	4,617	4,992	480	913	985	159	4	(A)	40	(A)	**23,297**	**35.96%**
Functional Verif of Flt/Grd Systems for Launch	5.03	B	842	2,962	800	116	987	2,015	16	88	35	223	(A)	(A)	**8,084**	**12.48%**
Launch Facility and Equipment Periodic Maintenance	5.11	C	1,655	2,216	1,320	494	587	600	96	336	(A)	(A)	(A)	(A)	**7,304**	**11.27%**
Vehicle System Servicing at Launch Point	5.06	D	45	2,888	2,560	576	104	524	(A)	(A)	(A)	50	(A)	(A)	**6,747**	**10.41%**
Position Vehicle at Launch Point, Mate to Ground Facility and Services & Provide Access	5.02	E	4,224	157	0	90	202	176	(A)	(A)	(A)	(A)	(A)	(A)	**4,849**	**7.48%**
Post-Launch Securing	5.10	F	3,499	17	16	493	309	8	(A)	(A)	(A)	(A)	(A)	(A)	**4,342**	**6.70%**
Launch Systems Unplanned Troubleshooting/Repair	5.04	G	1,073	992	100	243	240	122	40	44	8	16	4	(A)	**2,882**	**4.45%**
Launch the Vehicle	5.09	H	125	(A)	16	200	2,348	18	(A)	(A)	17	(A)	(A)	(A)	**2,724**	**4.20%**
Access/Umbilical Removal & Closeout for Flight	5.07	I	4	128	(A)	1,666	(A)	156	452	(A)	(A)	(A)	(A)	(A)	**2,406**	**3.71%**
Crew Ingress, Late Payload Stowage & Functional Verification for Flight	5.05	J	(A)	(A)	(A)	(A)	32	(A)	1,252	(A)	346	(A)	(A)	(A)	**1,630**	**2.52%**
Remote Automated Servicing and Propellant Loading	5.08	K	(A)	419	(A)	(A)	(A)	106	(A)	(A)	(A)	(A)	(A)	(A)	**525**	**0.81%**
Total (Task-hours)			**19,388**	**12,965**	**9,429**	**8,870**	**5,289**	**4,638**	**2,841**	**627**	**410**	**289**	**44**	**(A)**	**64,790**	**100.00%**
Percent			**29.92%**	**20.01%**	**14.55%**	**13.69%**	**8.16%**	**7.16%**	**4.38%**	**0.97%**	**0.63%**	**0.45%**	**0.07%**	**(A)**	**100.00%**	

(A) System disciplines not showing dedicated concentrations of direct work. This should not be interpreted that no direct work occurs. Some of the tasks involving the design discipline may have been identified, for the purposes of this analysis, as being dominated by another discipline. Proper interpretation is that these items have no identifiable direct work concentrations relative to the total annual direct work content.

5.2.3.2 Launch Facility Refurbishment, Reservicing, and Preparation for Vehicle Arrival

This set of activities includes the restoration of Pad 39 and the Mobile Launcher Platform (MLP) and was the largest contributor to the launch operations function (FBS 5.0). The work content exceeded 29,000 task-hours or nearly 43 percent of the launch work content. Ground systems and equipment are verified and controlled for cleanliness, propellant compatibility, and functionality prior to use on flight systems for launch operations. High-maintenance areas include the MLP Tail Service Masts, or TSMs (Figure 26); Water Deluge System elements; SRB holddown posts; and data acquisition systems. Background on the MLPs is provided below.[23]

Three MLPs are part of the Space Shuttle system (MLP-1, MLP-2, and MLP-3). These three units were used in the Apollo–Saturn operations and were modified for use in Space Shuttle operations. With cranes, umbilical towers, and swing arms removed, the redesigned Saturn/Apollo-era Launch Umbilical Towers (LUTs) were redesignated as MLPs. In place of one large exhaust hole in the platform, three smaller openings accommodate the hot exhaust gases from the Solid Rocket Boosters and the Orbiter. Segments of the dismantled Saturn-era LUTs became part of the permanent installation at the launch pad, where they serve as sections of the Fixed Service Structure (FSS). (With only two launch pads, the third tower was cut into 20-ft sections and currently is mothballed in the KSC Industrial Area). Each platform is nearly identical and is a two-story steel structure measuring 7.6 m (25 ft) high, 49 m (160 ft) long, and 41 m (135 ft) wide. Each weighs approximately 4.19 kg (9.25 million lb). When carrying an unfueled Shuttle vehicle, the system weighs 5.45 million kg (12.02 million lb), and 6.22 million kg (13.72 million lb) with a fully fueled vehicle. The system is supported by six fixed steel pedestals 7 m (22 ft) high, whether in the VAB, on the launch pad, or at the MLP park sites.

---

[23] "Countdown! NASA Space Shuttles and Facilities," IS-2003-09-006-KSC, NASA Information Summaries, September 2003.

Figure 26. MLP at Pad (TSMs seen at top)

At the surface level (MLP zero-ft level), there are two TSMs, one on each side of the main engine's exhaust hole. The masts provide several umbilical connections to the Orbiter, including a liquid-oxygen line through one and a liquid-hydrogen line through another. These cryogenic propellants feed into the ET from the pad tanks via these connections. At launch, the umbilicals pull away from the Orbiter and retract into the TSMs, where protective hoods rotate closed to shield them from the exhaust flames. Each TSM assembly is 4.6-m (15-ft)-long and 2.7-m (9-ft)-wide and rises 9.4 m (31 ft) above the MLP deck. Other umbilicals carry helium and nitrogen, as well as ground electrical power and communications links.

The turnaround process for MLPs includes a postlaunch securing at the pad with a median stay time of 31 calendar days (minimum of 2 days and maximum of 61 days). This may be followed by MLP maintenance and overhaul at the MLP park site located north of the VAB at Launch Complex 39, or the MLP may be directly transported to the VAB for SRB stacking operations. The park site stays average 31 calendar days, with a minimum of 5 days and a maximum of 175 days. The MLP stacking preparations in the VAB average 17 calendar days, with a minimum of 8 days and a maximum of 45 days. The Mobile Launcher TSMs that feed propellants and other services to the T-0 umbilicals on the Orbiter require a great deal of refurbishment (Figure 27), as does the Permanent Measurement System (PMS).

The PMS takes approximately 800 measurements throughout two launch complexes and in the VAB. NASA uses the system prior to and during launch to ensure launch readiness and safety.

The PMS supports multiple parts of the launch process. During the stacking of the SRBs on the MLP, the system monitors the strain on the booster holddown posts. The posts bear the load of the entire vehicle until it is launched. The strain measurements allow engineers to verify that the bearing assembly—where the booster meets the post—does not bind as each successive booster segment is stacked. After stacking and testing are completed, the vehicle and MLP are moved to the launch pad for final preparations. During countdown, the system takes many different types of measurements. Some, such as liquid-oxygen pump vibrations, are monitored in real time. Others, including pressure, temperature, vibration, displacement, and strain, are recorded at launch time for later analysis.

Figure 27. MLP Propellant Line Repair and Routine Sandblasting

5.2.3.3 Launch Facility and Equipment Periodic Maintenance

Periodic maintenance of the launch pad is required over time. In fact, just as Orbiter vehicles periodically are taken out of the flight manifest for long-term inspections and modification, so too are the launch pads to ensure facility and equipment durability against repeated launch exhaust environment, moisture, and weather exposure.

The analysis revealed that 11.6 percent of the launch operations work (i.e., FBS 5.0) was devoted to this function and accumulated over 7,800 task-hours for the eight processing flows analyzed.

In addition to the water deluge system (Figure 28), labor-intensive periodic maintenance is required for the Orbiter Access Arm (OAA), changeout of flex hoses, maintenance on the Hydraulic Charger Unit (HCU), cryogenic pumps, fire protection systems, industrial power systems, cooling units, and so forth.

Figure 28. Launch System Undergoes an All-Up Water Deluge System Test

5.2.3.4 Vehicle System Servicing at Launch Point

In addition to the main cryogenic propellant loading, the Space Shuttle Vehicle requires a variety of toxic, hazardous propellants to be loaded and systems serviced for launch, and various systems require ordnance to be installed at the launch point. This type of work accumulated over 6,700 tasks-hours, or about 10 percent of the launch activities, and involved 275 separately scheduled actions across the eight flows analyzed.

The hypergolic fuel propellant loading accumulated the most work in this category. Not only is this work labor-intensive, but it also requires the launch pad to be cleared of personnel performing other prelaunch work (such as payload preparations and other vehicle closeouts). The work is coordinated and managed from the remote Launch Control Center (LCC) firing rooms, with local servicing being performed by technicians in special protective gear (Figure 29).

In addition, local manual servicing for auxiliary power units for the SRBs, gaseous oxygen and nitrogen high-pressure supply servicing for breathing air, and work involving ground servicing equipment at the launch pad are required.

Figure 29. Local Servicing of Shuttle Systems at Pad Is Often Hazardous

5.2.3.5 Functional Verification of Flight and Ground Systems for Launch

The next highest accumulation of direct work involves functionally verifying that the many services, supplies, and support systems, as well as the many safing, hazardous gas detection, emergency power, and egress systems, are all functional following the previous launch. This work amounted to 9.3 percent of the work, totally over 6,300 task-hours across 350 separately required tasks.

Some of this work includes on-pad, on-vehicle inspections, and remote checkouts performed from a Firing Room in the LCC.

5.2.3.6 Vehicle Positioning at Launch Point, Mate to Ground Facility and Services, and Provide Access

Also accumulating a fair amount of work at the pad are transporting the mated vehicle to the pad (Figure 30), mating the vehicle and its MLP to the pad facility, and providing the personnel access that is required to perform the other functions. This work amounted to 7 percent of the total launch activity, or about 4,800 task-hours across nearly 400 separate operations.

Figure 30. Space Shuttle Vehicle Transfer Operation

5.2.3.7 Launch Systems Unplanned Troubleshooting and Repair

As with all the processing functions, unplanned troubleshooting and repair work accumulates. Since this processing function is essentially at the "end of the line," it is not surprising that this only accumulated 5.5 percent of the total launch work. This includes flight and ground systems component troubleshooting, removal and replacement, and ground launch structural systems repair.

5.2.3.8 Postlaunch Securing

This activity, which secures the launch facility following liftoff, required 3,300 task-hours, across the eight launches analyzed, or nearly 5 percent of the total launch work content. This work included postlaunch system safing operations and postlaunch inspection of facilities and equipment.

5.2.3.9 Access and Umbilical Removal and Closeout for Launch

This work involves removal of access equipment, such as temporary access stands, lighting, and scaffolding, and also major equipment, such as the large Rotating Service Structure (RSS) and engine service platforms (ESPs). In addition, servicing equipment for various systems is disconnected and stowed (Figure 31). Finally, all "remove before flight" items are removed, and all flight panels and final vehicle and ground launch system "closeouts" are performed. This work amounted to 3.5 percent, with over 2,400 task-hours associated with nearly 120 separately scheduled items across the eight flows analyzed.

Figure 31. Closeout Crew Member in the White Room with STS–111 Crew Members

5.2.3.10 Crew Ingress, Late Payload Stowage, and Functional Verification for Flight

Ingress of the flight crew, late stowage of payload, final checks, and any unplanned troubleshooting and repair for the payload accumulated over 1,600 task-hours across eight launches, or about 2.4 percent of the launch work.

5.2.3.11 Launching the Vehicle

Performing the launch countdown procedure for the Space Shuttle amounted to less than 2 percent of the total launch work. A highly complex procedure requiring a number of days to perform, this function is, of course, the final milestone of spaceflight production.

5.2.3.12 Remote Automated Servicing and Propellant Loading

This activity includes the cryogenic loading of the Shuttle Vehicle's main propellant system. The cryogenic loading operation, conducted as part of launch countdown procedure, commences about 8 hours prior to liftoff. The ET is loaded with about 500,000 gallons of cryogenic propellant consumed by the Shuttle's main engines. The liquid oxygen is transferred by ground pumps capable of pumping 1,300 gallons per minute (i.e., is "pump-fed"). The liquid-hydrogen fuel vaporizes and is transferred to the ET using pressure created by the hydrogen itself, or in other words, is "pressure-fed." Pumps are not needed for the liquid-hydrogen loading Figure 32.

Figure 32. Complex 39 Liquid-Hydrogen Storage Facility

5.2.4 Overall Flight Element Assembly Work Content

The next highest category of direct work content, though not as visible and apparent, is the assembly work to build up a flight element, including the installation of various subsystems onto the flight elements (14 percent). To produce a year's worth of flights in 1997, this involved scheduling over 3,700 tasks, totaling nearly 70,800 task-hours.

Most of the assembly work involves the construction of the left and right SRBs, made up of four solid rocket motor segments, a forward and aft skirt assembly, and attachment and verification of various systems tunnels, range safety equipment, and separation ordnance. The SRB assembly work begins at the RPSF at Launch Complex 39 (Figure 33).

Also included were such operations as auxiliary propulsion pod removals and reinstallations with the Orbiter flight element, the routine removal and reinstallation of the Space Shuttle Main Engines, and the installation and removal of the Orbiter's Remote Manipulator System (RMS) robotic arm. In addition, some relatively minor assembly work was accumulated for the ET standalone checkout activities.

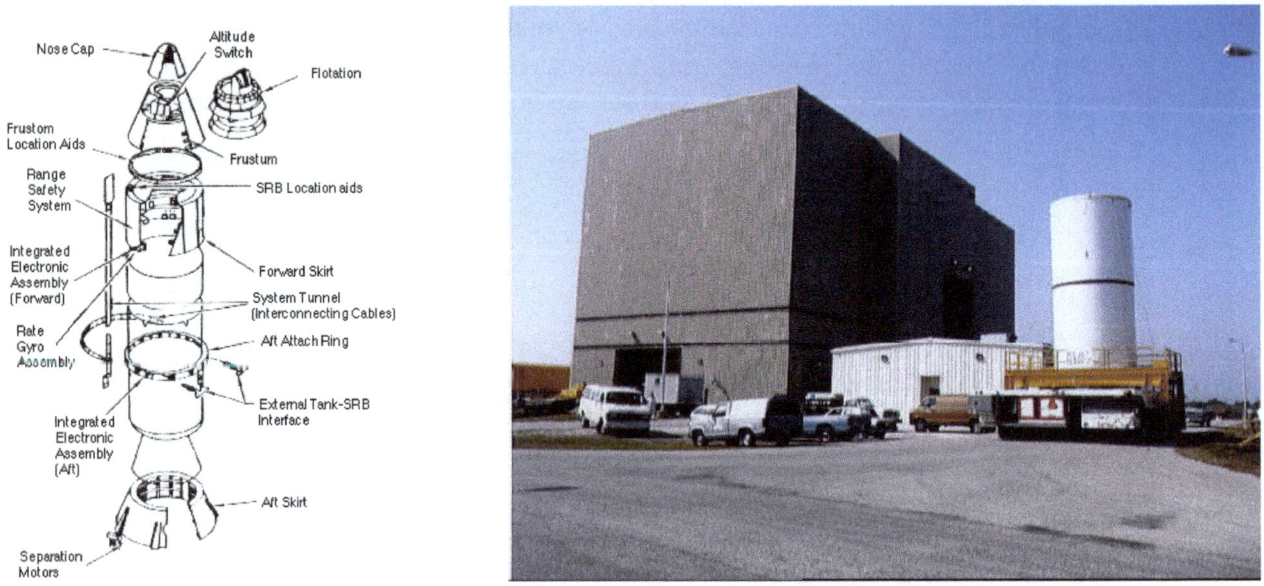

Figure 33. SRB Segment Assembly at the Rotation, Processing, and Surge Facility (RPSF)

5.2.4.1 Flight Element Assembly Work Content Summary

The results of the flight element assembly review are summarized in Table 10. Similar to previous tables, it shows the concentration of work required to assemble the various Space Shuttle flight elements by design discipline.

Table 10. Summary of Flight Element Assembly for Eight Shuttle Processing Flows

Concentrtion of Space Shuttle Flight Element Assembly Work Content for 1997 Launches (As measured by cumulative task-hours for eight processing flows)	FBS	ID	Generic Design Functions												Grand Total	Percentage
			Structures , Mechanisms, Veh Handling 1	Liquid Propulsion 2	Cmd, Ctl & Health Mngmt 3	Thermal Management 4	Ground Systems & Facilities 5	Power Management 6	Safety Management & Control 7	Guid, Nav & Ctl 8	Communications 9	Environmental Ctl & Life Spt 10	Payload Accommodations 11	Cockpit & Crew Cabin 12		
Generic Operations Functions																
Subsystem / Component Installations	3.03	A	29,175	4,082	2,497	1,890	52	271	227	301	285	221	115	--	**39,116**	**68.41%**
Element Structural Assembly	3.02	B	5,681	372	(A)	(A)	(A)	(A)	(A)	(A)	(A)	(A)	(A)	--	**6,053**	**10.59%**
Flight Element Assembly Functional Verifications	3.04	C	1,038	1,586	593	(A)	(A)	156	(A)	(A)	7	(A)	(A)	--	**3,380**	**5.91%**
Element Transport, Lift and Handling for Assembly	3.01	D	1,252	1,259	(A)	(A)	272	(A)	(A)	(A)	(A)	(A)	(A)	--	**2,783**	**4.87%**
Assembly Unplanned Troubleshooting & Repair	3.05	E	1,197	16	253	708	62	(A)	90	(A)	(A)	(A)	(A)	--	**2,326**	**4.07%**
Routine Reusable Flight Element Module/Component Disassembly	3.08	F	163	1,730	(A)	5	(A)	(A)	(A)	(A)	(A)	(A)	58	--	**1,956**	**3.42%**
Post-Assembly Operations	3.06	G	1,212	40	(A)	88	227	(A)	(A)	(A)	(A)	(A)	(A)	--	**1,567**	**2.74%**
Total (Task-Hours)			**39,718**	**9,085**	**3,343**	**2,691**	**613**	**427**	**317**	**301**	**292**	**221**	**173**	**(A)**	**57,181**	**100.00%**
Percent			**69.46%**	**15.89%**	**5.85%**	**4.71%**	**1.07%**	**0.75%**	**0.55%**	**0.53%**	**0.51%**	**0.39%**	**0.30%**	**(A)**	**100.00%**	

(A) System disciplines not showing dedicated concentrations of direct work. This should not be interpreted that no direct work occurs. Some of the tasks involving the design discipline may have been identified, for the purposes of this analysis, as being

5.2.4.2 Subsystem/Component Installations

The installation of various subsystems and components, particularly for the SRBs, was the most labor-intensive flight element assembly function. Subsystem/component installations led to over 39,100 identified task-hours, or 55 percent of the flight element assembly work. For example, in the left-hand photo in Figure 34, SRB technicians work on the cover of the SRB systems tunnel. The systems tunnel contains about 200 electrical/electronic cables carrying power, command, and instrumentation signals.

5.2.4.3 Element Structural Assembly

The most visible function associated with flight element assembly is the mating of various segments and modules. The work accumulated during the eight flows analyzed amounted to 26 percent of the flight element assembly work, or over 18,400 task-hours. This analysis included the mating of SRB segments, installation of OMS/RCS pods, SSMEs and the RMS robotic arms on the Orbiter.

5.2.4.4 Flight Element Design Modification

The design modification and process control change work amounted to about 5 percent of the flight element assembly work. This type of work can be an indication of an unstable design searching for a stable production process and can also represent continual improvement changes. Stability in process cycle time and general downward trends in total work content, however, were not analyzed in this effort.

Figure 34. Example SRB Subsystem Installations and Unplanned Work

5.2.4.5 Assembly Unplanned Troubleshooting and Repair

The complexity of the assembly operations can lead to some unplanned troubleshooting, rework, refit, repair, and reverification. This flight element assembly function contributed about 4.2 percent, or nearly 3,000 task-hours, of the assembly work. Again referencing Figure 34, the right-hand photo shows United Space Alliance (USA) and NASA personnel inspecting a range safety cable between the ET and SRBs on Space Shuttle *Discovery* in 1999. The cable, which relays a redundant emergency destruction signal between the SRBs in the unlikely event of a contingency, was damaged during closeout operations and was being replaced.

5.2.4.6 Element Transport, Lift, and Handling for Assembly

In support of the assembly activity, transporter operations, lifting, sling handling, crane operations, and so forth are required (Figure 35). This type of work accounted for 4.2 percent of the assembly activity.

Figure 35. SRB Forward Skirt Assembly and External Tank Transfer

5.2.4.7 Routine Reusable Flight Element Module/Component Disassembly

A category was needed for routine disassembly of flight elements, such as SSMEs, OMS/RCS pods, and even occasional SRB destacking, which was encountered in the analysis of STS–81 data (Figure 36). This activity accounted for nearly 3 percent of the overall flight element assembly work.

Figure 36. Routine Orbiter Assembly/Disassembly Operations (OMS Pod, Remote Arm, and SSME)

5.2.4.8 Postassembly Operations

The work required following assembly operations, such as stowing lifting and handling equipment and repositioning access stands and equipment, made up the balance of the assembly operations at 2.2 percent.

5.2.5 Overall Vehicle Integration Work Content

Space Shuttle Vehicle integration takes up about 7.4 percent of the total work content. Shuttle Vehicle integration work accumulated about 29,400 task-hours, involving approximately 1,600 separately scheduled operations over the eight processing flows examined. The integration processes include integration of the boosters to the MLP, the tank to the SRBs, and the Orbiter to the

ET. It also includes the integration and mating of the payload to the Orbiter, whether it occurs in the OPF or at the pad.

5.2.5.1 Vehicle Integration Work Summary

The results of the Space Shuttle Vehicle integration processing review is summarized in Table 11. Similar to previous tables, it shows the concentration of work required at the launch pad facility and for activities leading to launch, as well as postlaunch activities and prelaunch preparations by design discipline.

5.2.5.2 Element Transportation, Handling, Lifting, and Mating

This function accumulated over 19,400 task-hours, for 66 percent of the vehicle integration activity across eight processing flows. This work includes facility preparations for flight element integration operations, which involves setting up personnel access platforms, spotting support equipment, and ground interface alignments. Also included is the work involved with the ground transportation of the flight element, including operation of ground transporters and lifting, positioning, and mating operations (Figure 38). This includes crane operations and attachment of lifting slings, hydrasets, and other handling hardware. Finally, this activity includes postmate servicing and closeout of element-to-element interfaces and interfaces between the MLP and the vehicle stack.

5.2.5.3 Configure Systems for Vehicle Transfer to Launch Point

The next highest accumulation of work for vehicle integration involves disconnecting and stowing assembly and mating ground hardware and configuring the vehicle stack and MLP for transfer to the pad. This work accounted for nearly 14 percent of the total vehicle integration work volume, with nearly 4,000 task-hours accumulated across the eight processing flows analyzed.

5.2.5.4 Postmating Functional Verification

The inspections and checkout of the newly mated flight elements accounted for almost 11 percent of the integration work content, accumulating 3,100 task-hours in the eight flows analyzed.

This work includes verifying the interfaces of the flight element to ground electrical and fluid system, as well as structural and mechanical integrity. Also included is the verification of interfaces between flight elements. These verifications encompass electrical, fluid, and structural integrity checks and inspections.

5.2.5.5 Integration of Payload and Cargo

This category encompasses physically attaching and mating the payload element to the flight element. While most of the work in the vehicle integration category takes place in the VAB, the payload integration work may be performed either in the horizontal orientation in the OPF, or vertically at the launch pad (Figure 37 and Figure 38). This work amounted to about 5.7 percent of the integration work content.

This work includes facility preparations for payload installation (normally similar to clean-room operations), lifting and handling operations for payload installation into the vehicle, ground handling and transport operations, and payload inspections and testing.

It should be noted that the work involved with the Orbiter vehicle's payload accommodations (including the payload bay and payload flight deck) are accounted for in the turnaround (FBS 2.0) functions.

Figure 37. Payload Canister Installation at Pad-A and Space Station Cargo Installation Into *Endeavour* With Payload Ground Handling Mechanism (PGHM)

5.2.5.6 Integration Unplanned Troubleshooting and Repair

Unplanned troubleshooting and repair accounted for 3.9 percent of the vehicle integration work, accumulating over 1,100 task-hours across the eight processing flows analyzed.

5.2.5.7 Integration Facility and Equipment Periodic Maintenance

Very little integration facility and equipment periodic maintenance was observed as interfering with the direct work data set. Unlike other critical-path facilities, such as the OPF, most of the VAB facility maintenance work does not appear on the daily schedules for direct vehicle processing work and was not included in the data set. Eight task-hours proof-load testing a crossover platform was noted, however.

Table 11. Summary of Eight Flows of Vehicle Integration Work

Concentrtion of Space Shuttle Vehicle Integration Work Content for 1997 Launches (As measured by cumulative task-hours for eight processing flows)		FBS	ID	Generic Design Functions													
				1 Structures , Mechanisms, Veh Handling	2 Safety Management & Control	3 Liquid Propulsion	4 Power Management	5 Ground Systems & Facilities	6 Payload Accommodations	7 Cmd, Ctl & Health Mngmt	8 Thermal Management	9 Environmental Ctl & Life Spt	10 Cockpit & Crew Cabin	11 Guidance, Navigation and Control	12 Communications	**Grand Total**	**Percentage**
Generic Operations Functions	Element Transportation, Handling, Lift and Mate	4.01	A	20,202	5,993	3,898	4,293	2,989	88	569	1,020	57	(A)	(A)	(A)	**39,109**	**85.14%**
	Post-Mating Functional Verification	4.02	B	402	58	1,464	300	(A)	(A)	772	(A)	(A)	(A)	(A)	(A)	**2,996**	**6.52%**
	Perform Integration of Payload/Cargo	4.04	C	268	(A)	(A)	(A)	(A)	1,260	(A)	(A)	(A)	40	(A)	(A)	**1,568**	**3.41%**
	Integration Unplanned Troubleshooting & Repair	4.03	D	1,297	(A)	8	136	38	(A)	(A)	23	(A)	(A)	(A)	(A)	**1,502**	**3.27%**
	Configure Systems for Vehicle Transfer to Launch Point	4.05	E	354	(A)	(A)	198	8	(A)	(A)	201	(A)	(A)	(A)	(A)	**761**	**1.66%**
	Total (Task-hours)			**22,523**	**6,051**	**5,370**	**4,927**	**3,035**	**1,348**	**1,341**	**1,244**	**57**	**40**	**(A)**	**(A)**	**45,936**	**100.00%**
	Percent			**49.03%**	**13.17%**	**11.69%**	**10.73%**	**6.61%**	**2.93%**	**2.92%**	**2.71%**	**0.12%**	**0.09%**	**(A)**	**(A)**	**100.00%**	

Figure 38. Orbiter *Atlantis* Mated in the VAB for Its STS–112 Launch

5.3 Work Concentrations Across Space Shuttle Vehicle and Ground Systems

This section describes the direct work content distributed across various Space Shuttle vehicle and ground system design elements. Table 12 and Table 13 are similar to Table 6 in that the direct work is allocated to generic design disciplines and generic operations functions, but also provide insight as to which vehicle element or ground system element the work is devoted.

Table 12. Direct Work Content By Shuttle System Element and Design Discipline

Concentration of Space Shuttle Direct Operations by System Element for 1997 Launches (As measured by cumulative task-hours for eight processing flows)			GENERIC DESIGN DISCIPLINE													
			Structures, Mechanisms and Vehicle Handling	Liquid Propulsion	Thermal Management	Power Management	Safety Management & Control	Ground Interfacing Systems and Facilities	Payload Accommodations	Environmental Control and Life Support	Command, Control & Health Management	Communications	Guid, Nav & Ctl	Cockpit & Crew Cabin	Total (Task-hours)	Percentage
		ID	1	2	3	4	5	6	7	8	9	10	11	12		
STS FLIGHT SYSTEM TURNAROUND OPS	ORBITER OPS	A	62,266	44,973	42,825	32,727	17,486	1297	12,361	13,921	6,047	3,451	2,362	2,104	241,820	**60.67%**
	SOLID ROCKET BOOSTER ASSY/OPS	B	45,583	385	294	1,223	(A)	(A)	(A)	(A)	3,412	(A)	116	(A)	51,013	**12.80%**
	EXTERNAL TANK OPS	C	3,558	2,725	3,288	156	261	(A)	(A)	(A)	40	(A)	(A)	(A)	10,028	**2.52%**
	SPACE SHUTTLE VEHICLE OPS	D	1,066	2,275	(A)	377	246	140	28	(A)	2,593	(A)	(A)	35	6,760	**1.70%**
	PAYLOAD INTEGRATION & LAUNCH	E	294	(A)	(A)	(A)	(A)	725	2,414	(A)	32	(A)	(A)	40	3,505	**0.88%**
	FLT SYS Sub Total		112,767	50,358	46,407	34,483	17,993	2,162	14,803	13,921	12,124	3,451	2,478	2,179	**313,126**	**78.56%**
	FLT SYS Percent		36.01%	16.08%	14.82%	11.01%	5.75%	0.69%	4.73%	4.45%	3.87%	1.10%	0.79%	0.70%		
STS GROUND FACILITY/ELEMENT OPS	MOBILE LAUNCH PLATFORM (MLP) TURNAROUND/PREPS	F	21,529	5,094	186	4,182	8,600	12,221	272	393	1,307	(A)	(A)	(A)	53,784	**13.49%**
	PAD TURNAROUND	G	16	7,132	(A)	1,406	6,522	7,581	1,474	223	277	(A)	(A)	(A)	24,631	**6.18%**
	OPF FACILITY OPS	H	(A)	(A)	(A)	(A)	(A)	5,825	(A)	(A)	(A)	(A)	(A)	(A)	5,825	**1.46%**
	VAB FACILITYOPS	I	(A)	(A)	(A)	(A)	(A)	1,026	(A)	(A)	(A)	(A)	(A)	(A)	1,026	**0.26%**
	CT/CRAWLERWAY OPS	J	26	(A)	(A)	(A)	(A)	118	(A)	(A)	(A)	(A)	(A)	(A)	144	**0.04%**
	SHUTTLE LANDING FACILITY	K	(A)	(A)	(A)	(A)	(A)	28	(A)	(A)	(A)	(A)	(A)	(A)	28	**0.01%**
	GND FAC/ELEM Sub Total		21,571	12,226	186	5,588	15,122	26,799	1,746	616	1,584	--	--	--	**85,438**	**21.44%**
	GND FAC/ELEM Percent		25.25%	14.31%	0.22%	6.54%	17.70%	31.37%	2.04%	0.72%	1.85%	--	--	--		
Total (Task-hours)			134,338	62,584	46,593	40,071	33,115	28,961	16,549	14,537	13,708	3,451	2,478	2,179	**398,564**	**100.00%**
Percent			33.71%	15.70%	11.69%	10.05%	8.31%	7.27%	4.15%	3.65%	3.44%	0.87%	0.62%	0.55%	**100.00%**	

(A) System disciplines not showing dedicated concentrations of direct work. This should not be interpreted that no direct work occurs. Some of the tasks involving the design discipline may have been identified, for the purposes of this analysis, as being dominated by another discipline. Proper interpretation is that these items have no identifiable direct work concentrations relative to the total annual direct work content.

Table 13. Direct Work Content By Shuttle System Element and Generic Ops Function

Concentration of Space Shuttle Direct Operations by System Element for 1997 Launches (As measured by cumulative task-hours for eight processing flows)		ID	GENERIC OPERATIONS FUNCTION: Flight Element Turnaround/Checkout (FBS 2.0)	Launch (FBS 5.0)	Flight Element Assembly (FBS 3.0)	Vehicle Integration (FBS 4.0)	Flight Element Shipping, Receiving & Acceptance (FBS 6.0)	Landing & Recovery (FBS 1.0)	Total (Task-hours)	Percentage
			1	2	3	4	5	6		
STS FLIGHT SYSTEM TURNAROUND OPS	ORBITER OPS	A	217,287	10,919	9,909	2,697	953	55	241,820	**60.67%**
	SOLID ROCKET BOOSTER ASSY/OPS	B	(A)	590	41,358	5,406	3,659	(A)	51,013	**12.80%**
	EXTERNAL TANK OPS	C	(A)	351	5,301	1,941	2,435	(A)	10,028	**2.52%**
	SPACE SHUTTLE VEHICLE OPS	D	(A)	5,350	(A)	1,410	(A)	(A)	6,760	**1.70%**
	PAYLOAD INTEGRATION & LAUNCH	E	(A)	1,954	(A)	1,232	(A)	319	3,505	**0.88%**
	FLT SYS Sub Total		217,287	19,164	56,568	12,686	7,047	374	**313,126**	**78.56%**
	FLT SYS Percent		69.39%	6.12%	18.07%	4.05%	2.25%	0.12%		
STS GROUND FACILITY/ELEMENT OPS	MOBILE LAUNCH PLATFORM (MLP) TURNAROUND/PREPS	F	(A)	20,991	(A)	32,793	(A)	(A)	53,784	**13.49%**
	PAD TURNAROUND	G	24	24,363	(A)	244	(A)	(A)	24,631	**6.18%**
	OPF FACILITY OPS	H	5,825	(A)	(A)	(A)	(A)	(A)	5,825	**1.46%**
	VAB FACILITYOPS	I	(A)	128	613	213	72	(A)	1,026	**0.26%**
	CT/CRAWLERWAY OPS	J	(A)	144	(A)	(A)	(A)	(A)	144	**0.04%**
	SHUTTLE LANDING FACILITY	K	(A)	(A)	(A)	(A)	(A)	28	28	**0.01%**
	GND FAC/ELEM Sub Total		5,849	45,626	613	33,250	72	28	**85,438**	**21.44%**
	GND FAC/ELEM Percent		6.85%	53.40%	0.72%	38.92%	0.08%	0.03%		
	Total (Task-hours)		223,136	64,790	57,181	45,936	7,119	402	**398,564**	**100.00%**
	Percent		**55.98%**	**16.26%**	**14.35%**	**11.53%**	**1.79%**	**0.10%**	**100.00%**	

(A) System disciplines not showing dedicated concentrations of direct work. This should not be interpreted that no direct work occurs. Some of the tasks involving the design discipline may have been identified, for the purposes of this analysis, as being dominated by another discipline. Proper interpretation is that these items have no identifiable direct work concentrations relative to the total annual direct work content.

5.4 High-Concentration Operations Functions

It can be seen from the foregoing description of work that certain common types of work recur throughout the process. In particular, unplanned troubleshooting and repair, assembly, servicing, functional verification, and to a lesser extent, modifications and special tests were identified as major, common contributors to the overall workload. As a next step toward determining design root causes, these common functions are quantified from the analysis data and presented in detailed tables at one functional level down from the previous presentation. (While *flight element assembly* was covered in 5.2.4, this subsection will also explore assembly as a generic, cross-cutting operations function.) These high-concentration operations functions are presented in order of their direct work content contribution.

5.4.1 Unplanned Troubleshooting and Repair

In general, this class of work can be defined as the nonroutine activity necessary to recover from the discovery of out-of-specification conditions. The unplanned troubleshooting and repair activity totaled nearly 73,500 task-hours across the eight flows reviewed in the analysis. A summary of this work is provided in Table 14. The work includes initial problem isolation and troubleshooting, which may or may not lead to nonroutine access, component repair or replacement, and retest. Much of the unplanned work is normally concentrated on the Orbiter's turnaround operations. This is an important finding for all potential designers and operators of reusable spaceflight systems.

Unplanned work can involve major operations, such as destacking flight elements, or removing major flight element modules (e.g., SSME, OMS, and Forward Reaction Control System [FRCS]) pods. For example, in reviewing the results of the STS–81 flow, some unplanned disassembly and assembly work resulting from an SRB destack and restack was encountered.

More often encountered, though, are troubleshooting and repair of delicate thermal protection tiles, airframe structural repairs, and replacements of various flight components or line replaceable units (LRUs). These activities typically involve isolating the problem within a functional path of flight hardware and often require (1) gaining access to and deconfiguring previously certified flight connections and (2) installing and removing special protective "breakout boxes" for electrical functions and/or regulation panels (reg panels) for fluid and gas functions. The In-Flight Anomaly (IFA) List, normally cited as an indicator of component failure rates, was not used in this analysis. While the criteria for IFAs have fluctuated throughout the history of the program, the actual component replacements performed on the ground, as tracked by the Problem Reports in the Problem Reporting and Corrective Action (PRACA) System, are more indicative of component replacement rates and unplanned work.

Modern aircraft typically encounter this situation less than once per turnaround. A typical trend of unplanned Space Shuttle Orbiter component replacements is displayed in Figure 39 . It averages approximately 100 flight components per processing flow as measured by the number of LRU PRACA System. It is important to note that the trend data does not include the removal and replacement of limited-life components (of which there is an average of 300 per processing flow), nor the SSMEs, nor the acreage tile (Figure 39).

Table 14. Detailed Summary of Unplanned Troubleshooting and Repair Work

Concentrtion of Space Shuttle Unplanned Troubleshooting and Repair Work Content
for 1997 Launches
(As measured by cumulative task-hours for eight processing flows)

Generic Operations Functions	FBS	ID	1 Thermal Management	2 Structures , Mechanisms, Veh Handling	3 Power Management	4 Propulsion	5 Cmd, Ctl & Health Mngmt	6 Ground Systems & Facilities	7 Guid, Nav & Ctl	8 Environmental Ctl & Life Spt	9 Communications	10 Safety Management & Control	11 Cockpit & Crew Cabin	12 Payload Accommodations	Grand Total	Percentage
Repair or replacement of TPS hardware	2.08.05	A	19,937	(A)	(A)	(A)	(A)	(A)	(A)	(A)	(A)	(A)	(A)	(A)	**19,937**	**27.13%**
Troubleshooting, replacement, and disposition of failed/suspect LRUs	2.08.01	B	258	791	7,193	3,264	1,063	(A)	1,055	304	366	24	239	6	**14,563**	**19.82%**
Structural repair/refurbishment	2.08.06	C	96	12,031	16	220	(A)	(A)	9	83	8	21	80	(A)	**12,564**	**17.10%**
Troubleshooting, retest and repair of electrical cables, connectors	2.08.04	D	8	550	5,405	192	925	(A)	61	98	430	(A)	(A)	(A)	**7,669**	**10.44%**
Troubleshooting and repair of mechanisms and thermal/pressure seals	2.08.08	E	3,075	4,170	40	59	8	(A)	40	8	(A)	29	(A)	(A)	**7,429**	**10.11%**
Component troubleshooting, removal, and replacements	5.04.01	F	4	80	122	992	240	52	16	44	(A)	100	8	40	**1,698**	**2.31%**
Flight element troubleshooting and repair	3.05.02	G	624	577	(A)	16	241	62	(A)	(A)	(A)	90	(A)	(A)	**1,610**	**2.19%**
Troubleshooting and repair of ground equipment	2.08.09	H	(A)	(A)	8	521	8	908	(A)	(A)	(A)	32	(A)	(A)	**1,477**	**2.01%**
Troubleshooting and repair of leaks	2.08.02	I	50	51	976	257	(A)	(A)	(A)	92	(A)	(A)	(A)	(A)	**1,426**	**1.94%**
Ground Launch Structural Systems Repair	5.04.02	J	(A)	163	(A)	(A)	(A)	1,021	(A)	(A)	(A)	(A)	(A)	(A)	**1,184**	**1.61%**
Disconnection of flight elements	4.03.01	K	(A)	868	(A)	(A)	(A)	(A)	(A)	(A)	(A)	(A)	(A)	(A)	**868**	**1.18%**
Element Repair and Rework	6.06.01	L	(A)	812	(A)	(A)	(A)	(A)	(A)	(A)	(A)	(A)	(A)	(A)	**812**	**1.11%**
Repair of damaged ducts, tubes, and hoses	2.08.07	M	(A)	20	85	77	(A)	8	(A)	242	(A)	247	8	(A)	**687**	**0.93%**
Unplanned element disassembly/reassembly	3.05.01	N	(A)	596	(A)	(A)	(A)	(A)	(A)	(A)	(A)	(A)	(A)	(A)	**596**	**0.81%**
Postintegration ground element/equipment troubleshooting and repair	4.03.04	O	(A)	292	136	8	(A)	26	(A)	(A)	(A)	(A)	(A)	(A)	**462**	**0.63%**
Fluid/pneumatic system decontamination/cleaning	2.08.03	P	24	46	68	62	(A)	(A)	(A)	6	(A)	(A)	(A)	(A)	**206**	**0.28%**
Postintegration flight element troubleshooting and repair	4.03.03	Q	17	109	(A)	(A)	(A)	(A)	(A)	(A)	(A)	(A)	(A)	(A)	**126**	**0.17%**
Flight element modifications and process changes	3.05.03	R	84	24	(A)	(A)	12	(A)	(A)	(A)	(A)	(A)	(A)	(A)	**120**	**0.16%**
Interface troubleshooting and repair	4.03.02	S	6	28	(A)	(A)	(A)	12	(A)	(A)	(A)	(A)	(A)	(A)	**46**	**0.06%**
Total Task-hours			**24,183**	**21,208**	**14,049**	**5,668**	**2,497**	**2,089**	**1,181**	**877**	**804**	**543**	**335**	**46**	**73,480**	**100.00%**
Percent			**32.91%**	**28.86%**	**19.12%**	**7.71%**	**3.40%**	**2.84%**	**1.61%**	**1.19%**	**1.09%**	**0.74%**	**0.46%**	**0.06%**	**100.00%**	

Columns 1–12: Generic Design Functions.

(A) System disciplines not showing dedicated concentrations of direct work. This should not be interpreted that no direct work occurs. Some of the tasks involving the design discipline may have been identified, for the purposes of this analysis, as being dominated by another discipline. Proper interpretation is that these items have no identifiable direct work concentrations relative to the total annual direct work content.

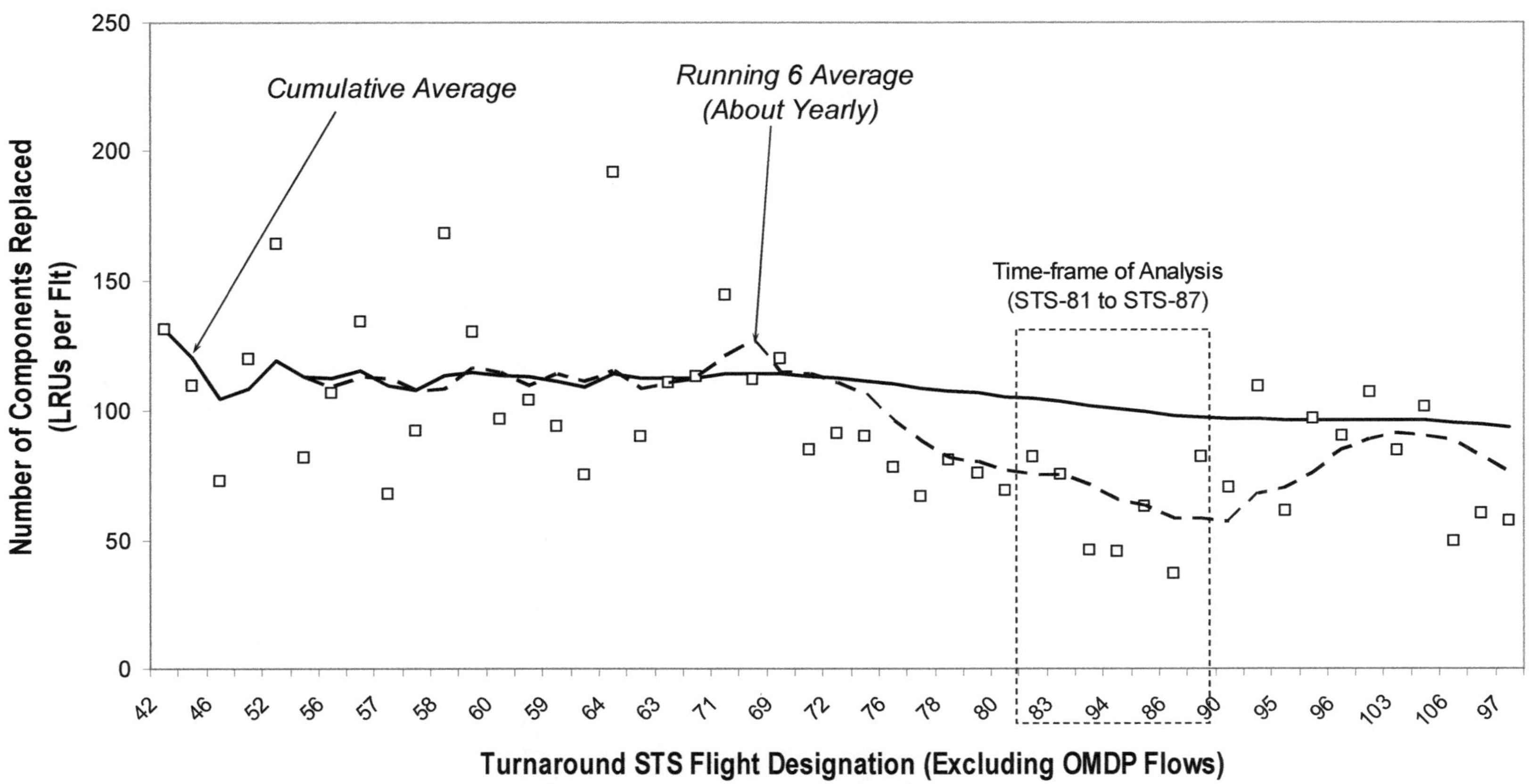

Figure 39. Typical Trend of Space Shuttle Orbiter Component Replacements (Exclusive of Tiles and SSMEs)

In terms of repair work to maintain electrical system integrity (Figure 40), it is instructive to look at processing data. The Shuttle Program has maintained management and control of electrical system integrity through an evolving set of engineering management methods.

Initially, the method used was to manually account, track, and sign off every pin function in each connector for demating, remating, and retest status. This highly labor-intensive method was maintained by the manufacturer's Connector Integrity Group (CIG) in paper form contained in bookcases full of ring binders in trailer facilities at the launch site for engineers to review and disposition. When the Shuttle Processing Contract (SPC) began an effort to consolidate and streamline the work content, the Shuttle Connector Analysis Network (SCAN) was created from knowledge of the vehicle's wire list connected with a functional database. While still somewhat labor-intensive in engineering, SCAN has dramatically reduced the amount of engineering management labor required to maintain the vehicle's electrical integrity, and also provided the operations team the ability to status and identify specific operational constraints to vehicle or system power-up. **Error! Reference source not found.** provides an overview of connector integrity from SCAN for the processing flows under analysis. No similar data was uncovered for plumbing systems, even though numerous critical fluid and propellant transfer lines often go through unplanned disassembly, troubleshooting, repair and reassembly (Figure 41).

Figure 40. Orbiter Wiring in Payload Bay (left) and Typical Short-Circuit Source

Table 15. Summary of Orbiter Electrical Connector Integrity

	Number of Orbiter Connector Demates Tracked in SCAN										
Design Discipline	STS-81	STS-82 (OMDP)	STS-83	STS-84	STS-94	STS-85	STS-86	STS-87	6-Flow Total	6-Flow Avg	Percent
Power Management	177	*787*	167	NA	169	164	118	132	927	155	**37.90%**
Liquid Propulsion	86	*296*	98	NA	101	92	59	75	511	85	**20.89%**
Cmd, Ctl & Health Mngmt	75	*361*	40	NA	39	63	25	142	384	64	**15.70%**
Payloads	13	*148*	63	NA	20	99	13	92	300	50	**12.26%**
Guid, Nav & Ctl	28	*241*	16	NA	18	24	14	16	116	19	**4.74%**
Communications	19	*309*	17	NA	2	19	18	11	86	14	**3.52%**
Environmental Ctl & Life Spt	22	*205*	8	NA	2	6	11	10	59	10	**2.41%**
Structures, Mechanisms & Veh Handling	12	*96*	13	NA	9	7	6	5	52	9	**2.13%**
Safety Management & Control	1	*34*		NA	1	6		3	11	2	**0.45%**
TOTAL	**433**	**2,477**	**422**	**0**	**361**	**480**	**264**	**486**	**2,446**	**408**	

Unplanned work also includes repair to structurally damaged ductwork, vacuum-jacketed propellant feed lines, tubes, and hoses. Also requiring repair work are critical structural items found to have excessive corrosion or cracks. Repair may include stop drilling, attachment of structural doublers, and reapplication of coatings.

The IFA List, normally cited as an indicator of flight system integrity, was not used in this analysis. While the criteria for "in-flight anomalies" have fluctuated throughout the history of the program, the component replacements, as tracked by the PRACA system, is more indicative of significant unplanned work. The PRACA data in Figure 39 reveals a chronic high level of component replacements for the Orbiter (excluding tile and main engines), which indicates that the "corrective actions" taken are those necessary only to restore flight system integrity prior to the next flight, but have not corrected the overall design to significantly lower the level of unplanned work. Discovering the design root causes of this type of work and implementing controls to prevent them throughout the design process will be vital to the creation of highly efficient space transportation.

Figure 41. Technicians with Special Viewing Apparatus Use Fiber-Optics for Unplanned Inspections of Propulsion System Flow Liner on Orbiter *Endeavour*

5.4.2 Servicing

In general, this class of work can be defined as the routine activity necessary to restore vehicle systems to a flight-worthy/flight-ready state. A summary of this work is provided inTable 16. This planned work may involve routine fluid system drains, fills, and pressurizations; filter inspections and replacements; replacement of single-use or limited-life items; and mechanical calibrations. This often-neglected operations function accumulated a dramatic amount of work as the Space Shuttle design progressed. Over 57,500 task-hours of accumulated servicing work were identified for the processing flows leading to 1997 launches.

During Orbiter processing, the high degree of maintenance traffic in sensitive areas has led to routine, labor-intensive cleaning to maintain contamination-free zones in the crew cabin, payload bay, and aft compartment. While this type of servicing work is a function of the overall duration of the Orbiter maintenance process, it nevertheless emerged, along with filter maintenance, as the highest concentration of servicing work, with over 12,900 task-hours associated with the eight processing flows analyzed.

As mentioned, the many different fluid systems that were designed into the Orbiter have led to numerous fluid-line filter inspections and replacements. Many of these filters have design agency requirements imposed with as little as 90 days of service before replacement, such as the Window Cavity Conditioning System (WCCS) filter-desiccant assembly. This operation is very intrusive and labor-intensive and subsequently requires time-consuming functional verification testing. This operation may occur during Orbiter processing in the OPF or could occur at the launch pad (Figure 42).

Another set of servicing activities that accumulates is the routine attention paid to the many thermal seals and aerodynamic pressure seals. The Orbiter has by design many mold line penetrations (i.e., areas where the thermal protection tile and aluminum skin are interrupted by access panels, windows, payload bay doors, landing gear doors, antennas, air data probes, star tracker doors, and propellant feed lines from the ET). These areas have sensitive thermal/pressure seals that are highly critical, often reveal signs of wear and tear, and have to be inspected and replaced frequently.

Other sensitive thermal/pressure seals are located at the aerosurface hinge lines. These areas require that the upper and lower surfaces be isolated from each other because of the pressure differential in the presence of hot plasma. Otherwise, hot gas jets would form between the airframe and the surface causing catastrophic structural failure. Between the wing and the elevon aerosurfaces, a complex mechanical "flipper door" and cove seal arrangement requires a great deal of routine maintenance work. Similarly, the hinge line seals at the body flap and the left and right rudder speed brake panels require routine servicing.

Fluid systems servicing is another area accumulating a lot of direct work. In particular, fluid-systems servicing and fill operations to support electrical and mechanical power generation and distribution functions are quite high (over 4,000 task-hours accumulated in the analyzed data set). This includes the fuel cell Power Reactant Storage and Distribution (PRSD) System (oxygen and hydrogen) and its supporting gas systems; and the hydrazine, hydraulics, and steam supply systems for the auxiliary power unit and hydraulic power distribution—and its supporting gas systems.

Table 16. Detailed Summary of Servicing Work Content

Generic Operations Functions: Concentration of Space Shuttle Servcing Work Content for 1997 Launches (As measured by cumulative task-hours for eight processing flows)	FBS	ID	Propulsion 1	Thermal Management 2	Structures , Mechanisms, Veh Handling 3	Power Management 4	Safety Management & Control 5	Environmental Ctl & Life Spt 6	Cmd, Ctl & Health Mngmt 7	Cockpit & Crew Cabin 8	Payload Accommodations 9	Guid, Nav & Ctl 10	Communications 11	Ground Systems & Facilities 12	Grand Total	Percentage
			Generic Design Functions													
Routine Vehicle Cleaning and Filter Inspections	2.04.09	A	6,758	(A)	6,016	(A)	(A)	46	(A)	92	(A)	(A)	4	(A)	12,916	22.44%
TPS Mold Line Penetration, Leading Edge/Nose, and Aerosurface Hinge Line Seal Servicing	2.04.06	B	(A)	7,377	585	16	(A)	(A)	(A)	(A)	(A)	(A)	(A)	(A)	7,978	13.86%
Vehicle Electrical Power Application, Data Processing, Monitoring, and Control Operations	2.04.13	C	14	(A)	1,446	5,661	(A)	(A)	412	(A)	(A)	(A)	4	(A)	7,537	13.09%
Acreage Tile and Thermal Protection System (TPS) Blanket Servicing	2.04.05	D	8	4,411	478	(A)	(A)	(A)	(A)	(A)	(A)	8	(A)	(A)	4,905	8.52%
Fluid Servicing and Fill	2.04.08	F	827	756	(A)	2,104	(A)	373	(A)	(A)	(A)	9	(A)	(A)	4,069	7.07%
Local Servicing with Facility/GSE by Remote Control Room	5.06.01	G	1,171	(A)	(A)	91	2,560	(A)	(A)	(A)	(A)	(A)	(A)	13	3,835	6.66%
Flight Crew Systems Servicing	2.04.04	H	(A)	(A)	16	(A)	(A)	1,851	(A)	835	408	(A)	(A)	(A)	3,110	5.40%
Local Manual Servicing at the Launch Facility	5.06.02	I	1,717	(A)	(A)	433	(A)	(A)	24	(A)	(A)	50	(A)	32	2,256	3.92%
Servicing of Ground Systems Prior to Flight Vehicle Mating	5.01.04	J	1,402	(A)	(A)	132	491	18	(A)	(A)	(A)	(A)	(A)	62	2,105	3.66%
Post-Mate Servicing and Interface Closeout	4.01.04	K	91	522	671	(A)	100	(A)	230	(A)	(A)	(A)	(A)	(A)	1,614	2.80%
Ordnance Handling and Installation	2.04.02	L	(A)	(A)	1,518	(A)	(A)	(A)	(A)	(A)	(A)	(A)	(A)	(A)	1,518	2.64%
Landing Gear Mechanical Servicing	2.04.01	M	(A)	32	915	190	(A)	(A)	284	(A)	(A)	(A)	(A)	(A)	1,421	2.47%
Fluid Drain and Deservicing	2.04.07	N	686	(A)	(A)	498	18	181	(A)	(A)	16	(A)	(A)	(A)	1,399	2.43%
Routine Replacement of Expendable and Limited-Life Items	2.04.12	O	122	(A)	155	384	8	205	59	(A)	(A)	138	(A)	(A)	1,071	1.86%
Ordnance Hookups	5.06.03	P	(A)	(A)	576	(A)	(A)	(A)	80	(A)	(A)	(A)	(A)	(A)	656	1.14%
Tank Gaseous Pressurization	5.08.03	Q	388	(A)	(A)	(A)	(A)	(A)	(A)	(A)	(A)	(A)	(A)	(A)	388	0.67%
Maintenance of Fuselage/PLB Vent filters and Window Cavity Purge	2.04.03	R	(A)	(A)	(A)	(A)	257	(A)	(A)	(A)	(A)	(A)	(A)	(A)	257	0.45%
Flight Pressurizations	2.04.10	S	14	(A)	155	27	(A)	(A)	58	(A)	(A)	(A)	(A)	(A)	254	0.44%
Navigation and Instrumentation Component Servicing	2.04.11	T	(A)	3	(A)	(A)	(A)	(A)	16	4	(A)	(A)	104	4	131	0.23%
Remote Fuel Cell Gas Supply	5.08.07	U	(A)	(A)	(A)	106	(A)	(A)	(A)	(A)	(A)	(A)	(A)	(A)	106	0.18%
Flight Vehicle Press	5.08.04	V	31	(A)	(A)	(A)	(A)	(A)	(A)	(A)	(A)	(A)	(A)	(A)	31	0.05%
Total (Task-Hours)			13,229	13,101	12,531	9,642	3,434	2,674	1,163	931	424	205	112	111	57,557	100.00%
Percent			22.98%	22.76%	21.77%	16.75%	5.97%	4.65%	2.02%	1.62%	0.74%	0.36%	0.19%	0.19%	100.00%	

(A) System disciplines not showing dedicated concentrations of direct work. This should not be interpreted that no direct work occurs. Some of the tasks involving the design discipline may have been identified, for the purposes of this analysis, as being dominated by another discipline. Proper interpretation is that these items have no identifiable direct work concentrations relative to the total annual direct work content.

Figure 42. Design Requirements Force Frequent, Intrusive, and Labor-Intensive Window Cavity Conditioning System (WCCS) Filter/Desiccant Inspection and Replacement

Routine replacement of expendable and limited-life items is another servicing function creating over 1,000 task-hours of work. This work includes major items, such as limited operating hours for fuel cells and auxiliary power units, to smaller items such as filters, to such hazardous items as pyrotechnic discharge devices. Typical trends were analyzed and are shown in Figure 43.

Another type of work classified as launch vehicle servicing was ordnance installation. Some of the preparatory work is performed during the vehicle stack buildup, whereas the hazardous installation operations are performed at the launch point while the pad area is cleared of personnel and all other work. This work accumulated over 1,500 task-hours of installation and over 650 task-hours of hookups across the eight flows analyzed.

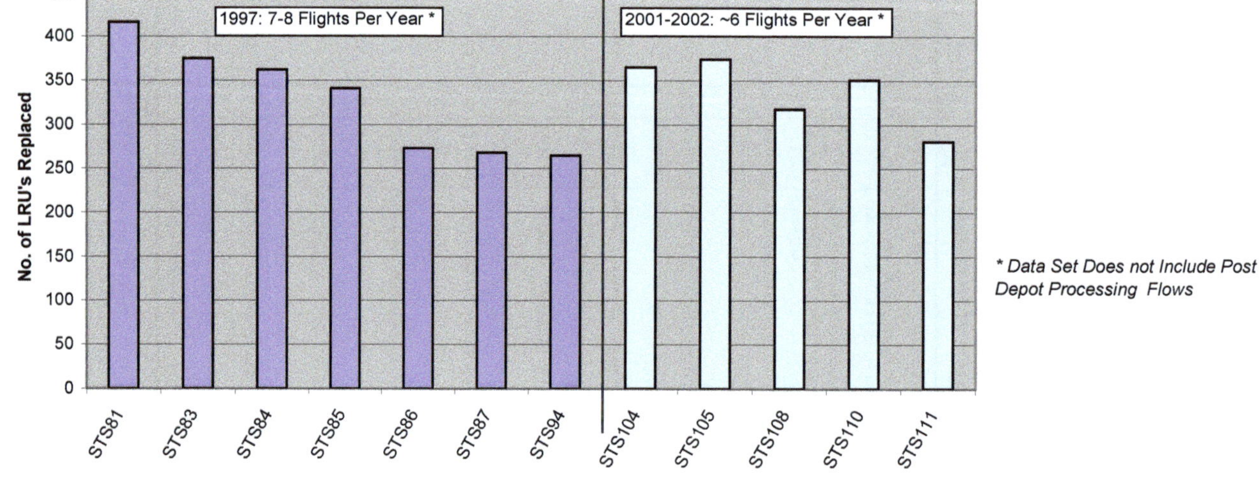

Figure 43. Number of Limited-Life Items Replaced Per Flow

5.4.3 Assembly

This high-concentration operations function accounted for nearly 57,100 task-hours of work. This function is summarized in Table 17. This table goes one level deeper in functional breakdown than the assembly functions presented in Table 10 (see 5.2.4).

Nearly three-quarters of the assembly work involved the following areas: (1) closeout work following the installation of various SRB components and subsystems, (2) the SRB segment connections and field joint assembly, and (3) the actual installation of various components and subsystems (as opposed to the installation "closeouts"). The overwhelming majority of this work was concentrated in the SRB, OMS/RCS, and SSME systems—the propulsion systems.

Table 17. Summary of Assembly Work Content

Concentration of Space Shuttle Assembly Work Content for 1997 Launches (As measured by cumulative task-hours for eight processing flows) — Generic Operations Functions	FBS	ID	Generic Design Functions: Structures, Mechanisms, Veh Handling 1	Propulsion 2	Cmd, Ctl & Health Mngmt 3	Thermal Management 4	Ground Systems & Facilities 5	Power Management 6	Safety Management & Control 7	Guid, Nav & Ctl 8	Communications 9	Environmental Ctl & Life Spt 10	Payload Accommodations 11	Cockpit & Crew Cabin 12	Grand Total	Percentage
Flight Component Installation Closeouts	3.03.05	A	18,666	2,223	338	1,890	(A)	159	2	(A)	51	145	25	(A)	23,499	41.10%
Flight Component Installation onto Flight Element	3.03.03	B	7,907	1,511	2,159	(A)	(A)	96	225	301	182	76	39	(A)	12,496	21.85%
Flight Segment Structural Connections	3.02.02	C	2,651	372	(A)	(A)	(A)	(A)	(A)	(A)	(A)	(A)	(A)	(A)	3,023	5.29%
Non-flight Component Installations	3.03.04	D	1,954	14	(A)	(A)	(A)	(A)	(A)	(A)	(A)	(A)	(A)	(A)	1,968	3.44%
Passive System Verifications	3.04.01	E	780	599	492	(A)	(A)	(A)	(A)	(A)	(A)	(A)	(A)	(A)	1,871	3.27%
Flight Element Troubleshooting and Repair	3.05.02	F	577	16	241	624	62	(A)	90	(A)	(A)	(A)	(A)	(A)	1,610	2.82%
Structural Assembly Verification and Closeout	3.02.03	G	1,537	(A)	(A)	(A)	(A)	(A)	(A)	(A)	(A)	(A)	(A)	(A)	1,537	2.69%
Flight Segment Handling Operations	3.02.01	H	1,493	(A)	(A)	(A)	(A)	(A)	(A)	(A)	(A)	(A)	(A)	(A)	1,493	2.61%
Facility Preps for Element Assembly Operations	3.01.01	I	194	851	(A)	(A)	272	(A)	(A)	(A)	(A)	(A)	(A)	(A)	1,317	2.30%
Active System Verifications	3.04.02	J	(A)	987	101	(A)	(A)	156	(A)	(A)	7	(A)	(A)	(A)	1,251	2.19%
Flight Element Component Disassembly, Handling, and Transport	3.08.02	K	(A)	1,121	(A)	(A)	(A)	(A)	(A)	(A)	(A)	(A)	22	(A)	1,143	2.00%
Flight Element Preparations for Integration	3.06.02	L	886	(A)	(A)	88	(A)	(A)	(A)	(A)	(A)	(A)	(A)	(A)	974	1.70%
Lift, Positioning, and Connection to Assembly Facility Support Fixtures	3.01.03	M	857	100	(A)	(A)	(A)	(A)	(A)	(A)	(A)	(A)	(A)	(A)	957	1.67%
Disassembly Preps and Attach/Closeout Hardware Removal	3.08.01	N	163	529	(A)	5	(A)	(A)	(A)	(A)	(A)	(A)	32	(A)	729	1.27%
Subsystem/Component Handling Ops	3.03.01	O	261	330	(A)	(A)	(A)	(A)	(A)	(A)	16	(A)	25	(A)	632	1.11%
Unplanned Element Disassembly/Reassembly	3.05.01	P	596	(A)	(A)	(A)	(A)	(A)	(A)	(A)	(A)	(A)	(A)	(A)	596	1.04%
Reconfigure/Relocate Facility and Ground Equipment	3.06.01	Q	326	40	(A)	(A)	227	(A)	(A)	(A)	(A)	(A)	(A)	(A)	593	1.04%
Facility Preps for Installations	3.03.02	R	387	4	(A)	(A)	52	16	(A)	(A)	36	(A)	26	(A)	521	0.91%
Ground Transport of Major Element Segments	3.01.02	S	(A)	308	(A)	(A)	(A)	(A)	(A)	(A)	(A)	(A)	(A)	(A)	308	0.54%
Inspections and Alignment Verifications	3.04.03	T	258	(A)	(A)	(A)	(A)	(A)	(A)	(A)	(A)	(A)	(A)	(A)	258	0.45%
Activate Element Eenvironmental and Safety Controls	3.01.04	U	201	(A)	(A)	(A)	(A)	(A)	(A)	(A)	(A)	(A)	(A)	(A)	201	0.35%
Flight Element Modifications and Process Changes	3.05.03	V	24	(A)	12	84	(A)	(A)	(A)	(A)	(A)	(A)	(A)	(A)	120	0.21%
Post-Disassembly	3.08.03	W	(A)	80	(A)	(A)	(A)	(A)	(A)	(A)	(A)	(A)	4	(A)	84	0.15%
Total (Task-Hours)			39,718	9,085	3,343	2,691	613	427	317	301	292	221	173	(A)	57,181	100.00%
Percent			69.46%	15.89%	5.85%	4.71%	1.07%	0.75%	0.55%	0.53%	0.51%	0.39%	0.30%	(A)	100.00%	

(A) System disciplines not showing dedicated concentrations of direct work. This should not be interpreted that no direct work occurs. Some of the tasks involving the design discipline may have been identified, for the purposes of this analysis, as being dominated by another discipline. Proper interpretation is that these items have no identifiable direct work concentrations relative to the total annual direct work content.

5.4.4 Functional Verification (Inspection and Checkout)

Another high-concentration work classification is the routine, planned functional verification of the launch vehicle, flight elements and critical systems, and critical ground elements and systems. This work accumulated to about 51,600 task-hours across the eight flight processing flows analyzed. A summary of this work is provided in Table 18. This planned work may involve routine physical inspections, cockpit-initiated checkout routines, remote firing-room test sequences, or portable equipment that executes testing.

The inspection and checkout work was concentrated in the Orbiter turnaround and the integrated Space Shuttle Vehicle processing activities in the VAB and at the launch pad.

The highest accumulation of functional verification work was in the area of structural integrity inspections and in the various Orbiter mechanisms (landing gear, payload bay doors, deployable air data probes, and so forth). This area amounted to about 24 percent of the inspection and checkout work, or about 12,300 task-hours accumulated across the eight analyzed processing flows. The thermal management/tile design discipline accounted for 2,500 of these task-hours.

The next highest functional verification concentration of work was devoted to Orbiter component and subsystem functional verification (about 17.6 percent). Much of this work requires vehicle power application and is most often performed remotely from the same firing-room infrastructure used for the Shuttle vehicle's launch (Figure 44). Execution of automated test sequences is managed by a Launch Processing System (LPS) that not only allows technicians and engineers the ability to monitor system test activity, but also has built-in capability for safely handling hazardous and critical timing commands required by the flight system design.

Orbiter propellant, fluid, and gas system leakage checks and other inspections and verifications also added more than 6,800 task-hours (13.2 percent). The Orbiter is designed to function with many different fluid systems—literally dozens of separately serviced operational fluids. With a well-known history of fluid system leakage (sometimes from toxic and hazardous designs) combined with the high level of required intrusive system servicing and unplanned maintenance, there has never been enough engineering confidence from the design agency and spaceport system engineers for Orbiter processing to occur without conducting thorough inspections and leak checks of these critical and hazardous systems.

Amongst the other many contributors, the functional verification of ground equipment prior to mating to critical vehicle systems amounted to 11.7 percent of the functional verification work, or over 6,000 task-hours uniquely identified in the analysis data. Most of the maintenance actions in the analysis have off-line procedures that are called out by the controlling on-line procedure. These prerequisite checks of the GSE, or "pre-ops," are often required of the literally thousands and thousands of Shuttle GSE items prior to conducting the maintenance action on the flight systems.

Table 18. Summary of Functional Verification Work

Concentration of Space Shuttle Functional Verification Work Content for 1997 Launches (As measured by cumulative task-hours for eight processing flows)

Generic Operations Functions	FBS	ID	Propulsion (1)	Structures, Mechanisms, Veh Handling (2)	Power Management (3)	Safety Management & Control (4)	Cmd, Ctl & Health Mngmt (5)	Thermal Management (6)	Ground Systems & Facilities (7)	Environmental Ctl & Life Spt (8)	Communications (9)	Payload Accommodations (10)	Guid, Nav & Ctl (11)	Cockpit & Crew Cabin (12)	Grand Total	Percentage
Structural integrity inspections and mechanism functional verifications	2.05.02	A	1,090	7,842	(A)	542	4	2,524	(A)	14	68	57	91	52	**12,284**	**23.81%**
Component/subsystem functional verification	2.05.01	B	494	562	2,789	2,011	488	78	(A)	969	1,120	68	403	99	**9,081**	**17.60%**
Propellant/fluid/gas system integrity verifications and inspections	2.05.03	C	4,916	322	1,073	27	2	57	(A)	413	(A)	(A)	(A)	(A)	**6,810**	**13.20%**
Functional verifications of ground systems prior to flight vehicle mating	5.01.03	D	1,289	320	225	2,776	(A)	(A)	1,085	141	(A)	201	(A)	(A)	**6,037**	**11.70%**
System integrity verification	5.03.02	E	1,150	100	1,695	140	433	(A)	10	(A)	(A)	(A)	29	(A)	**3,557**	**6.90%**
Periodic maintenance/calibrations	5.03.01	F	1,046	(A)	220	8	(A)	(A)	824	88	(A)	16	(A)	(A)	**2,202**	**4.27%**
Passive system verifications	3.04.01	G	599	780	(A)	(A)	492	(A)	(A)	(A)	(A)	(A)	(A)	(A)	**1,871**	**3.63%**
System functionality verification	5.03.03	H	766	16	100	596	54	(A)	8	(A)	(A)	(A)	194	(A)	**1,734**	**3.36%**
Element-to-element interfaces	4.02.02	I	535	32	284	(A)	735	(A)	(A)	(A)	(A)	(A)	(A)	(A)	**1,586**	**3.07%**
Structural assembly verification and closeout	3.02.03	J	(A)	1,537	(A)	(A)	(A)	(A)	(A)	(A)	(A)	(A)	(A)	(A)	**1,537**	**2.98%**
Element-to-ground-system interfaces	4.02.01	K	929	370	16	58	37	(A)	(A)	(A)	(A)	(A)	(A)	(A)	**1,410**	**2.73%**
Active system verifications	3.04.02	L	987	(A)	156	(A)	101	(A)	(A)	(A)	7	(A)	(A)	(A)	**1,251**	**2.43%**
Vehicle power-up/down, switch lists, functional checks, and onboard s/w configuration	2.05.04	M	187	36	45	(A)	169	(A)	(A)	54	88	(A)	(A)	141	**720**	**1.40%**
Payload functional verification for flight	5.05.03	N	(A)	(A)	(A)	(A)	32	(A)	(A)	(A)	(A)	528	(A)	32	**592**	**1.15%**
Human systems verification and training	5.03.04	O	(A)	(A)	(A)	56	500	(A)	(A)	(A)	(A)	(A)	(A)	35	**591**	**1.15%**
Inspections and alignment verifications	3.04.03	P	(A)	258	(A)	(A)	(A)	(A)	(A)	(A)	(A)	(A)	(A)	(A)	**258**	**0.50%**
Functional verification of ground systems prior to vehicle arrival	2.01.01	Q	(A)	(A)	(A)	16	(A)	(A)	48	(A)	(A)	(A)	(A)	(A)	**64**	**0.12%**
Total (Task-Hours)			**13,988**	**12,175**	**6,603**	**6,230**	**3,047**	**2,659**	**1,975**	**1,679**	**1,283**	**870**	**717**	**359**	**51,585**	**100.00%**
Percent			**27.12%**	**23.60%**	**12.80%**	**12.08%**	**5.91%**	**5.15%**	**3.83%**	**3.25%**	**2.49%**	**1.69%**	**1.39%**	**0.70%**	**100.00%**	

(Columns 1–12 are grouped under the heading **Generic Design Functions**.)

(A) System disciplines not showing dedicated concentrations of direct work. This should not be interpreted that no direct work occurs. Some of the tasks involving the design discipline may have been identified, for the purposes of this analysis, as being dominated by another discipline. Proper interpretation is that these items have no identifiable direct work concentrations relative to the total annual direct work content.

Figure 44. Complex Shuttle Operations Are Conducted From Several Firing Rooms Throughout the Turnaround Process

6 DESIGN CAUSES OF OPERATIONS AND INFRASTRUCTURE

6.1 Top-Level Design Root Cause Results

After documenting numerous design causes associated with various work content groupings, an effort was made to analyze common themes and recurring causes associated with the major generic operations functions. These are described below in decreasing order of direct work volume.

6.1.1 Excessive Unplanned Troubleshooting and Repair

Unplanned troubleshooting and repair accounted for approximately 24 percent of the annual direct work in processing the Space Shuttle at Launch Complex 39. Preliminary analysis revealed the following design root causes:

a. Flight and ground system dependability (i.e., the hardware design life and flight element reliability in relation to total parts count).

b. Flight and ground system complexity manifested through high parts count in components, subsystems, and redundancy management techniques.

c. Lack of an effective means for balancing ground safety and maintainability while achieving flight safety objectives.

d. Lack of an effective means for implementing design corrective actions and continuous improvement.

6.1.2 Complex Assembly, Handling, Access, and Mating

The operationally complex assembly, handling, and access and vehicle element mating accounted for approximately 19 percent of the annual direct work in processing the Space Shuttle at Launch Complex 39. Preliminary analysis revealed the following design root causes:

a. Number of launch-site-installed subsystems and components per flight element.

b. Number and complexity of interfaces across assembled flight elements.

c. Number and frequency of reusable flight element modules and components requiring disassembly and/or reassembly (SSMEs, OMS/RCS pods, RMS, etc.).

d. Number of single-use, limited-life components requiring assembly and access.

e. Decomposition of design functions discourages well-integrated design processes and resulting products and duplicates and proliferates hardware without regard to recurring assembly operations, ground assembly, and handling infrastructure impacts.

6.1.3 Excessive Flight System Servicing

The excessive amount of vehicle fluid and mechanical servicing accounted for approximately 18 percent of the annual direct work in processing the Space Shuttle at Launch Complex 39. Preliminary analysis revealed the following design root causes:

a. Number and type of different fluid commodities requiring routine servicing and separate flight-ground interfaces.

b. Number of limited-life items requiring routine removal, replacement, and functional verification (relates to Root Cause 6.1.1a, design life issue).

c. Decomposition of design functions discourages well-integrated design processes and resulting products, duplicating and proliferating hardware without regard to recurring flight system servicing operations and ground servicing infrastructure and logistics impacts.

6.1.4 Lack of Demonstrated System Dependability and Resulting Functional Verification

The functional verification work (i.e., the systems checkout and physical inspection work) accounted for approximately 14 percent of the annual direct work in processing the Space Shuttle at Launch Complex (LC)-39. Preliminary analysis revealed the following design root causes:

a. Lack of demonstrated operational reliability (i.e., routine system operation without need for functional restoration between flights).

b. Overredundancy in design to overcome lack of inherent reliability relative to the space transportation system operating environments.

c. Number of critical flight functions requiring functional verification.

d. Amount of reusable flight element disassembly and reassembly, and amount of expendable flight item assembly resulting in the requirement for recertification prior to each flight commitment.

e. Amount of automation employed in the system—both flight and ground elements.

6.1.5 Excessive Facility and Equipment Preparation and Refurbishment

a. Ground system dependability (i.e., design life and hardware reliability in relation to total parts count).

b. Lack of ground launch system design life to withstand the induced launch environment for frequent, routine launches.

c. Number of separate GSE items to support routine operations, driven in turn by flight systems design complexity.

d. Lack of an effective means for balancing ground safety and maintainability.

6.1.6 Complex, Customized Payload Integration With Flight Vehicle

a. Unique vehicle payload accommodations that are customized for each flight during routine on-line operations—by design.

b. Large number of flight-unique, vehicle-provided services to the payload that drives on-line operations (e.g., lack of payload center-of-gravity and weight margin).

6.2 Detailed Turnaround Work Causes

6.2.1 Turnaround Facility Preparations Work

This activity may arise from the requirement to configure and/or verify that facility support systems are functional and prepared to receive the flight vehicle for turnaround operations. This includes:

6.2.1.1 Functional verification of ground systems prior to vehicle arrival. Examples include toxic-vapor scrubber configuration, ground support system calibrations, and functional checks. This work in turn arises from design requirements to:

a. Protect personnel and high-value vehicle assets.

b. Provide functional checks of ground systems that directly interface to flight elements (as a result of 6.2.1.1a).

6.2.1.2 Servicing and staging of ground systems prior to vehicle arrival. This can include filter changeouts, painting, positioning, and spotting of GSE. This work in turn arises from design requirements to:

a. Support total number of accumulated ground support services and their equipment for subsequent vehicle turnaround functions (safing, servicing, repair, and so forth).

b. Prepare for removal of major flight element assemblies, subsystems, and components (reaction control modules, main engines, remote manipulators, payload elements, etc.).

c. Ready safety equipment and protective devices for immediate hookup at vehicle arrival.

6.2.1.3 Contamination control. This may involve hangar bay cleaning, foreign-object-debris (FOD) control, access equipment cleaning, and so forth. This work in turn arises from requirements driven by:

a. Sensitivity of vehicle to debris damage.

b. Payloads needing clean-room environment.

6.2.2 Flight Element Positioning, Connection to Services, Gaining Access, and Protection

This activity arises from a requirement to provide personnel access to the vehicle for subsequent flight system maintenance, repair, servicing, etc. This work in turn arises from requirements driven by:

6.2.2.1 Vehicle transport and alignment. This work is required for operational concepts using separate turnaround facilities for post- and preflight element safing, servicing, repair, and manual inspections. The work may involve tug transports and spotting, jack stand configuration/reconfiguration, and facility door operations. It may also include transfer of purge, power and cooling functions from mobile convoy to facility-provided services, if vehicle design requires it. Drivers of this work include requirements for:

a. Interconnected convoy vehicles that complicate the vehicle transfer operations (return from flight).

b. Alignment of vehicle into complex service and access stands and critical alignment for connection to fixed facility services.

6.2.2.2 Ground access equipment positioning. This work includes personnel access platform extension/configuration, positioning of bridge buckets, flipdowns, flipups, ladders, rollups, and so forth. Drivers of this work include requirements for:

a. Manual elevated (or otherwise out-of-reach) connection of mold line vehicle interfaces.

b. Internal vehicle access without aid of vehicle-provided means of ingress/egress for connection of ground services.

6.2.2.3 Connection to facility services. This activity performs the actual transfer of purge, power, and thermal conditioning functions from onboard or mobile equipment-provided (i.e., equipment convoy) to facility- and/or GSE-supplied services. This may include scrubbers, conditioned air purges, cooling loops, and electrical power. Drivers of this work include requirements for:

a. Power on the vehicle from landing until internal access for full safing of internal systems can be completed.

b. The accumulated number of ground-connected services (umbilicals, drag-on hose sets, drag-on ducts, electrical connections, and so forth) in the flight element design for postlanding safing and preflight element servicing (subsystem drain, fill, purge, etc.).

6.2.2.4 Establish protective enclosures and install/remove flight element protective covers. This activity may involve setup of protective white-room tents, small clean rooms, payload bay curtains, etc. Drivers of this work include requirements for:

a. Flight-element-provided payload protection in a payload bay or shroud, rather than payload-provided protective packaging.

b. Design utilization of contamination-sensitive optical navigation systems (e.g., star-trackers).

c. Zero- or microgravity environment of crew or passenger module driving contamination control operations.

d. Ground protection against inherent leak potential of auxiliary propulsion and other fluid-based subsystem designs. Gives rise, for example, to propulsion system throat plugs, and engine nozzle clamshell covers.

e. Protection of flight hardware against operator-intensive designs and maintenance concepts.

6.2.2.5 Removal of flight equipment covers and/or panels to gain design-required personnel access. This work may include avionics bay panel and flight cover/access door removal, and payload bay liners. Drivers of this work include design requirements for:

a. Intrusive personnel access (i.e., design requires destroying previously certified flight configuration) for routine ground maintenance of flight equipment.

b. Confined spaces in turn driving need for complex remove-before-flight internal access kits (e.g., ladders, lighting, oxygen monitoring, uniquely designed platforms, scaffolding, handholds, footholds, and flight equipment covers).

6.2.2.6 Opening and closing of vehicle hatches and hinged/articulated doors. This may include crew module hatches, airlock hatches, payload bay doors, landing gear doors, element disconnect doors (e.g., ET disconnect doors), and star tracker doors. Drivers of this work include requirements for:

a. Door and hatch locations that are sensitive to the high-temperature environment of atmospheric reentry.

b. Crew internal and external hatch design for maintaining pressurization.

6.2.3 Flight Element Safing Work

This activity arises from a requirement to configure flight systems to protect personnel and equipment, as effectively as possible, from hazardous return-from-flight conditions. This does not include initial precautions that may have been performed immediately prior to launch or immediately after vehicle return at the landing point. This type of work also tends to arise if the operations concept and the vehicle's inherent design qualities call for extensive routine servicing and repair operations in a dedicated servicing and maintenance facility where personnel-intensive activities may be concentrated.

6.2.3.1 Hazardous fluid system safing. These activities may involve the off-loading of toxic, hazardous hypergolic bipropellants (such as that used in the Orbiter's OMS/RCS systems) and toxic, hazardous monopropellants, such as that used for APUs. In addition, this work involves off-loading of stored oxygen and hydrogen from cryogenic fuel cell power reactant storage and distribution (PRSD) systems. Work of this type is driven by:

a. Overaccumulation of required turnaround labor and concentration of vehicle turnaround personnel, in turn driving personnel and equipment hazards, in turn driving safing and deservicing work.

b. Technology limitations caused by seal material properties.

c. Technology limitations leading the designer to specify unique, hazardous fluid commodities.

d. Intrusiveness of vehicle maintenance activities routinely exposing maintenance personnel to various hazardous-fluid systems and risks.

e. No or poor accessibility for basic, planned turnaround tasks, such as fluid replenishment.

f. Lack of subsystem and system design integration into a simple-to-operate design characterized by few parts, few ground interfaces, and few commodities.

6.2.3.2 Pyrotechnic and/or ordnance safing. This work may be required to disconnect (and otherwise "safe") explosive ordnance and pyrotechnic devices not quickly and effectively performed upon return to the landing point. For the Shuttle Orbiter, this includes pyro/ordnance safing in the aft propulsion compartment for the Orbiter/ET umbilical separation function, safing of the landing drag-chute deployment ordnance, and mid-body fuselage/landing gear deployment pyros. Work of this type is driven by:

a. Lack of simplicity in flight element design (i.e., undesirable configuration ["architecture"] from design trade outcome in turn driving added systems and/or explosive components).

b. Inadequate component/system design reliability (i.e., design life or mean time between failures [MTBF] relative to operational environment).

6.2.3.3 Vehicle purges for personnel/flight element safety. This work may involve establishing flight-ground system purge circuits for temperature and humidity, as well as hazardous-gas accumulation and flammability. These operational functions help prevent inherent design risks for fire, explosion, and toxic-vapor exposure to personnel and prevent corrosion from developing on sensitive, critical, and/or high-value components. Work of this type is driven by:

a. Routine, designed-in, internal vehicle access requirements. These requirements are in turn driven by vehicle safing, servicing, and functional verification (such as required physical inspection).

b. Nonroutine internal vehicle access requirements. These requirements are driven by the inherent design life and dependability of vehicle hardware, in turn producing unplanned troubleshooting, repair, reservicing, and retest work.

c. Aerostructure configuration. Configuration can create (1) closed compartments requiring routine personnel access for maintenance and repair, (2) hazardous-gas confinement, and (3) safety monitoring and control of hardware and operations.

6.2.4 Flight Element Systems Servicing Work

This major class of ground activity arises from flight system requirements to routinely return on-board systems to acceptable design criteria. Possible requirements may include fluid systems servicing and mechanical systems servicing, as well as electrical and information systems servicing.

6.2.4.1 Landing gear mechanical servicing. This activity may involve brake calibrations and hydraulic and/or pneumatic strut checks, refill, and repressurization. Work of this type is driven by:

a. Excessive vehicle landing energy (i.e., combination of landing speed and vehicle landing mass) and poorly distributed landing loads.

b. Inadequacy of tire and brake material for landing loads.

c. Landing facility (e.g., runway) incompatibility with landing scheme (for example, runway surface that is too abrasive for loads on tires).

6.2.4.2 Ordnance handling and installation. This activity may involve replacement of single-use ordnance devices (e.g., pyrotechnic flight element separations and parachute deployments), or any designs employing explosive devices and charges. Work of this type is driven by:

a. Lack of simplicity in flight element design (i.e., undesirable configuration ["architecture"] from design trade outcome), in turn driving added systems and/or explosive components.

b. Inadequate component/system design reliability (i.e., design life or MTBF relative to operational environment).

6.2.4.3 Maintenance of fuselage/payload bay/shrouded payload compartment vent filters and crew module payload cavity purges. These may involve routine replacement of filters, filter modules, and desiccants. This type of activity may be driven by:

a. Inadequate design life of filters/desiccants (e.g., 90-day life on Orbiter window cavity purge "racetrack" assembly).

b. Lack of structural margin in crew cabin windows.

c. Lack of fault tolerance and cleanliness tolerance in vent paths, which results in time/labor-consuming maintenance.

6.2.4.4 Flight crew systems servicing. This activity may involve removing cabin-stowed items from the previous flight and performing initial stowage of flight crew cabin equipment prior to rotation to the vehicle integration/launch attitude. This type of activity may be driven by:

a. Number of unique mission preparation items.

b. Colocation of flight crew system items and stowage lockers with vehicle equipment needing routine access, such as avionics and associated thermal management systems (blankets, cold plates, and so forth).

6.2.4.5 Acreage tile and blanket Thermal Protection System (TPS) servicing. This activity may include waterproofing of tile and blankets to prevent water and moisture absorption. This type of activity may be driven by:

a. Use of thermal protection material that requires application of special coating to prevent water or moisture intrusion that itself can weight down the vehicle's TPS layer.

b. Use of customized tiles with time-consuming means of vehicle attachment.

6.2.4.6 TPS moldline penetration, leading-edge/nose and aerosurface hinge line seal servicing. This set of activities conducts routine wear-and-tear adjustments and replacements around the high-temperature mold line penetrations and seals. This type of activity may be driven by:

a. Excessive number of mold line penetrations located in thermally sensitive regions.

b. Lack of materials that are inherently tolerant of routine heating cycles around mechanized components (landing gear doors, hinge lines, etc.).

6.2.4.7 Fluid drain and deservicing. This set of activities includes the removal of various fluid system commodities, either in their entirety or to a safe and/or serviceable quantity. This includes installation of post-deservicing protective devices (e.g., plugs, caps, and covers). This type of activity may be driven by:

a. Use of hazardous, corrosive, and toxic fluids.

b. Design constraints associated with the working pressure of containers and vessels (pressurization cycle limits).

c. Location of hazardous or high-pressure fluid systems too near required personnel or equipment access or within confined spaces.

d. Extended vehicle turnaround durations.

6.2.4.8 Fluid servicing and fill. This set of activities includes any routine filling or replenishment of various fluid commodities to quantify for either test, standby, or flight levels, as well as installation of fluid system elements at off-line facilities. This type of activity may be driven by:

a. Use of hazardous, corrosive, and toxic fluids.

b. Design constraints associated with the working pressure of containers and vessels (pressurization cycle limits).

c. Location of hazardous or high-pressure fluid systems too near required personnel or equipment access or within confined spaces.

d. Extended vehicle turnaround durations.

6.2.4.9 Routine vehicle cleaning and filter inspections. This set of activities includes any routine vehicle cleaning operations and inspection of fluid system filters or sensors for contamination and integrity. This type of activity may be driven by:

a. System susceptibility to particulate contamination.

b. System susceptibility to moisture intrusion.

6.2.4.10 Flight pressurizations. This set of activities includes the fluid system pressurizations to a test, standby, or flight pressure level. This type of activity may be driven by:

a. Intrusive servicing or disassembly requirements that lead to extensive system integrity reverification.

b. Subsystem design that requires deservicing system pressures to provide safety.

6.2.4.11 Navigation and instrumentation component servicing. This set of activities includes the calibration work necessary for maintaining accurate navigation and instrumentation operation. This type of activity may be driven by:

a. Use of air data sensors that routinely drift and require external pneumatic sources for recalibration.

b. Use of sensor processing techniques that require extremely tight calibration tolerances.

c. Use of navigation techniques that require overly precise sensor data.

d. Use of mechanically sensitive optical star trackers—too many moving parts requiring precise alignment and recalibration.

6.2.4.12 Routine replacement of expendable and limited-life items. This activity may include routine replacement of limited-age-life items and limited-operating-life items (active time or operating cycles). *Limited-age-life items* are items that deteriorate with the passage of time and thus require periodic service, operation, replacement, refurbishment, or retesting to ensure thattheir operating characteristics have not degraded beyond acceptable design performance limits. These include installed items. *Limited-operating-life items* are items that deteriorate with increased accumulation of operating time or cycles and thus require periodic replacement or refurbishment by design. Common items include main gear tire (one flight), nose landing gear tire (two flights), and filters or filter elements (often every flight), but also include major components such as main engines (every flight), APUs (47 hours), fuel cells (less than 2,600 hrs), and wing leading-edge panel and seal assemblies (approximately 18 flights). Not included in this category are nonroutine replacements or refurbishments because of unanticipated failures. This type of activity may be driven by:

a. Lack of adequate design life and/or robustness relative to routinely encountered operating environments.

b. Lack of operationally relevant environments for adequately testing and demonstrating design life objectives prior to operational use.

c. Lack of continuous-improvement capability for increasing component dependability and validating design life objectives subsequent to operational deployment.

6.2.4.13 Vehicle electrical power application and data processing system software loading, unloading, and updating. This activity includes any electrical power system calibrations or voltage, current, or impedance adjustments. It also includes routine flight software updates and dumps. (This does not include routine data processing system test and checkout or nonroutine repair and retest). While not generally a major directwork contributor, the flight-relative labor intensity of the flight software management infrastructure and responsiveness is a major design issue.

6.2.5 Flight Element Inspection and Checkout Work

This major class of ground activity is required to gain confidence in the integrity of critical flights systems functions for next flight not obtained through previous flight certification procedures or invalidated since previous flight certification procedure. These may include physical inspections on or within the vehicle, onboard checkout routines (e.g., cockpit checks), or remote command and control from the maintenance facility or from a remote site (control center). This work in turn arises from requirements driven by:

6.2.5.1 Component or subsystem functional verification. This activity includes routine operations required to verify the functional integrity of various components and system functional paths. Confidence in system functional integrity can be compromised by (1) component failure or degradation in performance detected during flight or ground operations, (2) compromise of functional path integrity as a result of flight operations, ground servicing or maintenance (e.g., flight panel removals, disconnection of cables and wiring, plumbing, structural/mechanical hardware), and (3) discovery of fluid system leakage, electrical shorts and opens, cracks, dents, corrosion, and other signs of wear and tear. Verifications may have various turnaround *effectivities,* or required intervals (e.g., every flight, every third flight, on depot maintenance, or on contingency event). The required verification interval depends largely on demonstrated system reliability and functional criticality. This type of activity may be driven by:

a. Lack of demonstrated reliability of function and/or functional integrity.

b. Lack of any redundancy (criticality category 1, or "Crit 1" failure mode).

c. Overreliance on redundancy, accumulating an excessive number of functional paths to verify.

d. Turnaround servicing and access requirements designed into the architecture force the operator to break the functional integrity and consequently require recertification to flight-worthy condition.

e. Lack of dependability (resulting in continual troubleshooting and repair) designed into the architecture, which forces the operator to break the functional integrity and consequently requires recertification to flight-worthy condition.

6.2.5.2 Structural integrity inspections and mechanism functional verifications. A large variety of activities is included in this category, a sampling of which is provided here. For the cockpit and crew cabin, this type of work includes crew cabin equipment fit checks; stowage and destowage checks; pilot seat functional tests; and cabin air recirculation, inspection, and maintenance. In the mid-body fuselage area, the fuselage purge, vent, and drain checkout includes window purge system functional testing, air-to-GN_2 Orbiter purge system functional testing, ground T-0 umbilical quick-disconnect functional testing, and vehicle structural leakage/positive-pressure tests. Also in the mid-body are vehicle jacking and leveling checks; landing gear wheel and wheel well inspections (and supporting access platform installations and fit checks); payload bay door latch functional checks; cryogenic propellant feed line disconnect doors inspections; fuselage vent door functional checks; ultrasound inspections of element separation bolts; vehicle drain hole inspections; inspections of critical areas, including areas inside the propulsion compartment and docking mechanisms inside the payload bay. Other critical areas of inspection around the external portions of the vehicle are the vertical tail/stabilizer; left-hand and right-hand wing structures; element mating attach bolts; thrust structures; and drag-chute compartment. Installation of ordnance devices also requires subsequent resistance checks. Other articulated mechanisms requiring functional checkout are antenna mechanisms checks; left-hand and right-hand air data probe deployment checks; star tracker door operation tests; and airlock and crew cabin hatch checks. The design causes for all this checkout and inspection activity may be driven by:

a. Lack of demonstrated immunity of structure/mechanism design to corrosion, cracks, and other structural defects.

b. Inadequate or unknown and undemonstrated design margins.

c. Inspections that are required to restore confidence in flight-critical structures/mechanisms and certify flight worthiness; some required after vehicle reassembly to flight configuration.

d. Overly complex system design with many active parts and subsystems and undemonstrated reliability requiring functional checkout.

6.2.5.3 Propellant, fluid, and gas system integrity verifications and inspections. This set of processes includes functional verification of fluid subsystem operation and integrity (no leaks), as well as physical inspection of components. This work may include required periodic leak checks or pressurization decay tests. It may also require connection of ground-supplied pressurization sources, as well as reconfiguration of the system under test outside the normal ground test or flight condition. This type of work uses detection equipment or materials, such as optical inspection devices, mass spectrometers, liquid leak detection solutions, ultrasonic leak detectors, or other materials. Functional retest operations with reservicing of the affected system or subsystems are also often required. This type of activity may be driven by:

a. Lack of demonstrated reliability of function and/or functional integrity; reliance on redundancy.

b. Lack of any redundancy (Crit 1 failure mode).

c. Turnaround servicing and access requirements, designed into the architecture, force operator to break the fluid system integrity and consequently require fluid system reservicing and recertification to flight-worthy condition.

d. Lack of dependability and continual troubleshooting and repair, designed into the architecture, force operator to break the functional integrity and consequently require recertification to flight-worthy condition.

6.2.5.4 Vehicle power-up/down, switch lists, functional checks, and onboard software checks. This type of activity includes routine avionics and communications checks upon vehicle power-up. However, the process of powering up the vehicle can be both time-consuming and labor-intensive depending on the level of operability built into the design. Vehicle power application may possess many unwanted system-to-system interactions and constraints created by fluid or propellant system safety issues, pyrotechnic and ordnance implications, complex thermal management constraints for ground ops, and potential for personnel injury and hardware damage caused by mechanism motion upon power up (position versus command offsets not compensated for in the design). In addition, power application and maintenance may require many ground services be readied and sustained for environmental control and thermal management, as well as multiple electrical power services. Many functions require manual cockpit or crew compartment switch-throws that cannot be accomplished remotely or automatically (if not designed into the architecture, as is the case by and large for many Orbiter functions). This type of activity may be driven by:

a. Excessive number of ground-supplied vehicle services (thermal, environmental, power, data, and radio frequency [RF]) required for ground power application and monitoring.

b. Lack of system automation; overaccumulation of manual switches to operate the system.

c. Multiple vehicle telemetry formats/modes needed to conduct ground operations are the cause of excessive processing time and manual procedural steps to repeatedly reconfigure flight software and vehicle system functions.

6.2.6 Vehicle Payload Accommodation Work

This class of flight element turnaround work involves the operations functions required for disconnection and removal of any special airborne support equipment (ASE) not needed or required for next flight; relocation devices for next flight's payload; installation and/or connection of any special ASE to the vehicle for next flight; integrity verification of any newly mated fluid, power, and/or data connections; installation of any needed thermal protection; and other payload closeout operations. This work in turn arises from requirements driven by:

6.2.6.1 Reconfiguration of unique payload thermal protection and insulation. This activity involves the replacement/relocation of special or mission-unique thermal blankets, covers, etc.

a. Lowest-mass payload packaging design. That is, no payload pre-packaging "tare weight" accounted for in design. Perception that cargo carriers and other transportation system packaging solutions impact vehicle performance. Disregards positive impact to vehicle utilization, cycle time, and throughput capability optimization. Design focus tends to be on optimizing mass-to-orbit per flight, rather than throughput capability.

b. Reliance on transportation vehicle for in-space operations.

6.2.6.2 Repositioning of vehicle-fixed mechanisms. This type of work involves actuation and repositioning operations associated with permanently installed vehicle mechanisms, such as payload bay doors and antennas. This type of activity may be driven by:

a. Lowest-mass payload packaging design. That is, no payload pre-packaging "tare weight" accounted for in design. Perception that cargo carriers and other transportation system packaging solutions impact vehicle performance. Disregards positive impact to vehicle utilization, cycle time, and throughput capability optimization. Design focus tends to be on optimizing mass-to-orbit per flight, rather than throughput capability.

b. Reliance on transportation vehicle for in-space operations.

6.2.6.3 Removal and/or installation of mission-unique mechanisms. Examples include remote manipulators, docking ports and adapters, tunnels, bridges, and latches. This work in turn arises from requirements driven by:

a. Lowest-mass payload packaging design. That is, no payload pre-packaging "tare weight" accounted for in design. Perception that cargo carriers and other transportation system packaging solutions impact vehicle performance. Disregards positive impact to vehicle utilization, cycle time, and throughput capability optimization. Design focus tends to be on optimizing mass-to-orbit per flight, rather than throughput capability.

b. Reliance on transportation vehicle for in-space operations.

6.2.6.4 Reconfiguration of vehicle-supplied fluid services. This activity reworks the plumbing interfaces between the vehicle and the payload. It may include purge ducts, water supply, water cooling, ammonia loops, Freon loops, gaseous nitrogen supply, gaseous helium supply, and cryogenic fluids. This type of activity may be driven by:

a. Lowest-mass payload packaging design. That is, no payload pre-packaging "tare weight" accounted for in design. Perception that cargo carriers with embedded fluid services impact vehicle performance. Disregards positive impact to vehicle utilization, cycle time, and throughput capability optimization. Design focus tends to be on optimizing mass-to-orbit per flight, rather than throughput capability.

b. Reliance on transportation vehicle for in-space operations.

c. Excessive number of unique fluid interfaces between the flight vehicle and the payload, requiring removal, installation, and functional verification.

6.2.6.5 Reconfiguration of vehicle-supplied electrical and information services. This activity reworks the electrical/electronic interfaces between the vehicle and the payload. This may include (1) connector interface panels, (2) payload external signal conditioners or switchgear, as required, (3) numerous cables and harnesses, and (4) software. This type of activity may be driven by:

a. Design focus that tends to be on optimizing payload mass-to-orbit per flight, rather than overall payload throughput capability (see Figure 7). This leads to lowest-mass payload integration design without regard to system availability. That is, no payload prepackaging "tare weight" is accounted for in the design process. There is a perception that cargo carriers with embedded power and data services impact vehicle weight. Positive impact on vehicle utilization, cycle time, and throughput capability should, however, be considered in system payload delivery design optimization.

b. Reliance on Earth-to-orbit (ETO) transportation vehicle for in-space mission operations.

c. Excessive number of unique electrical interfaces between the flight vehicle and the payload requiring removal, installation, and functional verification.

6.2.6.6 Access to vehicle payload accommodations. This activity provides maintenance personnel external, mold line access to the payload and internal access to payload items and locations (such as through a crew compartment or access hatch). It may also include intrusive access (requiring disassembly of vehicle structure, panel removal, blanket removal, etc.) to gain access to payload items. This type of activity may be driven by:

a. Lowest-mass payload packaging design. That is, no payload pre-packaging "tare weight" accounted for in design. Perception that cargo carriers with embedded power and data services impact vehicle performance. Disregards positive impact to vehicle utilization, cycle time, and throughput capability optimization. Design focus tends to be on optimizing mass-to-orbit per flight, rather than throughput capability.

b. Reliance on transportation vehicle for in-space operations.

c. Excessive number of unique electrical interfaces between the flight vehicle and the payload requiring removal, installation, and functional verification.

6.2.6.7 Payload accommodations cleaning and closeout. This set of activities may be required for contamination-sensitive payloads and final installation of thermal protection and other final closeout items. This type of activity may be driven by:

a. Lowest-mass payload packaging design. That is, no payload pre-packaging "tare weight" accounted for in design. Perception that cargo carriers with embedded power and data services impact vehicle performance. Disregards positive impact to vehicle utilization, cycle time, and throughput capability optimization. Design focus tends to be on optimizing mass-to-orbit per flight, rather than throughput capability.

b. Reliance on transportation vehicle for in-space operations.

c. Excessive number of unique fluid, mechanical, and electrical interfaces between the flight vehicle and the payload requiring removal, installation, and functional verification.

6.2.6.8 Payload accommodations turnaround unplanned troubleshooting and repair. This set of activities may be required to ensure proper payload operation. This type of activity may be driven by:

a. Lack of payload item and/or payload accommodations system dependability and design life.

b. Excessive number of unique fluid, mechanical, and electrical interfaces between the flight vehicle and the payload, requiring troubleshooting, repair, retest, and re-servicing.

6.2.6.9 Payload accommodations inspections and testing. This set of activities may be required to ensure proper payload operation. This type of activity may be driven by:

a. Payload accommodation design approach requires too many first-time connections and assemblies that in turn require subsequent verification to gain confidence in functionality, alignment, etc.

b. Lack of demonstrated reliability in payload item and/or payload accommodations.

c. Excessive number of unique fluid, mechanical, and electrical interfaces between the flight vehicle and the payload requiring troubleshooting, repair, retest, and reservicing.

6.2.7 Processing Support Work

This class of flight element turnaround work involves routine operations required to configure flight systems to support another system's operation for safing, servicing, or functional verification or to obtain access for troubleshooting and repair. This work in turn arises from requirements driven by the following.

6.2.7.1 Purge operations and monitoring. This set of activity services the various vehicle purge circuits and specialized vehicle compartment purges (payload, propulsion, etc.). This type of activity may be driven by:

a. Excessive number of vehicle compartments that are inherently susceptible to temperature and humidity. These compartments require active ground control of their environment or control of flammability and hazardous-gas accumulation.

b. Susceptibility of metallic heat exchangers to corrosion.

6.2.7.2 Hazard monitoring systems operation. This set of activities involves setup or hookup of monitoring systems such as oxygen analyzers, hazardous-gas detectors, and other means of safety hazard detection.

6.2.7.3 Vehicle system reconfigurations. This set of activities can involve a variety of mechanical systems, such as landing gear movements, aerosurface movements, engine movements, antenna movements, payload bay door movements, hatch openings and closings, software moding and reconfiguration, and jack transfers.

6.2.8 Turnaround Unplanned Troubleshooting and Repair Work

This work includes the nonroutine activity required to recover from discovery of out-of-specification conditions. Initial actions include problem isolation and troubleshooting, which may or may not lead to nonroutine gaining of access, component repair, component removal and replacement, closeout, and functional recertification. This may include (1) items that failed (or were discovered to have failed) during flight, (2) items that failed because of collateral damage sustained during required, intrusive maintenance, and (3) failed items discovered during routine inspection, functional verification, or other routine turnaround operations. This work in turn arises from requirements driven by the following.

6.2.8.1 Troubleshooting, replacement, and disposition of failed or line replaceable units (LRUs). Troubleshooting is the process of isolating a problem to its source. This first requires gaining access to a suspect area and may entail hooking up fluid system reg panels; using drag-on electrical "breakout boxes;" employing "cross-strapping" techniques, where a suspect end item's upstream functional path is reconnected to a similar good end item to observe whether the anomaly follows to help isolate the problem; using special equipment such as meggars, time delay reflectometers (TDRs) to isolate open circuits to a particular cable segment in very long cable runs; using mass spectrometers and various leak detection methods and equipment; using x-ray, ultrasound, capacitive field detection and other nondestructive evaluation (NDE) techniques and devices. Replacement may require further teardown of the flight-certified configuration (protective panels, covers, thermal blankets, structural brackets, mechanical rigging, etc.) and special access kits, handling equipment, and other special hardware, such as device installers. Functional retest operations are required and may also call for a combination of reservicing, recalibration, rerigging, or realignment, depending on the complexity of the device. Final engineering acceptance of replaced LRUs also includes documenting and approving discrepant conditions and anomaly closeout rationale. This type of activity may be driven by:

a. Inadequate design life requirements.

b. Excessive power-on time and cycles created by other time-consuming ground functions (e.g., servicing and checkout).

c. Overaccumulation of intrusive personnel traffic required for routine vehicle turnaround—a second-order processing effect.

d. Inadequate reliability requirements.

e. Inadequate seal material.

f. Inadequate electrical insulation materials and connector designs.

6.2.8.2 Troubleshooting and repair of leaks. This type of activity includes troubleshooting and repair of leaking ducts, tubes, hoses, valves, or other fluid containment equipment. This often requires connection of GSE pressurization sources and reconfiguration of the affected system beyond the normal ground test or flight condition. It also often includes the use of detection equipment or materials, such as mass spectrometers, liquid lead detection solutions, or ultrasonic leak detectors. This type of activity may also involve hazardous operations or extended-duration pressure decay tests and evaluations. Functional retest operations are required, as well as reservicing of the affected system. This type of activity may be driven by:

a. Design choices using damage-sensitive mechanical fittings.

b. Use of limited-life or marginal-design components that require frequent removal or intrusive inspection (frequent joint demates).

c. Location of leak-sensitive equipment in areas requiring high personnel or equipment traffic.

6.2.8.3 **Fluid or pneumatic system decontamination and cleaning.** This work involves the removal of foreign material or contaminants from the internal or external surfaces of fluid lines, ducts, reservoirs, and other components, and reverification of acceptable cleanliness in the exposed areas. It also often requires subsequent functional testing, re-servicing, and closeout activities. This type of activity may be driven by:

a. Insufficient performance insight or sampling required to detect contamination.

b. Improper choices for compatible lubricants and other materials inside fluid flow streams.

c. Fluid system intrusions that introduce contaminants.

d. Inadequate designs or materials used for seals.

e. Use of toxic or hazardous materials that could be liberated or expose adjacent hardware.

6.2.8.4 **Troubleshooting, repairing, and retesting electrical cables and connectors.** This type of activity includes locating wire or connector damage or other interfering electrical problem. This may involve added operations for system power-up, vehicle access to wire and cable trays, installation of breakout or breakthrough boxes (with appropriate fuse configurations and location and installation of clamp-on ammeters), and examination of connectors for recessed pins or foreign contamination. This activity may also include cable splicing repairs, replacement of entire cable segments and/or connectors and feed-throughs. It often involves the inspection of wiring insulation for wear and tear from brittleness or vibration causing cracks and other failures of the insulation material. Operations often involve relocating cables away from nearby sharp or abrasive fraying sources. This type of activity may be driven by:

a. Excessive number of wire and cable paths.

b. Excessive number of connectors.

c. Lack of electrical/electronic equipment design integration into a minimum number of LRUs with minimum number of electrical paths.

d. Inaccessible location of avionics equipment.

e. Electrical wiring material and connector designs inadequate for repeated exposure to flight environment.

6.2.8.5 Repair and replacement of Thermal Protection System (TPS) hardware. This activity includes the unplanned repair of thermal protection tiles damaged during launch, flight, landing, and ground processing operations. Unplanned tile removals are performed for access to other discrepant hardware, fluid contamination, resolving gap-filler or room-temperature vulcanization (RTV), strain isolation pad (SIP) damage, or instrumentation anomalies (for instrumented TPS). This type of activity may be driven by:

a. Susceptibility to damage from launch, landing, and routine ground turnaround environment.

b. Requirement to remove TPS for access to resolve problems in other areas.

6.2.8.6 Structural repair and refurbishment. This type of activity includes repair of corrosion damage, damaged captive fasteners, rivets, impact damage, and damaged or delaminated composite structures. This damage often occurs during launch, landing, and ground turnaround activities. There are repairs of scratches, gouges, and dents caused by high traffic and labor-intense servicing and inspection activities. Installation of structural doublers is occasionally required. There are also unplanned structural repairs of failed thermal protection systems. Some of the structural repairs result from discrepancies observed during closeout inspections prior to major flight element moves or launch. This type of activity may be driven by:

a. Use of materials in the design that are susceptible to corrosion.

b. Lack of structural margins to tolerate reasonable corrosion over the lifetime of the airframe.

c. Inadvertent design of moisture traps. Inadequate design testing to locate moisture traps.

d. Requirement for routine ground personnel activity near damage-susceptible structures.

6.2.8.7 Repair of damaged ducts, tubes, and hoses. This type of activity involves Kevlar/epoxy purge ducts, which are repaired by patching and/or touching up damaged conductive coating or sometimes remanufacturing the duct section. It also includes repair or replacement of damaged tubes, hoses, or other fluid system interconnection equipment. This type of activity may be driven by:

a. Weakness in compression of Kevlar/epoxy ducts (though they are stron in tension) and their susceptibility to crush damage.

b. Location of sensitive Kevlar/epoxy ducts in areas requiring routine ground personnel traffic.

c. Use of fragile duct/tube material in areas requiring routine vehicle ground maintenance.

d. Inadequate definition of acceptable damage criteria.

e. Inadequate design of hardware, requiring difficult or cumbersome activity to gain access for inspection, repair, or replacement.

6.2.8.8 Troubleshooting and repair of mechanisms and thermal/pressure seals. This work includes repairing damage caused by routine personnel traffic, excessive binding of mechanisms, misadjustment, and interference with other structures. Repair consists of replacing damaged seals and bearings/bushings, correcting misalignments, eliminating interferences, and readjusting mechanisms or actuators. This type of activity may be driven by:

a. Overaccumulation of intrusive personnel traffic required for routine vehicle turnaround—a second-order processing effect.

b. Mechanism/seal wear—inadequate design life.

c. Mechanism design that is conducive to misalignment and interference.

d. Seals that are by design not adequately protected from damage susceptibility.

6.2.8.9 Troubleshooting and repair of ground equipment. This involves the work encountered during the course of vehicle turnaround to troubleshoot, repair, and retest faulty GSE. This type of activity may be driven by:

a. Excessive amount of required ground equipment to support vehicle design for turnaround.

b. Excessive number of ground interfaces.

c. Program reuse of old GSE (use of GSE beyond its original design intent).

d. Inadequate design life of GSE. Underestimation of required GSE utilization per turnaround.

6.2.9 Modifications and Special Testing

This type of activity includes modification of flight components and subsystems between flights or special engineering tests between flights. This excludes similar work performed during periodic, long-term downtimes for depot maintenance. This work in turn arises from requirements driven by the following.

6.2.9.1 Flight element modifications and process changes. This entails relatively urgent design changes and upgrades that are of a safety nature that should not to wait until depot.

6.2.9.2 Special tests and fleet system checks. This work involves suspect design problems, design checks and reports, and other items similar to FAA circulars and airworthiness directives.

6.2.9.3 Payload-driven flight element weight and center-of-gravity alterations. This involves ballasting alterations, component removals, etc., related to payload margin limits and center-of-gravity adjustments.

6.2.10 Flight Element Closeout Work and Removal of Access and Umbilicals

This work involves functions required to remove equipment previously located on, in, or connected to the flight vehicle. This work in turn arises from requirements driven by the following.

6.2.10.1 Removal of umbilical and access equipment. This work is associated with disconnects, service disconnection, internal access kit removals, platform retraction. This type of activity may be driven by:

a. Number of ground interfaces.

b. Number of internal ground servicing connections.

c. Level of vehicle system, subsystem, and component reliability driving component removals needing access platforms and special access kits.

d. Amount of limited-life items needing internal access to replace.

6.2.10.2 Flight element system closeouts for vehicle and ground launch system integration. This work involves final installation of vehicle areas with a high degree of multisystem maintenance (such as avionics bays, payload bay, or aft system compartment) or highly complex mechanisms (such as aerosurface hinge line seals) near TPS that requires tile work completion before proceeding with the final rigging. The closeout work involves installation of flight doors, panels, thermal protection carrier panels, closeout seals, and so forth. This type of activity may be driven by:

a. Interior-located fluid service ports.

b. Dependability and design life of systems, subsystems, and components, in turn driving unplanned deconfiguration of previously flight-certified areas within the vehicle for unplanned troubleshooting and repair.

c. Overly-complex mechanical rigging schemes for aerosurfaces and landing gear.

6.2.10.3 Preps for vehicle element transfer. This work involves weighing and locating the vehicle's centerofgravity jack transfers and preparing the vehicle and its transporter for move to the next facility. This type of activity may be driven by:

a. Requirement for the vehicle to be in different jacking positions for routine maintenance.

b. Use of specialized vehicle transporters in the operations concept.

c. Very small vehicle weight and center-of-gravity margins.

6.2.11 Turnaround Facility and Equipment Periodic Maintenance

This work includes required calibrations; functional verifications of facility services and associated facility service outages; lifting-device proof-loading; and hazardous waste removal. This work in turn may arise from requirements driven by the following.

6.2.11.1 Interval maintenance of ground systems. This includes maintenance of a variety of electrical supplies required by the flight system, air and nitrogen purges, hazardous-gas scrubbers, emergency alarm system testing, cranes, articulated access structures, powered facility doors, communications infrastructure, radio frequency (RF) equipment, etc. This type of activity may be driven by:

a. Number of separate electrical power services required (e.g., single-phase, dual-phase, three-phase alternating-current (ac) power; separately supplied frequencies; separately supplied direct-current (dc) voltage and current supplies).

b. Number of separately supplied fluid and gas commodities.

c. Number of toxic and hazardous materials required by the vehicle for routine maintenance.

6.2.11.2 Ground system and facility modifications and process changes. This type of activity may be driven by:

a. Number of separate electrical power services required (e.g., single-phase, dual-phase, three-phase ac power; separately supplied frequencies; separately supplied dc voltage and current supplies).

b. Ease of upgrading facilities for communications and data transfer systems.

c. Number of separately supplied fluid and gas commodities.

d. Number of toxic and hazardous materials required by the vehicle for routine maintenance.

6.3 Flight Element Assembly Work Causes

This major classification of work is required to structurally assemble and mate components and segments of a flight element as received at the operational launch area or spaceport. This classification of work also includes the transfer of assembled flight elements to the flight vehicle integration point. Thus, this work includes the assembly activities associated with the SRB, the assembly and disassembly activity associated with the Shuttle Orbiter, such installation and removal tasks for the SSMEs, the forward and aft auxiliary propulsion pods (FRCS and OMS pods), and the assembly and removal of the Remote Manipulator System (RMS) arm. This class of work does not include the mating of the Orbiter to the ET, nor the payload lift and mate operations into the Orbiter flight element. This work in turn arises from requirements driven by the following.

6.3.1 Element Transport, Lift, and Handling for Assembly

This type of work involves receiving the primary flight element segments, and the necessary ground handling hardware into the assembly facility. This work in turn arises from requirements driven by the following.

6.3.1.1 Facility preparations for element assembly operations. This work stages and sets up handling equipment, fixtures, and access platforms. This type of activity may be driven by:

a. Deferred element manufacturing and assembly functions required to be performed by the operator by design.

b. Internal and/or elevated access required to perform assembly operations.

c. Element size and dry weight (fully loaded) (driven by choice of solid propulsion), which in turn drive separate, large-segment element assembly by the operator.

6.3.1.2 Ground transport of major element segments. This type of activity may be driven by:

a. Deferred element manufacturing and assembly functions required to be performed by the operator by design.

b. Element size and dry weight (fully loaded) (driven by choice of solid propulsion), which in turn drive separate, large-segment element assembly by the operator.

6.3.1.3 Lift, positioning, and connection to assembly facility support fixtures. This type of activity may be driven by:

a. Deferred element manufacturing and assembly functions required to be performed by the operator by design.

b. Internal and/or elevated access required to perform assembly operations.

c. Element size and dry weight (fully loaded) (driven by choice of solid propulsion), which in turn drive separate, large-segment element assembly by the operator.

6.3.1.4 Activation of element environmental and safety controls. This type of activity includes ground conditioning of the flight element, blanket purges, electrical grounding, safety ropes, etc. This type of activity may be driven by:

a. Choice of solid propulsion, which drives facility clears of all personnel not essential to lifting tasks.

b. Choice of low-margin, lightweight propellant pressure vessels for element, which drives need to maintain blanket pressure in element during element assembly and handling.

c. Blanket purges, which may also be required for moisture protection.

6.3.2 Element Structural Assembly

This type of work includes such functions as Solid Rocket Motor/Booster segment stacking and payload fairing assembly (for Expendable Launch Vehicles [ELVs]). This work in turn arises from requirements driven by the following.

6.3.2.1 Flight segment handling operations. This work is required for preparing flight element segments, lifting fixture attachment, and positioning for assembly. This type of activity may be driven by:

a. Deferred element manufacturing and assembly functions required to be performed by the operator by design.

b. Use of segmented solid-propulsion concepts.

6.3.2.2 Flight segment structural connections. This work is required for physical alignment and connection of the element segments, critical seal installation, and pinning. This type of activity may be driven by:

a. Deferred element manufacturing and assembly functions required to be performed by the operator by design.

b. Use of segmented solid-propulsion concepts.

6.3.2.3 Structural assembly, verification, and closeout. This work is required solid motor band installation; seal leak checks, gap checks, and port plug installations. This type of activity may be driven by:

a. Choice of solid-propulsion field joint design, which drives time-consuming, labor-intensive functional verification of critical primary, secondary, and tertiary seals and subsequent closeout of numerous test ports.

b. Use of liquid-propulsion engines requiring intensive turbopump and other internal, intrusive inspections to gain confidence for next use. Imposes either sever *in situ* inspection on or in the vehicle or disassembly of the engine modules from the vehicle and subsequent reinstallation of an engine.

6.3.3 Subsystem and Component Installations

This type of activity is needed when the design requires various subsystems and components be installed by the operator at the assembly and launch area. Examples include battery installation, Range Safety Subsystem hardware and ordnance, interconnecting cables, wiring, cable trays, systems tunnels, fairings, closeout panels, and thermal protection material application. This work in turn arises from requirements driven by the following.

6.3.3.1 Subsystem and component handling operations. This work involves installation of handling fixtures and equipment; logistics kitting; component and equipment spotting; and handling gear configurations, adjustments, field calibrations and any other necessary preparatory tasks. This type of activity may be driven by:

a. Deferred element manufacturing and assembly functions required to be performed by the operator by design.

b. Choice of segmented solid-propulsion element design, which drives numerous field closeouts of the integrated, fully assembled segments (electrical, thermal, and structural/aerodynamic functions).

c. Use of liquid-propulsion engines requiring intensive turbopump and other internal, intrusive inspections to gain confidence for next use. Imposes either sever *in situ* inspection on or in the vehicle or disassembly of the engine modules from the vehicle and subsequent reinstallation of an engine.

6.3.3.2 Facility preparations for installation. This work includes tasks involved in setting up any necessary environmental enclosures and access platforms and stands. This type of activity may be driven by:

a. Deferred element manufacturing and assembly functions required to be performed by the operator by design.

b. Choice of segmented solid-propulsion element design, which drives numerous field closeouts of the integrated, fully assembled segments (electrical, thermal, and structural/aerodynamic functions).

c. Use of liquid-propulsion engines requiring intensive turbopump and other internal, intrusive inspections to gain confidence for next use. Imposes either sever *in situ* inspection on or in the vehicle or disassembly of the engine modules from the vehicle and subsequent reinstallation of an engine.

6.3.3.3 Flight component installation onto flight element. This work involves the installation of various vehicle or flight element modules/components and major assemblies, such as the SSMEs, the forward RCS pod, the aft OMS/RCS pods, and the RMS onto the Orbiter. It also includes cable harnesses, heater assemblies, attach struts, and batteries on the SRBs, and the ground umbilical carrier panel (GUCP) on the ET. This type of activity may be driven by:

a. Lack of design life and inherent system dependability, causing routine liquid-propulsion engine and auxiliary propulsion pod assembly/disassembly from the vehicle required between depot maintenance intervals.

b. Lack of adequate operational payload and center-of-gravity margin, requiring frequent removal and reinstallation of the RMS arm and its equipment by the operator (rather than just leaving it permanently installed on one vehicle or leaving it permanently removed on another).

c. Deferred element manufacturing and assembly functions required to be performed by the operator by design.

d. Choice of segmented solid-propulsion element design, which drives numerous field closeouts of the integrated, fully assembled segments (electrical, thermal, and structural/aerodynamic functions).

6.3.3.4 Nonflight component installations. This work involves installation of crew module window covers, actuator locks, ground doors, and other remove-before-flight items. This type of activity may be driven by:

a. Overaccumulation of intrusive personnel traffic required for routine vehicle turn-around—a second-order processing effect.

b. Sensitive flight-critical hardware exposure time to environmental and required ground handling operations, which drives installation of numerous remove-before-flight items.

c. Control system actuators whose design tendency is to "drift" under gravity loads away from position at power-off, thus requiring operator installation of external actuator locks to prevent engine collision or undesirable/uncontrolled aerosurface motion or drift (e.g., during ferry flight).

6.3.3.5 Flight component installation closeouts. This work may involve application of thermal protection material (foams, coatings, etc.) and aerodynamic fairings and protective covers. This type of activity may be driven by design requiring element aerodynamic fairings and thermal protection application due to multiple-element, field-assembled vehicle architecture.

6.3.4 Flight Element Assembly Functional Verifications

This type of activity is needed following assembly of functional hardware. This may include physical inspections, systems checks of newly assembled functional electronic hardware paths, leak checks for any assembled fluid or gas lines, and load-bearing checks for assembled structural/mechanical assemblies (e.g., element attach struts). This work in turn arises from requirements driven by the following.

6.3.4.1 Passive system verifications. This work may involve passive fluid and gas system leak checks and electrical polarity and continuity. This type of activity may be driven by multiple, field-installed systems requiring subsequent functional validation to verify functional path integrity and that safety-critical hardware has been assembled according to design.

6.3.4.2 Active system verifications. This work may involve active command, control, and actuation of assembled active systems. This may involve thrust-vectoring flight control subsystems or purge system valve actuation, as examples. This type of activity may be driven by a multiple, field-installed systems requiring subsequent functional validation to verify functional path integrity and that safety-critical hardware has been assembled according to design.

6.3.4.3 Inspections and alignment verifications. This work may involve safety walkdowns, closeout photography, alignment checks, and other visual inspections. This type of activity may be driven by:

a. Need for documenting as-built assembly for reconstruction of field-manufactured process.

b. Multiple, field-installed systems requiring subsequent functional validation to verify functional path integrity and that safety-critical hardware has been assembled according to design.

6.3.5 Assembly Unplanned Troubleshooting and Repair

This work arises from requirements driven by the following.

6.3.5.1 Unplanned element disassembly and reassembly. This may involve destacking of SRBs, unplanned SSME removals, unplanned pod removal, disassembling previously closed-out areas (removal of fairings and TPS); removals and disconnections (system tunnels, struts, associated nonflight supports). This type of activity may be driven by:

a. Late, untimely discovery of manufacturing process concern, which drive element disassembly, verification, corrective action, and reassembly.

b. Numerous flight element component malfunctions caused by a lack of adequate design life/dependability and failure rate exposure created by high-parts-count design.

6.3.5.2 Flight element troubleshooting and repair. This may involve troubleshooting heaters, cables and wiring, and seals, for example. This type of activity may be driven by:

a. Late, untimely discovery of manufacturing process concerns.

b. Numerous flight element component malfunctions caused by a lack of adequate design life/dependability and failure rate exposure created by high-parts-count design.

6.3.5.3 Flight element modifications and process changes. This activity involves changes of materials, replacements of LRUs, and new assembly processes changing the level or work. This type of activity may be driven by a lagging corrective action process on critical items (hardware and functions) causing remanufacturing in the field rather than at the manufacturing source.

6.3.6 Postassembly Operations

This includes disconnection and stowage of assembly and access ground hardware. This work arises from requirements driven by the following.

6.3.6.1 Refiguration and/or relocation of facility and ground equipment. This set of activities involves safety ropes and rail teardown; access platform reconfigurations (retraction, flipup, etc.); and stowage of special handling equipment, lifting sling sets, etc. This type of activity may be driven by:

a. Deferred element manufacturing and assembly functions required to be performed by the operator by design.

b. Choice of segmented solid-propulsion element design, which drives numerous field closeouts of the integrated, fully assembled segments (electrical, thermal, and structural/aerodynamic functions).

c. Use of liquid-propulsion engines requiring intensive turbopump and other internal, intrusive inspections to gain confidence for next use. This imposes either sever *in situ* inspection on or in the vehicle or disassembly of the engine modules from the vehicle and subsequent reinstallation of an engine.

6.3.6.2 Flight element preparations for integration. This activity involves configuring interface hardware for integration and temporary ground support installations. This type of activity may be driven by the choice of multielement, field-assembled vehicle or architecture.

6.3.7 Assembly Facility and Equipment Periodic Maintenance

This work arises from requirements driven by the following.

6.3.7.1 Interval maintenance of assembly ground systems. This type of activity may be driven by a multitude of flight and ground elements and/or segments and associated operational requirements that drive numerous ground support systems and assembly facilities needing wear-and-tear maintenance and upgrade to support changes in the flight element design and regulation compliance.

6.3.7.2 Assembly ground system and facility modifications and process changes. This type of activity may be driven by a multitude of flight and ground elements and/or segments and associated operational requirements that drive numerous ground support systems and assembly facilities needing wear-and-tear maintenance and upgrade to support changes in the flight element design and regulation compliance.

6.3.8 Routine Reusable Flight Element Module/Component Disassembly

This work involves the routine removals of SSMEs, aft and forward OMS/RCS pods, and Remote Manipulator System (RMS) robotic arm, and Extended-Duration Orbiter (EDO) pallet, and so forth. This work arises from requirements driven by the following.

6.3.8.1 Disassembly preparations and attachment and/or closeout hardware removal. This work includes the staging of disassembly handling equipment and the removal of flight carrier panels and other thermal protection hardware, such as engine heat shields. This type of activity may be driven by:

a. Lack of design life and inherent system dependability requiring routine liquid-propulsion engine and auxiliary propulsion pod assembly/disassembly from the vehicle between depot maintenance intervals.

b. Lack of adequate operational payload and CG margin requiring frequent removal and reinstallation of the RMS arm and its equipment by the operator (rather than just leaving it permanently installed on one vehicle or leaving it permanently removed on another).

6.3.8.2 Flight element module/component disassembly, handling, and transport. This work includes lifting and positioning operations, installation on transporter, and removal to offline storage and maintenance areas or facilities. This type of activity may be driven by:

a. Lack of design life and inherent system dependability requiring routine liquid-propulsion engine and auxiliary propulsion pod assembly/disassembly from the vehicle required between depot maintenance intervals.

b. Lack of adequate operational payload and CG margin requiring frequent removal and reinstallation of the RMS arm and its equipment by the operator (rather than just leaving it permanently installed on one vehicle or leaving it permanently removed on another).

6.3.8.3 Post-disassembly operations. This work includes operations to remove and store ground handling fixtures and transporters. This type of activity may be driven by:

a. Lack of design life and inherent system dependability requiring routine liquid-propulsion engine and auxiliary propulsion pod assembly/disassembly from the vehicle required between depot maintenance intervals.

b. Lack of adequate operational payload and CG margin requiring frequent removal and reinstallation of the RMS arm and its equipment by the operator (rather than just leaving it permanently installed on one vehicle or leaving it permanently removed on another).

6.4 Detailed Vehicle Integration Work Causes

This major classification of work is required to structurally assemble and mate the flight elements into an integrated spaceflight vehicle and mate to the ground support systems. This classification of work does not include the transfer of the assembled vehicle to the launch point but does include the preparation at the integration point for subsequent transfer, if required by the concept. Thus, this work includes the assembly activities associated with mating the ET to the left- and right-hand SRBs and the mating activities associated with attaching the Shuttle Orbiter to the ET. While the Space Shuttle operational concept performs most of these functions in the VAB, this general class of work does include the payload lift and mate operations into the Orbiter flight elementsince this is considered (functionally) to be a vehicle integration task, regardless of where it is performed in a specific operations concept. This work in turn arises from requirements driven by the following.

6.4.1 Vehicle Element Transportation, Handling, Lifting, and Mating

This work involves receiving the various flight elements and ground handling hardware at the vehicle integration point and mating to the ground launch support element. This work arises from requirements driven by the following.

6.4.1.1 Facility preparations for flight element integration operations. This activity includes personnel access platform operations, spotting support equipment, ground interface alignment, access configuration, and rotational equipment operations. This type of activity may be driven by:

a. Position and asymmetric cross-section geometry of element arrangement for maintenance and assembly-intensive designs that force complex access platform design, setups, and operation.

b. Complexity of vehicle holddown design requiring time-consuming, manually intensive assembly operations; and alignments, tensioning, attachment, and assembly of complex separation and ordnance hardware.

c. Need for separate and numerous ground umbilicals and element interfaces that create extra covers and operations to remove and install separate covers and interface protection devices (flight and ground equipment).

6.4.1.2 Ground transportation of flight element. This work involves moving and positioning the various flight elements for subsequent lift and mate operations. This work arises from requirements driven by the number of different elements and stages that in turn drive separate element movement operations, reorientations, and often uniquely designed transporters that are themselves complex to operate and support.

6.4.1.3 Element lift, position, and mate. This work involves receiving the various flight elements and ground handling hardware at the vehicle integration point and mating to the ground launch support element. This work arises from requirements driven by number and complexity (i.e., manual operations) of separately mated:

a. Flight-element-to-flight-element structural connections.

b. Ground-element-to-flight-element structural connections.

c. Flight-element-to-flight-element propellant and fluid/gas connections.

d. Ground-element-to-flight-element propellant and fluid/gas connections.

e. Flight-element-to-flight-element electrical/electronic connections.

f. Ground-element-to-flight-element electrical/electronic connections.

6.4.1.4 Postmate servicing and interface closeouts. This work involves fairing installations, thermal protection closeouts, and establishment of vehicle servicing (initiate purges, etc.). This work arises from requirements driven by the number and complexity of manual operations required to:

a. Structurally close out the element-to-element attachments and umbilicals (flight and ground), including attachment of aerodynamic fairings, access way covers, etc.

b. Thermally close out the element-to-element attachments and umbilicals (flight and ground), including installation of thermal curtains, protective foam/ablator application, etc.

6.4.2 Postmating Functional Verification Work

This work involves checking newly assembled interfaces via lead checks, continuity and connector integrity, etc. This work arises from requirements driven by the following.

6.4.2.1 Element-to-ground-system interfaces. This activity includes electrical checks, leak checks, and purge verifications. This type of activity may be driven by:

a. Functional verification of flight-element-to-ground-element electrical/electronic integrity (continuity and system load functionality).

b. Functional verification of flight-element-to-ground-element propellant, fluid, and gas integrity (continuity and system load functionality).

c. Functional verification of flight-element-to-ground-element structural integrity (continuity and system load functionality).

6.4.2.2 Element-to-element interfaces. This activity includes functional path verification of flight-element-to-flight-element umbilicals. This type of activity may be driven by:

a. Functional verification of flight-element-to- flight-element electrical/electronic integrity (continuity and system load functionality).

b. Functional verification of flight-element-to-flight-element propellant, fluid, and gas integrity (continuity and system load functionality).

c. Functional verification of flight-element-to-flight-element structural integrity (continuity and system load functionality).

6.4.3 Vehicle Integration Unplanned Troubleshooting and Repair

6.4.3.1 Disconnection of flight elements. This activity includes operational situations where unplanned demating of vehicle flight elements occurs because of an anomaly or other concern. For the Space Shuttle, this would include disconnecting the Orbiter from the ET, disconnecting the ET from the SRB, and destacking the SRBs. This type of activity may be driven by late, untimely discovery of anomaly or manufacturing process concern requiring vehicle disassembly.

6.4.3.2 Interface troubleshooting and repair.

6.4.3.3 Postintegration flight element troubleshooting and repair.

6.4.3.4 Postintegration ground element/equipment troubleshooting and repair.

6.4.4 Payload Integration Work

6.4.4.1 Facility preparations for payload installation. This activity includes preparation of any special payload integration clean rooms (such as the LC-39 Payload Changeout Room [PCR] and handling of containers and/or canisters). This activity also includes hoisting and transferring payload structural load to any specialized payload installers (such as the LC-39 Payload Ground Handling Mechanism [PGHM]) and demating and removing canisters or containers.

6.4.4.2 Payload installation into vehicle. This activity includes rotating and/or positioning the payload into the flight vehicle. It also includes vehicle payload bay door operations, preinstallation payload servicing, and the payload mate to the vehicle.

6.4.4.3 Postinstallation operations. This set of activities involves establishing purges, if required; initial payload commodity servicing; trickle-charging of payload power sources, if required; and deconfiguring any ground handling and access equipment. This type of activity may be driven by:

a. Payload accommodation design approach requiring too many first-time connections and assemblies, in turn requiring subsequent verification to gain confidence in functionality, alignment, etc.

b. Excessive number of unique fluid, mechanical, and electrical interfaces between the flight vehicle and the payload requiring troubleshooting, repair, retesting, and reservicing.

6.4.4.4 Payload handling and transport operations. This set of activities involves payload shipping and receiving at the vehicle-payload integration site and includes moves, lifts, and transfers.

6.4.4.5 Payload inspections and testing. This set of activities involves payload functional path verification between the payload packages and the vehicle. It also includes sharp-edge inspection and crew familiarity activity.

6.4.5 Configuration of Vehicle for Transfer and Rotation to Launch Point

This work involves disconnecting and stowing assembly and mating ground hardware. This work arises from requirements driven by the following.

6.4.5.1 Disconnecting and stowing integration and mating ground hardware. This set of activities involves platform retractions and stowage, portable purge stowage, facility power disconnection, facility gas connections, etc.

6.4.5.2 Configuration of vehicle systems for launch functions. This set of activities includes power-down for transport, installation of protective covers, and securing of items for transport.

6.4.5.3 Preparations for integrated vehicle transfer/rotation to launch position. This set of activities includes preparation for transport roadway and transporter operations.

6.4.6 Vehicle Integration Facility and Equipment Periodic Maintenance

6.4.6.1 Interval maintenance of integration ground systems.

6.4.6.2 Integration ground system and facility modification and process changes.

6.5 Flight Vehicle Launch Operations Work Causes

This major classification of work is required to provide final preparation, servicing, payload installation, and/or crew ingress, as well as launching the flight vehicle into space. It also includes the postlaunch securing functions, as well as refurbishment of the facilities and equipment at the launch point. It encompasses replenishment of liquid and gas commodities for the next launch. This work in turn arises from requirements driven bythe following.

6.5.1 Launch Facility and Equipment Refurbishment, Reservicing, and Preparations for Vehicle Arrival

This work includes operations required to return the facility and its equipment to a serviceable condition. This work arises from requirements driven by the following.

6.5.1.1 Launch equipment refurbishment and reservicing. This set of activities includes restoration of ground hardware to flight-serviceable condition. Ground systems and equipment are verified and controlled for cleanliness and functionality prior to use on flight systems for launch operations. This type of activity may be driven by:

a. Manual systems with no or little automated health management capability in ground systems.

b. Lack of reliability and dependability for ground systems used in routine launch operations.

c. Multitude of separate ground systems inherently required by the vehicle's design (cumulative number of different fluid types, electrical power types, launch umbilical connections, and so forth).

6.5.1.2 Contamination control. This work involves fluid system cleanliness control of for sensitive ground, flight, and payload systems. It also includes work associated with debris and foreign object debris (FOD) prevention. This type of activity may be driven by:

a. Separate fluid system cleanliness requirements (LOX versus hypergols, etc.).

b. Payload clean-room requirements to accommodate late access/installation at the pad.

c. FOD impact prevention requirements.

d. Exposed payloads sensitive to contamination (no designed-in protective payload packaging scheme accounted for in the concept).

6.5.1.3 Functional verifications of ground systems prior to flight vehicle mating. This work involves inspections and checkout of critical ground systems at the launch facility (pad) and on mobile launch equipment (e.g., MLP). This type of activity may be driven by a multitude of separate ground systems inherently required by the vehicle's design (cumulative number of different fluid types, electrical power types, launch umbilical connections, and so forth).

6.5.1.4 Servicing of ground systems prior to flight vehicle mating. This work involves servicing ground systems at the pad and MLP. This type of activity may be driven by a multitude of separate ground systems inherently required by the vehicle's design (cumulative number of different fluid types, electrical power types, launch umbilical connections, and so forth).

6.5.2 Positioning Vehicle at Launch Point/Orientation, Mating to Ground Facility and Services, and Providing Access as Necessary

This work includes operations required to configure/position the mated flight vehicle into its launch/departure orientation. It also includes configuring launch servicing equipment, such as servicing towers, swing arms, access devices, and any weather/environmental protection equipment. This work arises from requirements driven by the following.

6.5.2.1 Vehicle transportation and alignment. This may include transporter operations, precision alignment, and leveling. This type of activity may be driven by:

a. Mobile launch concept that separates integration point from launch point.

b. Multitude of flight and ground interfaces

c. Lack of operational margin in dimensional design tolerances that "stack-up" and make alignment operations difficult.

6.5.2.2 Umbilical mating, environmental protection, and access positioning. This may include launch equipment operations, such as the LC-39 RSS, umbilical/swing arms, weather protection doors, etc. This type of activity may be driven by:

a. Location and distribution of numerous interfaces (propellant, power, data, crew/payload services, environmental purges, etc.).

b. Vehicle environmental protection (length of exposure time at launch point).

c. Providing access for manual operations required at the launch point (umbilical connections, inspections, vehicle entry, payload installation, etc.).

6.5.3 Functional Verification of Flight and Ground Systems for Launch

This may include activities associated with verifying integrity of umbilical mates and with functional verification of critical equipment for subsequent launch operations. This type of activity may be driven by the following.

6.5.3.1 System integrity verification. This set of activities includes various integrity checks of newly mated launch-facility-provided services. This may include leak checks, continuity checks, power quality verification, fluid flow rates and pressures, and so forth. This type of activity may be driven by:

a. Manually operated systems.

b. Multitude of unique systems.

c. Criticality of systems/interfaces.

d. Lack of demonstrated reliability that requires flight-by-flight certification of systems.

6.5.3.2 System functionality verification. This activity includes system confidence checks, torque checks, and hot-fires.

a. Criticality of flight-to-ground interface systems (e.g., vent arms, swing arms, and safing systems).

b. Hazardous-component validation because of late replacement/changeout (e.g., APU/hydrolic power unit [HPU] hot-fire runs).

c. Verification of safety management and control systems (e.g., hazardous-gas detection systems [HGDS], control room hardwire safing functions, personnel/hardware protection, and fire suppression).

6.5.3.3 Human systems verification and training. This activity includes crew system familiarization and simulations for the launch team. This type of activity may include:

a. Flight crew familiarization with actual flight and ground systems, emergency procedures, and launch team (e.g., S0017 – Terminal Countdown Demonstration Test).

b. Launch team training and simulations for contingency and corrective action.

6.5.3.4 Periodic maintenance and calibrations. This includes planned maintenance activities and/or periodic calibration of ground/flight system interface equipment. This type of activity may be driven by:

a. Degradation of functional integrity of systems over time.

b. Environmental exposure causing physical degradation.

c. Inadequate design life limitations of ground systems (operational use often exceeded by two to four times the original design certification limit).

6.5.4 Launch Systems Unplanned Troubleshooting and Repair

This includes nontoutine activity required to recover from discovery of out-of-spec conditions in flight or ground systems. Initial actions include problem location and troubleshooting, which may or may not lead to nonroutine gaining of access, component repair, component removal and replacement, closeout, and functional recertification. This type of activity may be driven by the following.

6.5.4.1 Component troubleshooting, removal and replacements. This set of activities includes GSE valves, tubing, actuators, sensors, and various covers. This type of activity may be driven by:

a. Launch-induced acoustic environment.

b. Launch-induced thermal environment.

6.5.4.2 Ground launch structural systems repair. This work includes welding of cracks, fractures, etc. It also includes replacement of railings and fixing brackets and activities associated with protective coating restoration. This type of activity may be driven by:

a. Launch-induced acoustic environment.

b. Launch-induced thermal environment.

c. Corrosive environment.

6.5.5 Crew Ingress, Late Payload Stowage, and Functional Verification for Flight

This includes launch and departure activities associated with flight crew and payload. This type of activity may be driven by the following.

6.5.5.1 Flight crew ingress/egress. This type of activity may be driven by:

a. Piloted vehicle (human flight vehicle).

b. Mixed orientations that lengthen ingress time (vertical ingress/horizontal seating).

c. Complexity and need for specialized flight suits.

d. Complex, manually intensive hatch closeout and verification.

e. Access arm reconfiguration for emergency egress and launch environment.

6.5.5.2 Late payload stowage. This includes vertical payload installation, personal items stowage, and locker closeout. This type of activity may be driven by:

a. Accommodation of certain biological/live-specimen science experiments.

b. Vertical payload installation at the launch pad.

c. Stowage of perishable items and supplies.

6.5.5.3 Payload functional verification for flight. This type of activity may be driven by:

a. Payload dependence on vehicle-supplied services (power, cooling, command and control functions, etc.).

b. Payload end-to-end interface testing (payload health insight during terminal count and ascent).

6.5.5.4 Unplanned payload troubleshooting and repair. This includes nonroutine troubleshooting and replacement of installed payload items. This type of activity may be driven by:

a. Low inherent dependability of mission-unique items and lack of design life requirements.

b. Excessive power-on time and cycles caused by other time-consuming launch/payload ground functions (e.g., servicing and checkout).

c. Overaccumulation of intrusive personnel traffic required for routine vehicle turnaround—a second-order processing effect.

d. Inadequate seal material.

6.5.6 Vehicle Systems Servicing at Launch Point

This set of activities includes manual system servicing operations at the launch/departure point. This type of activity may be driven by the following.

6.5.6.1 Local servicing with facility/GSE by remote control room. This includes fluid fill, drain, and pressurization; and system flight closeout at the departure point with both local and remote control room operations. This also includes local servicing of the Orbiter's fuel cell power reactants, storage, and distribution (PRSD) system, requiring servicing of such commodities as special-grade liquid oxygen (LOX) and hydrogen, which are stored in a supercritical state onboard the Orbiter .In addition, OMS/RCS nitrogen tetroxide, monomethylhydrazine (MMH), and helium are all locally serviced. This activity also includes left- and right-hand SRB and Orbiter hydraulic local fluid servicing. This type of activity may be driven by:

a. Propellant quantity measurement accuracy.

b. Zero-gravity environment requires elimination of entrained gases in liquid systems (e.g., OMS/RCS and feed systems).

c. Environmental control servicing (e.g., breathing air, extravehicular activity [EVA] suit coolant loop, and other self-contained systems).

d. Local, drag-on hydraulic system servicing for ground functions.

e. Hazardous, toxic-fluid transfer.

f. Flight system pressurant bottle charging (e.g., helium and nitrogen)

6.5.6.2 Local manual servicing at the launch facility. This includes fluid fill, drain, pressurization; and system flight closeout at the departure point through local manual operations only (no automated remote control). This includes left- and right-hand SRB thrust vector control (TVC) hydrazine and gaseous-nitrogen servicing at the booster's aft skirts. Also included are the Orbiter APU servicing of hydrazine (which is a different grade from the OMS/RCS hydrazine) and local servicing of the APU/hydraulic system's gaseous-nitrogen system. It also includes breathing air and high-pressure-supply nitrogen and oxygen bottle fills. This type of activity may be driven by:

a. Many different vehicle fluids required.

b. Different grades of same commodity specified for different functions.

6.5.6.3 Ordnance hookups. This includes final connections for separation systems, removal of safety caps, etc. This also includes range safety system arming. This type of activity may be driven by:

a. Manual connection and verification of pyrotechnic flight element separation systems.

b. Manual connection and verification of pyrotechnic landing/deceleration systems (e.g., gear deployment and drag-chute deployment functions).

c. Manual connection and verification of pyrotechnic flight-to-ground separation systems (e.g., explosiveholddown bolts).

d. Manual connection and verification of range safety destruct systems.

e. Vehicle mechanism emergency/contingency pyrotechnic functions (e.g., emergency escape, remote manipulator arm and payload bay antenna emergency separation on-orbit).

f. Manual connection and verification of payload and upper-stage pyrotechnic systems (e.g., igniters and separation chords).

6.5.7 Access/Umbilical Removal and Closeout for Flight

This set of activities includes system servicing at the launch/departure point (i.e., for fluids, ordnance hookups, and final flight closeouts). This type of activity may be driven by the following.

6.5.7.1 Final flight crew and payload closeout. This encompasses all work associated with final flight configuration of all payload items, including installation of covers, flight crew cue cards, payload locker closeout and securing, final payload bay closeouts, walkdown inspections, and so forth.

6.5.7.2 Access removal. This includes relocating large access structures, such as the Rotating Service Structure (RSS); lowering the SRB and SSME engine service platforms (ESPs); retracting and securing the gaseous hydrogen and gaseous oxygen (GOX) vent arms (GVAs); removing access platforms; stowing loose handrails, and so forth.

6.5.7.3 Umbilical removal and closeout. An example of this activity is retracting and securing the Orbiter Midbody Umbilical Unit (OMBUU).

6.5.7.4 Environmental cover removal and other remove-before-flight closeouts. This includes all work associated with reconfiguring the Orbiter Weather Protection System (OWPS) hardware on the pad facility and service structures, remove-before-flight cover removals, and RCS throat plug flyaway covers.

6.5.8 Remote Automated Servicing and Propellant Loading for Flight

This set of activities includes remote servicing of the integrated (i.e., mated) flight vehicle for launch. Remote servicing means control-room-operated servicing distanced from the launch point. This type of activity may be driven by the following.

6.5.8.1 Cryogenic fluid transfer. This includes the main propellant servicing and loading operations for the main propulsion fuel (liquid hydrogen) and main propulsion oxidizer (liquid oxygen).

6.5.8.2 Thermal management fluid servicing. This includes the servicing of Freon in the vehicle thermal management loops.

6.5.8.3 Tank gaseous pressurization. This includes the main propellant tank pressurization with helium.

6.5.8.4 Flight vehicle pressurization. This includes pressurizing vehicle propellant tanks with helium, for example.

6.5.8.5 Environmental control purges. This includes activating and establishing air and gaseous-nitrogen purges.

6.5.8.6 Environmental control purges. This includes activating and establishing heated nitrogen purges.

6.5.8.7 Remote fuel cell gas supply. This includes establishing ground supply of gaseous oxygen and hydrogen for the Fuel Cell/PRSD system.

6.5.9 Launching the Vehicle

This set of activities includes procedures required to execute the departure of the flight vehicle and its payload from the launch/departure point to its ascent/climb portion of flight operations. This type of activity may be driven by the following.

6.5.9.1 Coordinating the network of local and off-site support functions. For the Space Shuttle system this includes activities to coordinate the Eastern Range, NASA's Mission Control, Tracking and Data Relay networks with firing room launch operations. This type of activity may be driven by the following.

a. Need for ground-based, manual spaceflight tracking and control.

b. Requirement for separate launch and ascent flight control and monitoring.

c. Lack of demonstrated reliability and engineering confidence in achieving routine, successful spaceflight.

6.5.9.2 Activating flight vehicle and ground systems for launch. This includes call-to-stations activities to coordinate work teams and system power-ups and activations. The level of intensity of this type of activity may be driven by:

a. Number of separate subsystems (flight and ground).

b. Degree of subsystem automation onboard or from the ground.

c. Dependability and complexity of subsystems (flight and ground) requiring activation for launch.

6.5.9.3 Obtaining weather and range clearances for launch. This includes operations associated with obtaining all necessary launch range and contingency abort site weather clearances. This type of activity may be driven by:

a. Fragility of vehicle design relative to surface winds and wind shear.

b. Requirement for visual range observation. Lack of demonstrated reliability and engineering confidence in achieving routine, successful spaceflight.

6.5.9.4 Configuring flight vehicle for terminal (final) launch operations. This includes operations required for flight crew ingress and subsequent crew cabin securing, switch lists, navigation system updates, verification of communications system integrity, and hatch closeout. It also includes flight and ground preparations for propulsion system final launch sequencing. The level of intensity of this type of activity may be driven by:

a. Degree of manual, human control (primary and secondary, onboard and ground) designed into the space transportation system concept of operation.

b. Number of separatelyoperated subsystems (flight and ground).

c. Degree of subsystem automation (i.e., little or no manual operation required from either onboard the flight vehicle or on the ground).

6.5.9.5 Final launch sequencing. This includes operations involving propulsion system pressurization for flight; securing of prelaunch swing arms (e.g., vent arms and personnel access arms); onboard activation and transfer of power from ground to all active flight elements (e.g., fuel cells and Orbiter, left-hand and right hand SRB auxiliary power); initiation of main propulsion engine start sequences, final checks, release systems (e.g., ordnance-fired systems) for flight operation. The level of intensity of this type of activity may be driven by:

a. Degree of manual, human control (primary and secondary, onboard and ground) designed into the space transportation system concept of operation.

b. Number of separately operated subsystems (flight and ground).

c. Degree of subsystem automation (i.e., little or no manual operation required from either onboard the flight vehicle or on the ground).

d. Complexity of main propulsion conditioning requirements for start and emergency shutdown.

6.5.9.6 Configuring systems for launch/abort/recycle, as needed. This includes operations required to safe the crew/passengers, the flight vehicle, and the ground support systems and facilities for aborted launch attempts and emergencies. This includes such operations as propellant off-loading, repositioning of access and service arms/umbilicals, and shutdown of critical systems. The level of intensity of this type of activity may be driven by:

a. Degree of manual, human control (primary and secondary, onboard and ground) designed into the space transportation system concept of operation.

b. Number of separately operated subsystems (flight and ground).

c. Operational complexities in removing flight crew, passengers, and time-sensitive payload items from the launch vehicle in emergency and nonemergency contingencies.

d. Complexity of main propulsion conditioning requirements for emergency shutdown.

e. Complexity of design in off-loading of propellants and gases and emergency safing of hazardous-gas systems.

f. Designed-in hazards and operational complexities in managing the safety of confined spaces and compartments (flight and ground systems) containing flammable, hazardous fluids.

g. Hazards in managing armed and explosive pyrotechnic devices and ordnance designed into the architecture (flight and ground).

6.5.10 Postlaunch Securing

This set of activities includes operations associated with safing and securing the launch systems and facilities following flight vehicle departure. This type of activity may be driven by the following.

6.5.10.1 Postlaunch system safing operations. This set of activities includes remote washdowns, system purging, fluid system inerting, pyrotechnic safing, and so forth. The level of intensity of this type of activity may be driven by:

a. Number of flight and ground systems using pyrotechnics and ordnance in the design.

b. Number of flight and ground systems using toxic, hazardous commodities in the design.

c. Ground facility design sensitivity to launch acoustic/thermal environment.

d. Ground facility design sensitivity to exhaust environment (e.g., solid-propellant exhaust residuals).

6.5.10.2 Postlaunch inspection of facilities and equipment. This set of activities includes both remote video surveys and on-site walkdowns for damage assessment. The level of intensity of this type of activity may be driven by:

a. Ground facility design sensitivity to launch acoustic/thermal environment.

b. Ground facility design sensitivity to exhaust environment (e.g., solid-propellant exhaust residuals).

6.5.11 Launch Facility and Equipment Periodic Maintenance

6.5.11.1 Interval maintenance of launch ground systems. The level of intensity of this type of activity may be driven by:

a. Ground facility and equipment design life and durability against launch acoustic/thermal environment.

b. Ground facility and equipment design life durability against launch exhaust environment (e.g., solid-propellant exhaust residuals).

c. Ground facility and equipment design durability against solar, moisture, humidity and other long-term ambient weather exposure.

6.5.11.2 Launch ground system and facility modifications and process changes.

6.6 OMS/RCS: A "Core Drill" of System-Driven Infrastructure

The foregoing causal descriptions involved the direct work emanating from the space transportation architectural design. Since an overwhelming majority of the program recurring costs accumulate from the spaceport and national support infrastructure, an examination of a specific system function (the Orbiter's OMS and RCS auxiliary propulsion functions), its resulting direct work, and a top-level itemization of its support functions and infrastructure was undertaken. As was the case for the eight-flow direct work analysis, the results of this OMS/RCS support infrastructure "core drill" exercise were compiled and collected in the STS Design Root Cause Knowledge Capture Tool (the MS Access database).

6.6.1 OMS/RCS System Architecture and Direct Operations Overview

6.6.1.1 The Orbital Maneuvering System (OMS) and Aft OMS/RCS Pods

The OMS provides the thrust for orbit insertion, orbit circularization, orbit transfer, rendezvous, deorbit, abort to orbit, and abort once around and can provide up to 1,000 pounds of propellant to the aft Reaction Control System (RCS).[24] The OMS is housed in two independent pods located on each side of the Orbiter's aft fuselage (Figure 45). The pods also house the aft RCS and are referred to as the OMS/RCS pods. Each pod contains one OMS engine and the hardware needed to pressurize, store, and distribute the propellants to perform the velocity maneuvers. The two pods provide redundancy for the OMS. The vehicle velocity required for orbital adjustments is approximately 2 ft per second for each nautical mile of altitude change.

[24] "Shuttle Reference Manual," <http://spaceflight.nasa.gov/shuttle/reference/shutref/index.html>.

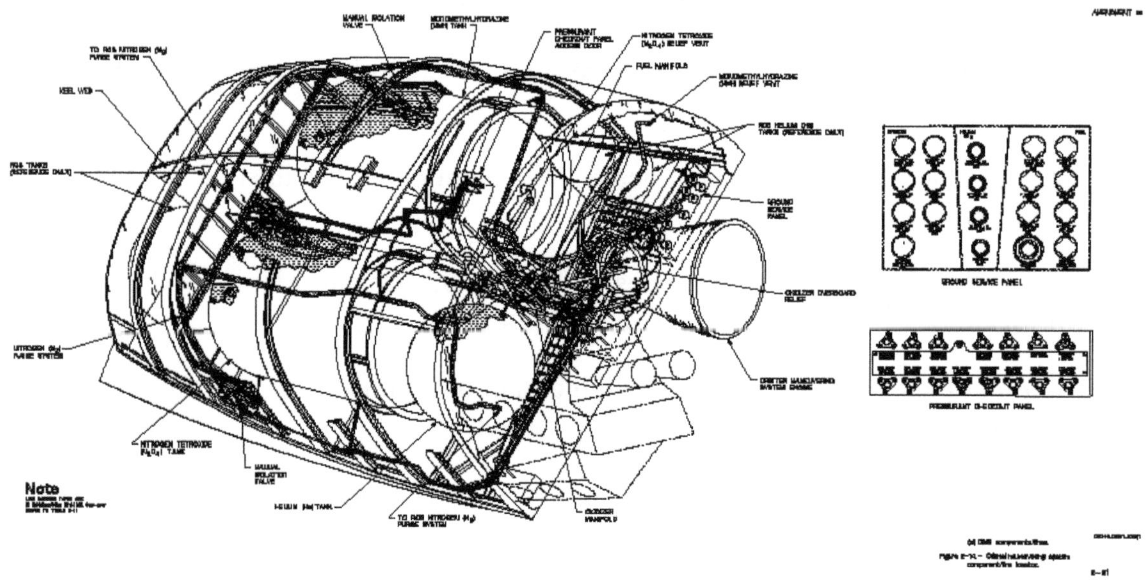

Figure 45. Orbital Maneuvering System/Reaction Control System (OMS/RCS) Pod

6.6.1.2 OMS System Design and Construction

The OMS in each pod consists of a high-pressure gaseous-helium storage tank, helium isolation valves, dual-pressure regulation systems, vapor isolation valves for only the oxidizer-regulated helium pressure path, quad check valves, a fuel tank, an oxidizer tank, and a propellant distribution system consisting of tank isolation valves, cross-feed valves, and an OMS engine. Each OMS engine also has a gaseous-nitrogen storage tank, gaseous-nitrogen pressure isolation valve, gaseous-nitrogen accumulator, bipropellant solenoid control valves and actuators that control bipropellant ball valves, and purge valves.

In each of the OMS pods, gaseous-helium pressure is supplied to helium isolation valves and dual-pressure regulators, which supply regulated helium pressure to the fuel and oxidizer tanks. The fuel is monomethylhydrazine, and the oxidizer is nitrogen tetroxide. The propellants are Earth-storable liquids at normal temperatures. They are pressure-fed to the propellant distribution system through tank isolation valves to the OMS engines. The OMS engine propellant ball valves are positioned by the gaseous-nitrogen system and control the flow of propellants into the engine. The fuel is directed first through the engine combustion chamber walls and provides regenerative cooling of the chamber walls; it then flows into the engine injector. The oxidizer goes directly to the engine injector. The propellants are sprayed into the combustion chamber, where they atomize and ignite upon contact with each other (hypergolic), producing a hot gas and, thus, thrust.

The gaseous-nitrogen system is also used after the OMS engines are shut down to purge residual fuel from the injector and combustion chamber, permitting safe restarting of the engines. The nozzle extension of each OMS engine is radiation-cooled and is constructed of columbium alloy.

Each OMS engine produces 6,000 pounds of thrust. The oxidizer-to-fuel ratio is 1.65:1. The expansion ratio of the nozzle exit to the throat is 55:1. The chamber pressure of the engine is 125 psia. The dry weight of each engine is 260 pounds.

Each OMS engine can be reused for 100 missions and is capable of 1,000 starts and 15 hours of cumulative firing. The minimum duration of an OMS engine firing is 2 seconds. The OMS may be utilized to provide thrust above 70,000 ft. For vehicle velocity changes of between 3 and 6 ft per second, normally only one OMS engine is used.

Each engine has two electromechanically gimbaled actuators, which control the OMS engine thrust direction in pitch and yaw (thrust vector control). The OMS engines can be used singularly by directing the thrust vector through the Orbiter center of gravity or together by directing the thrust vector of each engine parallel to the other. During a two-OMS-engine thrusting period, the RCS will come into operation only if the OMS gimbal rate or gimbal limits are exceeded and should not normally come into operation during the OMS thrust period. However, during a one-OMS-engine thrusting period, roll RCS control is required. The pitch and yaw actuators are identical except for the stroke length and contain redundant electrical channels (active and standby), which couple to a common mechanical drive assembly.

The OMS/RCS pods are designed to be reused for up to 100 missions with only minor repair, refurbishment, and maintenance. The pods are removable to facilitate Orbiter turnaround, if required.

The OMS/RCS left- and right-hand pods are attached to the upper aft fuselage left and right sides. Each pod is fabricated primarily of aluminum and graphite epoxy material. Each pod is 21.8 ft long and 11.37 ft wide at its aft end and 8.41 ft wide at its forward end, with a surface area of approximately 435 sq ft. Each pod is divided into two compartments: the OMS and the RCS housings. Each pod houses all the OMS and RCS propulsion components and is attached to the aft fuselage with 11 bolts. The pod skin panels are graphite epoxy honeycomb sandwich. The forward and aft bulkhead aft tank support bulkhead and floor truss beam are machined aluminum 2124. The centerline beam is 2024 aluminum sheet with titanium stiffeners and graphite epoxy frames. The OMS thrust structure is conventional 2124 aluminum construction. The cross braces are aluminum tubing, and the attach fittings at the forward and aft fittings are 2124 aluminum. The intermediate fittings are corrosion-resistant steel. The RCS housing, which attaches to the OMS pod structure, contains the RCS thrusters and associated propellant feed lines. The RCS housing is constructed of aluminum sheet metal, including flat outer skins. The curved outer skin panels are graphite epoxy honeycomb sandwich. Twenty-four doors in the skins provide access to the OMS and RCS and attach points.

The two graphite epoxy pods per spacecraft reduce the weight by 10 percent, approximately 450 pounds. The pods will withstand 162-decibel acoustic noise and a temperature range from –170 to +135 °F.

The exposed areas of the OMS/RCS pods are covered with a reusable thermal protection system, and a pressure and thermal seal is installed at the OMS/RCS pod aft fuselage interface. Thermal barriers are installed, and they interface with the RCS thrusters and reusable thermal protection system.

6.6.1.3 Reaction Control System (RCS) and Forward RCS Pod

The Orbiter's reaction control system comprises the forward and aft RCS. The forward RCS (FRCS) is located in the forward fuselage nose area. The aft (right and left) RCS is located with the OMS in the OMS/RCS pods (Figure 46).

Each RCS consists of high-pressure gaseous-helium storage tanks, pressure regulation and relief systems, a fuel and oxidizer tank, a system that distributes propellant to its engines, and thermal control systems (electrical heaters). The forward and aft RCS units provide the thrust for attitude (rotational) maneuvers (pitch, yaw, and roll) and for small velocity changes along the Orbiter axis (translation maneuvers). Two helium tanks supply gaseous-helium pressure to the oxidizer and fuel tanks. The oxidizer and fuel are then supplied under gaseous-helium pressure to the RCS engines. Nitrogen tetroxide is the oxidizer, and monomethylhydrazine is the fuel. The propellants are Earth-storable and hypergolic (they ignite upon contact with each other). The propellants are supplied to the engines, where they atomize, ignite, and produce a hot gas and thrust.

The FRCS has 14 primary and 2 vernier engines. The aft RCS has 12 primary and 2 vernier engines in each pod. The primary RCS engines provide 870 pounds of vacuum thrust each, and the vernier RCS engines provide 24 pounds of vacuum thrust each. The oxidizer-to-fuel ratio for each engine is 1.6:1. The nominal chamber pressure of the primary engines is 152 psia. For each vernier engine, it is 110 psia.

The primary engines are reusable for a minimum of 100 missions and are capable of sustaining 20,000 starts and 12,800 seconds of cumulative firing. The primary engines are operable in a maximum steady-state thrusting mode of 1 to 150 seconds, with a maximum single-mission contingency of 800 seconds for the aft RCS aft firing engines and 300 seconds (maximum) for the FRCS forward firing engines, as well as in a pulse mode with a minimum impulse thrusting time of 0.08 second above 125,000 ft. The expansion ratio (exit area to throat area) of the primary engines ranges from 22:1 to 30:1. The multiple primary thrusters provide redundancy. The vernier engines' reusability depends on chamber life. They are capable of sustaining 330,000 starts and 125,000 seconds of cumulative firings. The vernier engines are operable in a steady-state thrusting mode of 1 to 125 seconds maximum, as well as in a pulse mode with a minimum impulse time of 0.08 second. The vernier engines are used for finite maneuvers and stationkeeping (long-time attitude hold) and have an expansion ratio that ranges from 20:1 to 50:1. The vernier thrusters are not redundant.

The FRCS module is constructed of conventional 2024 aluminum alloy skin-stringer panels and frames. The panels are composed of single-curvature and stretch-formed skins with riveted stringers. The frames are riveted to the skin-stringer panels. The FRCS module is secured to the forward fuselage nose section and forward bulkhead of the forward fuselage with 16 fasteners, which permit the installation and removal of the module. The components of the FRCS are mounted and attached to the module, which will have a reusable thermal protection cover, in addition to thermal barriers installed around it and the RCS engine interfaces and the interface attachment area to the forward fuselage.

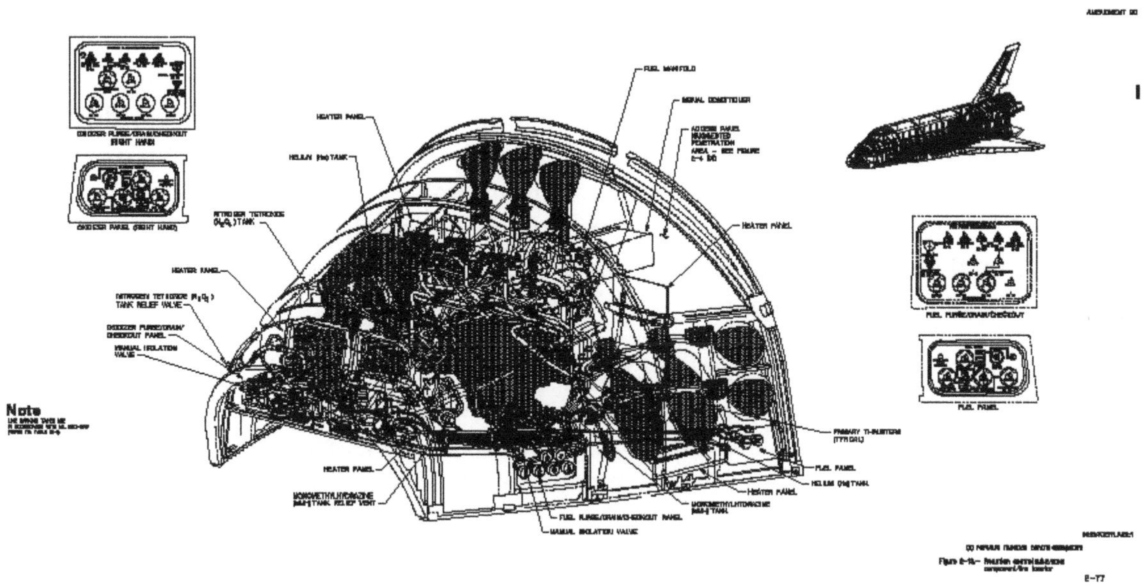

Figure 46. FRCS Pod

6.6.1.4 OMS/RCS Turnaround Performance

The various forward and aft pods in the Shuttle fleet are numbered and designated with the Orbiter it was originally delivered with (e.g., "RP02" is the Shuttle *Columbia,* right aft OMS pod). The pods are interchangeable with similar locations on other Orbiters in the fleet.

During a normal Orbiter Processing Facility (OPF) turnaround flow, both OMS pods and the FRCS are intended to remain installed on the Orbiter vehicle. All planned maintenance can be performed in the OPF. However, because of the overall system reliability characteristics, this is not always the case. Often the OMS pods are removed because of RCS thruster anomalies requiring complex, intrusive, and difficult repair and replacement operations, often with difficult access issues. For this reason OMS/RCS pods are often removed on an unplanned basis to the Hypergolic Maintenance Facility (HMF). The pods are also removed on a planned basis for periodic depot maintenance, at approximately eight flight intervals per pod.

For normal OPF turnaround processing flows (nondepot) in the time frame between 1992 and 1997, the FRCS was removed 12 of 34 flows analyzed (this time frame represented a stable flight processing tempo of 7 to 8 flights per year). This equates to a 35-percent FRCS pod removal rate. The average time between roll-in of the Orbiter vehicle to the OPF and the removal of the FRCS pod is approximately 16 calendar days, although this varied from a minimum of 2 days to a maximum of 45 days. The average duration of an FRCS repair in the HMF is 27 calendar days, with a minimum of 9 days and a maximum of 68 days for normal Orbiter turnaround flows. The time from the installation of the FRCS into the Orbiter until Orbiter roll-out from the OPF averaged 37 calendar days, with a minimum of 20 days and maximum of 54 days.

For normal OPF turnaround processing flows (nondepot) in the time frame between 1992 and 1997, the left OMS pod was removed 4 of 34 flows analyzed (this time frame represented a stable flight processing tempo of 7 to 8 flights per year). This equates to a 12-percent left OMS pod removal rate. The right OMS pod was also removed 4 of 34 flows for a 12-percent removal rate. The average time between roll-in of the Orbiter vehicle to the OPF and the removal of the aft OMS pods (both right- and left-hand pods) was approximately 25 calendar days, although this varied from a minimum of 16 days to a maximum of 56 days. The average duration of OMS pod repairs in the HMF was 47 calendar days, with a minimum of 15 days and a maximum of 166 days for normal Orbiter turnaround flows. The time from the installation of an OMS pod back into the Orbiter until Orbiter roll-out from the OPF averaged 29 calendar days, with a minimum of 19 days and maximum of 56 days.

The durations between the roll-in milestones and the pod removal and the durations between the pod installations and roll-out milestones are highly variable because of the numerous intersystem processing constraints for removing and installing OMS/RCS pods. Nevertheless, the processing minimums and maximums provide a relative quantification of these variables and complexities.

6.6.1.5 OMS/RCS Mission Control and Flight Monitoring

Not analyzed in this work are the functions associated with mission control and flight operations monitoring at NASA's Johnson Space Center (JSC). What follows is an overview of the OMS/RCS operation from ascent through landing.

The ascent profile of a mission determines if one or two OMS thrusting periods are used and the interactions of the RCS. After main engine cutoff, the RCS thrusters in the forward and aft RCS pods are used to provide attitude hold until ET separation. At ET separation, the RCS provides a minus (negative) Z translation maneuver of about minus 4 ft per second to maneuver the Orbiter away from the ET. Upon completion of the translation, the RCS provides Orbiter attitude hold until time to maneuver to the OMS-1 thrusting attitude. The targeting data for the OMS-1 thrusting period is selected before launch; however, the target data in the onboard general-purpose computers (GPCs) can be modified by the flight crew via the cathode ray tube (CRT) keyboard, if necessary, before the OMS thrusting period.

During the first OMS thrusting period, both OMS engines are used to raise the Orbiter to a predetermined elliptical orbit. During the thrusting period, vehicle attitude is maintained by gimbaling (swiveling) the OMS engines. The RCS will not normally come into operation during an OMS thrusting period. If, during an OMS thrusting period, the OMS gimbal rate or gimbal limits are exceeded, RCS attitude control is required. If only one OMS engine is used during an OMS thrusting period, RCS roll control is required.

During the OMS-1 thrusting period, the liquid oxygen and liquid hydrogen trapped in the main propulsion system ducts are dumped. The liquid oxygen is dumped out through the Space Shuttle main engines' combustion chambers, and the liquid hydrogen is dumped through the starboard (right) side T-0 umbilical overboard fill and drain. A "delta" velocity increment required to reach an initial orbital was precomputed in conjunction with the OMS-1 thrusting period.

Upon completion of the OMS-1 thrusting period, the RCS is used to null any residual velocities, if required. The flight crew uses the rotational hand controller and/or translational hand controller to command the applicable RCS thrusters to null the residual velocities. The RCS then provides attitude hold until time to maneuver to the OMS-2 thrusting attitude.

The second OMS thrusting period using both OMS engines occurs near the apogee of the orbit established by the OMS-1 thrusting period and is used to circularize the predetermined orbit for that mission. The targeting data for the OMS-2 thrusting period is selected before launch; however, the target data in the onboard GPCs can be modified by the flight crew via the CRT keyboard, if necessary, before the OMS thrusting period.

Upon completion of the OMS-2 thrusting period, the RCS is used to null any residual velocities, if required, in the same manner as during OMS-1. The RCS is then used to provide attitude hold and minor translation maneuvers as required for on-orbit operations. The flight crew can select primary or vernier RCS thrusters for attitude control on orbit. Normally, the vernier RCS thrusters are selected for on-orbit attitude hold.

If the ascent profile for a mission uses a single OMS thrusting maneuver, it is referred to as direct insertion. In a direct-insertion ascent profile, the OMS-1 thrusting period after main engine cutoff is eliminated and is replaced with a 5-ft-per-second RCS translation maneuver to facilitate the main propulsion system dump. The RCS provides attitude hold after the translation maneuver. The OMS-2 thrusting period is then used to achieve orbit insertion. The direct-insertion ascent profile allows the Main Propulsion System (MPS) to provide more energy to orbit insertion and permits easier use of onboard software.

Additional OMS thrusting periods using both or one OMS engine are performed on-orbit according to the mission's requirements to modify the orbit for rendezvous, payload deployment, or transfer to another orbit.

The two OMS engines are used to deorbit. Target data for the deorbit maneuver is computed by the ground and loaded in the onboard GPCs via uplink. This data is also voiced to the flight crew for verification of loaded values. After verification of the deorbit data, the flight crew initiates an OMS gimbal test on the CRT keyboard unit.

Before the deorbit thrusting period, the flight crew maneuvers the spacecraft to the desired deorbit thrusting attitude using the rotational hand controller and RCS thrusters. Upon completion of the OMS thrusting period, the RCS is used to null any residual velocities, if required. The spacecraft is then maneuvered to the proper entry interface attitude using the RCS. The remaining propellants aboard the FRCS are dumped by burning the propellants through the FRCS thrusters before the entry interface if it is necessary to control the Orbiter's center of gravity.

The RCS aft firing jets can be used to complete any planned OMS thrusting period in the event of an OMS engine failure. In this case, the OMS-to-aft-RCS interconnect would feed OMS propellants to the aft RCS.

From entry interface at 400,000 ft, the Orbiter is controlled in roll, pitch, and yaw with the aft RCS thrusters. The Orbiter's ailerons become effective at a dynamic pressure of 10 pounds per square foot, and the aft RCS roll jets are deactivated. At a dynamic pressure of 20 pounds per square foot, the Orbiter's elevons become effective, and the aft RCS pitch jets are deactivated.

The rudder is activated at Mach 3.5, and the aft RCS yaw jets are deactivated at Mach 1 and approximately 45,000 ft.

6.6.2 OMS/RCS Pod Shipping, Receiving, and Acceptance

This set of activities includes functions required to prepare, ship, and receive OMS/RCS pods to and from various OMS/RCS facilities throughout the national Shuttle program infrastructure. Facilities include the Hypergolic Maintenance Facilities (HMF) at NASA's John F. Kennedy Space Center and the White Sands Space Harbor Test Facility in New Mexico (e.g., Test Stand 401) (Figure 47). This type of activity may be driven by the following.

6.6.2.1 OMS/RCS Receiving and Handling Operations

a. OMS/RCS facility and equipment preparations and functional verification.

b. OMS/RCS pod transport receiving and positioning operations.

c. OMS/RCS pod off-load from transport system.

6.6.2.2 OMS/RCS Ground Transfer and Storage Operations

a. OMS/RCS pod ground handling equipment and fixtures.

b. OMS/RCS pod environmental shelter and conditioning.

c. Removal and stowage of OMS/RCS transportation hardware.

6.6.2.3 OMS/RCS Pod Inspection and Receiving Acceptance

a. OMS/RCS pod receiving inspection.

b. OMS/RCS pod hardware inventory.

c. OMS/RCS pod delivery soft-goods inventory.

6.6.2.4 OMS/RCS Pod Conditioning (Purging, Temperature, and Humidity Control)

a. Facility and GSE preparation for OMS/RCS pod storage.

b. Activation and monitoring of OMS/RCS pod conditioning systems.

6.6.2.5 OMS/RCS Design Modifications

a. OMS/RCS modifications.

b. Fleetwide OMS/RCS special tests and inspections.

6.6.2.6 OMS/RCS Pod Receiving Inspection Discrepancy Resolution and Repairs

a. OMS/RCS repair and rework.

b. OMS/RCS inventory discrepancy resolution.

6.6.2.7 OMS/RCS Pod Shipment

a. OMS/RCS pod facility and equipment preparation for flight element hardware shipment.

b. Shipment kit installation on the OMS/RCS pod.

c. OMS/RCS pod handling and connection with pod transport system.

d. OMS/RCS pod soft-goods preparation for shipment.

6.6.3 OMS/RCS Pod Depot Maintenance

This set of activities includes functions required to perform off-line periodic inspections, testing, checkout, and modifications/upgrades of the OMS/RCS pods. This type of activity may be driven by the following.

6.6.3.1 OMS/RCS pod overhaul, inspection, functional verification, and modification.

Figure 47. OMS/RCS Test Facilities at White Sands

6.6.3.2 OMS/RCS system/component test firings. This set of infrastructure deals with the off-line facilities, equipment, and workforce required to conduct the Shuttle OMS/RCS pod fleet leader test program. This program subjects flightlike test articles to mission duty or operating cycles, propellant exposure, maintenance operations, and component functional tests that simulate flight and launch site turnaround operations, at a more rapid rate than the actual flight hardware. This provides NASA insight into life-dependent failures and anomalies before they affect the actual fleet hardware. Specific infrastructure includes:

- White Sands Test Stand 301
- White Sands Test Stand 328
- White Sands Test Stand 403

This type of activity is driven by:

a. OMS engine qualification hot-fire test using Test Stand 403.

b. OMS tank pressure-volume-temperature (PVT) gauging using Test Stand 403.

c. OMS/RCS cross-feed line gas sweep test using Test Stand 301.

d. Forward interconnect system test using FRCS test article in Test Stand 328.

e. Understanding technology and design unknowns involved with the corrosion products that occur with the accumulation of hypergolic fuel-oxidizer reaction product (FORP) mixtures within the pressure chamber tubes of reaction jets.

f. Understanding technology and design unknowns involved with iron nitrate contamination of nitrogen tetroxide systems.

g. Other recurring OMS/RCS system operation issues discovered during maintenance.

6.6.3.3 OMS/RCS flight/ground software upgrades.

6.6.4 Spaceport Infrastructure Support for OMS/RCS

This set of activities includes the indirect services and functions that may be needed in support of OMS/RCS operations. This type of activity may be driven by the following.

6.6.4.1 Shop and lab support to OMS/RCS.

6.6.4.2 Photographic services to support OMS/RCS.

6.6.4.3 Fire protection and medical support to OMS/RCS operations.

6.6.4.4 Security support services (area access and facility control) dedicated to OMS/RCS infrastructure.

6.6.4.5 OMS/RCS technical documentation/library services.

6.6.4.6 Utility services provided to, dedicated to, or consumed by OMS/RCS infrastructure and operations (power, communications/info, water, and air product utilities provided by the spaceport and/or test complex).

6.6.4.7 Dedicated OMS/RCS facility road and grounds maintenance.

6.6.4.8 Heavy equipment dedicated and support to OMS/RCS.

6.6.4.9 Ground transportation services for OMS/RCS.

6.6.4.10 OMS/RCS environmental compatibility management.

6.6.4.11 Personal environmental protection equipment support for OMS/RCS.

6.6.5 OMS/RCS System Logistics

This set of activities includes functions required to ship, receive, and repair flight line replaceable units (LRUs) and fabrication of other parts and materials after removal from the Orbiter or OMS/RCS pod. It also includes logistics support for OMS/RCS propellants and gas consumables, off-line repair of OMS/RCS LRUs at various repair levels (depot and vendor-level repair). This type of activity may be driven by the following.

6.6.5.1 OMS/RCS propellant logistics (acquisition, storage, distribution, conditioning and verification).

6.6.5.2 Other OMS/RCS fluids and gas logistics and consumables.

6.6.5.3 OMS/RC LRU repair and replacement hardware logistics (flight and ground support systems).

6.6.6 OMS/RCS System Operations Planning and Management

This set of activities includes functions that may be needed to support OMS/RCS flight and ground systems processing (direct and indirect). These may include a host of management and engineering services needed to plan, manage, and control the OMS/RCS operations and the OMS/RCS infrastructure. This type of activity may be driven by the following.

6.6.6.1 Fleet manifesting of OMS/RCS pods and OMS/RCS task scheduling.

6.6.6.2 OMS/RCS flight/mission planning.

6.6.6.3 OMS/RCS work control support.

6.6.6.4 OMS/RCS ground systems and facilities scheduling (e.g., OMS/RCS GSE, HMF, JSC, and White Sands test rigs and test cells).

6.6.6.5 OMS/RCS flight software and ground software maintenance.

6.6.6.6 OMS/RCS sustaining operations engineering.

6.6.6.7 OMS/RCS continuous improvement. This includes systems engineering support for upgrades of the flight and ground Shuttle OMS/RCS systems. It also includes lessons learned and support for advanced auxiliary propulsion concepts and technologies.

6.6.6.8 OMS/RCS safety, reliability, and quality assurance operational support and management.

6.6.6.9 OMS/RCS Systems training.

6.6.6.10 Management and administrative support to OMS/RCS personnel.

7 CONCLUSIONS

The foregoing discussion of direct operations work (5.2) and its design causes (Section 6) revealed a pattern of design issues related to inherent operability, supportability, and dependability. The following subsections draw top-level, system-of-systems needs of the operator based on the design root causes summarized in 6.1 (these were in turn derived by the causes identified in the analysis and documented in 6.2 through 6.5). These primary system design and technology needs are explored in more detail below, and in a few cases, document subsequent analytical work.

7.1 Need for Operator-Driven Systems Engineering and Integration

Traditional space transportation conceptual design uses deterministic *outside-in* vehicle-configuration-driven approaches. Most of the unmet needs of the operator, however, are tied to the internal hardware characteristics. From the beginning of the design process, there needs to be a concept for managing and controlling the many components, assemblies, subsystems, systems, elements and stages—they all need to undergo radical changes from the *inside out,* rather than outfitting "advanced" concepts and configurations that continually reach for traditional off-the-shelf flight and ground systems, with traditional operability, dependability, and supportability.

Significant functional hardware changes are needed to significantly reduce physical interfaces to the ground support infrastructure such that the required amounts of direct and indirect support labor are likewise dramatically reduced. The inherent dependability of the hardware must be dramatically improved such that the chances of the operator encountering any indications of failure or nonconformance on a vehicle or GSE are rare, rather than the norm. In addition, a design process needs to be defined for system disciplines to combine like functions (e.g., thermal management, power management, communications, and heath management) into a reduced number of parts, subsystems, and ground interfaces. Examples of this might include designing common thermal transport systems and use of common working fluids. The design and manufacture of all systems, of course, should be composed of robust and dependable components (not just for flight, but for all phases of operation).

7.1.1 Functional System Breakdown Structure (SBS)

In an effort to determine where in the design sequence opportunities are being lost to perform early systems integration, the analysis team reviewed typical weight statements used by such popular weights and sizing programs as CONSIZ. The team found relatively detailed functional breakdowns for engine systems and primary structures. Not found was an equivalent level of definition and identification for the other design disciplines, such as thermal management, communications, guidance and flight control, and power management. While much of this functional hardware may have been "covered" at higher indentured levels in the weight breakdown structures, the generic design functions are not being treated homogeneously during the design process for attributes other than weight; such as total subsystem count, relative order of magnitude of ground interfaces, number of separate working fluids, propellants and gases, and so forth. Thus, the conceptual design process is likely to be left blind to these key design characteristics. Further, in some weight breakdown structures being used, functional hardware required for the integration of flight elements and stages was going unidentified (such as ordnance subsystems, attach hardware, and thermal protection). For the operations specialist participating in concept reviews or for the operations modeler, these nonhomogeneously defined system weight breakdown structures were the best available.

The question was raised as to whether one could generically define vehicle and flight element design functions homogeneously across all the design disciplines, similar to the way the ground functions were generically defined. This might provide a pathway to reform traditional weight statements into more uniformly defined system statements that would inherently possess more conceptual-level definition and insight than just mass, volume, and center of gravity. A cataloging of generic ground functions had already been created and refined several times (see Section 3 and the FBS in Appendix A). It seemed reasonable to perform a similar, start-from-scratch exercise for generic flight system definition.

Thus, a *functional* Systems Breakdown Structure (SBS) effort was organized in the Systems Engineering Office at KSC in 2003 in support of NASA's Next-Generation Launch Technologies (NGLT) program. By 2004, a Government, industry and academia Space Propulsion Synergy Team (SPST) began actual development work on an SBS. During the formative stages of the work, the SBS was significantly broadened in scope to support advanced technology life cycle analysis for NASA's Exploration Systems Research and Technology (ESR&T) effort as part of the Vision for Space Exploration.

7.1.2 Generic Spaceport Design Sequence

The flight vehicle design process has evolved and matured to a point where a generic "design structure matrix" can be used for "closing" a vehicle concept's performance. Still lacking, however, is a similar structured approach for ground process design, spaceport master planning, and launch complex, facility, and equipment design.

7.2 Need for Managing and Controlling Flight-to-Ground Interfaces

7.2.1 Preliminary Analysis of Orbiter-to-Orbiter Processing Facility (OPF) Interfaces

A recurring cause of direct work that emerged from the study of Space Shuttle systems related to ground interfaces with the flight systems.[25] Much needed work and attention has gone into automating the main propellant feed interfaces to address this general concern.[26,27] However, much labor was encountered in other technical disciplines and even propulsion functions *other* than main propellant fill and drain. An analysis, therefore, was undertaken to identify and roughly characterize specific functional interfaces between an existing flight element and its direct ground support infrastructure. For this effort, the Orbiter and its interfaces to the Orbiter Processing Facility (OPF) were chosen. The analysis concluded that there were literally hundreds of functional interfaces—402 summarized in the Orbiter-to-OPF Interface Control Document (ICD).

---

[25] C.M. McCleskey and C. Guidi, "Potential Strategies for Spaceport Systems Towards Airport-Like Operations," AIAA-RS2-2004-6003, *2nd Responsive Space Conference,* Los Angeles, CA, April 19-22, 2004.
[26] *John F. Kennedy Space Center Research and Technology 2003 Annual Report,* NASA-TM-2003-211190, Kennedy Space Center, FL, 2003.
[27] T. Gillespie, "Atlas V Evolved Expendable Launch Vehicle: The Evolution of Responsive Launch;" AIAA-RS2-2004-8001, *2nd Responsive Space Conference,* Los Angeles, CA, April 19-22, 2004.

As can be seen from Table 19, many of the interfaces relate to servicing of propulsion and power management systems. Contributing to the propulsion interface density is the RCS and OMS functions, and the 25 interfaces per SSME (Figure 48). Electrical and mechanical power management functions also drive a large number of dedicated fluid service interfaces, including the fuel cell's power reactant storage and distribution (PRSD) interfaces, and the many dedicated fluid service interfaces (some intrusive) for the APU/Hydraulic/Water Spray Boiler power distribution subsystems. A visual sense of the density of these ground interfaces can be seen from the Orbiter-to-OPF ICD (see Figure 49 and Figure 50). As important as the shear number of interfaces, the labor intensity for technicians connecting, disconnecting and functionally verifying the integrity of the interface is equally important to note.

Some of these ground interfaces are at the outer mold line of the vehicle. Many, however, are internal and require intrusive access that not only adds labor and risk to the operation, but also requires the vehicle's structural integrity to be routinely compromised for normal fluid and mechanical systems servicing. This characteristic of the design has had profound effects on the level of subsequent physical inspections, functional verification, and "close-out" activity.

Table 19. Generic Classification of Space Shuttle Orbiter Interfaces

MATRIX OF STS ORBITER TO ORBITER PROCESSING FACILITY (OPF) INTERFACES (Number of Interfaces)	Liquid Propulsion	Power Management (Mechanical and Electrical)	Structures, Mechanisms, and Vehicle Handling	Guidance, Navigation & Controls	Thermal Management (Passive & Active)	Environmental Control and Life Support	Payload Accommodations	Communications	Multiple Discipline, Integrated Interfaces	Command & Control	Common Avionics Interfaces	SubTotal	*Percentage*
Servicing	171	43			12	11	2		5	3		**247**	**61%**
Functional Verification	34	3	1	24	4	7	2	8				**83**	**21%**
Protective, Safety Purges	23		10		2		2					**37**	**9%**
Handling/Access	8		17	1			8				1	**35**	**9%**
Subtotal	**236**	**46**	**28**	**25**	**18**	**18**	**14**	**8**	**5**	**3**	**1**	**402**	**100%**
Percentage	**59%**	**11%**	**7%**	**6%**	**4%**	**4%**	**3%**	**2%**	**1%**	**1%**	**0%**	**100%**	

Figure 48. Technician Performs Maintenance on Aft Propulsion Subsystems

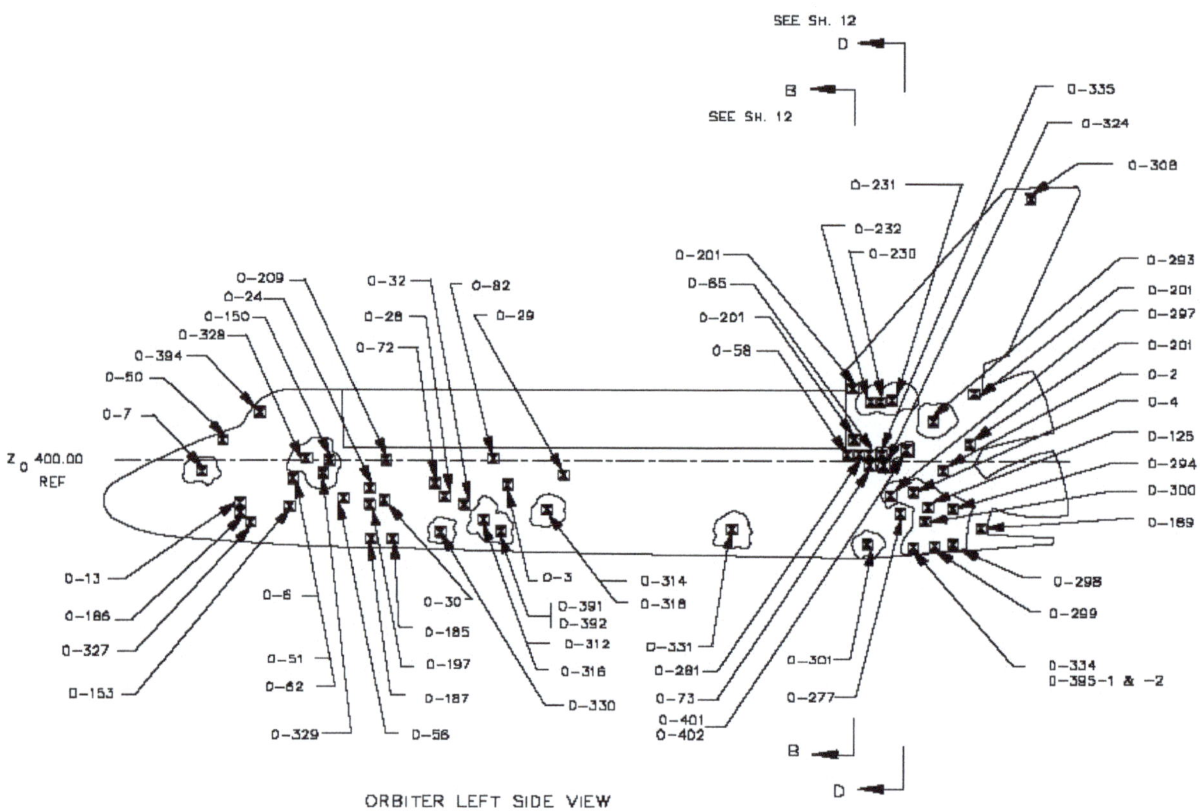

Figure 49. Side View—Orbiter Interfaces to Orbiter Processing Facility (OPF)

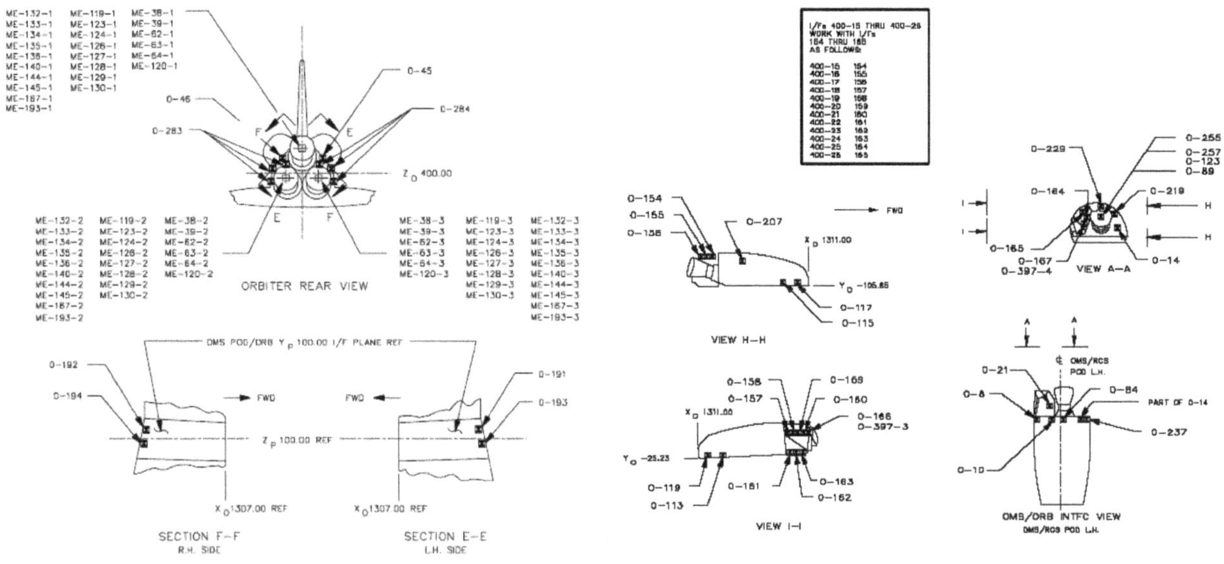

Figure 50. Orbiter Aft Propulsion Ground Interfaces (SSMEs, Left OMS Pod)

In addition to the fluid systems servicing interfaces, there are protective caps, covers, thruster throat plugs with desiccants and indicators, and numerous vehicle and flight element handling fixtures, transporters and jacks—all needing labor, equipment, and materials to operate, maintain, and support.

7.2.2 Conclusions Relating to Ground Interfaces

The primary conclusion that can be derived from the preliminary Orbiter interface analysis is that an accumulation of ground servicing, functional verification, and assembly and handling functions can drive an overaccumulation of ground interfaces, which in turn drives direct operations work and support infrastructure.

An unmanaged design process that is not focused on controlling ground interfaces will inevitably lead to an overaccumulation of ground interfaces in a design, thus becoming a root source of operational impediments and excessive support infrastructure. While the particulars of these items are not likely to be known, accumulation of functions that are likely to give rise to accumulation of functional hardware interfaces in a design can, nonetheless, be recognized and accounted for early in the conceptual design phases.

7.2.3 An Accounting of Space Shuttle Fluid Systems

In addition to the interface analysis, an accounting of the Shuttle's operational fluids was also prepared. In the course of producing a spaceflight, the operator has to deal with the following:

a. *Operational fluids*—fluids that are required onboard to operate the vehicle in flight.

b. *Assembly fluids*—fluids used for internal purging during installation or higher-level assembly.

c. *Cleaning, flushing, and testing fluids*—fluids that are not flown but are used for cleaning and flushing of various flight system hardware. In addition, *test fluids* are fluids that may be used as an alternative to the operational fluid for evaluating the performance of a system.

The Space Shuttle's Specification for Fluid Procurement, Use, and Control (SE-S-0073, rev. G, NASA Johnson Space Center, May 27, 1999) was reviewed with the results identified in Table 20. The assessment indicates that for the Space Shuttle system there are 31 operational fluids, 3 assembly fluids, and 29 vehicle cleaning fluids (not all test fluids, passivation fluids, and so forth were identified).

7.3 Need for Effective Balancing of Flight Reliability With Maintainability and Ground Safety

Regarding the level of unplanned work, there is a strong need in the design process to effectively balance flight reliability with maintainability and ground safety. The traditional approach to providing greater system-level flight reliability is inserting extra copies of hardware (i.e., triple, and even quaduple redundancy). This greatly increases the odds that an operator or maintainer will encounter unplanned component replacements, however (see Figure 51).

An alternative is to provide dual redundancy for critical systems but to aim for near-zero defects, or high component reliability. The advantage of this approach is to help reduce the total parts count across the vehicle systems and subsystems, while increasing the operability and supportability of the system by dramatically reducing exposure to unplanned troubleshooting and repair, as well as less complex ground equipment and few GSE items.

The traditional argument against the robust component design approach has been the perceived extra costs for building components of greater reliability. Consider, however, the design of an aerospace-quality primary flight control actuator, a critical component for passenger aircraft, expendable launch vehicles, and the Space Shuttle thrust vector control (TVC) and aerosurfaces. These devices are tested for an operating life of a certain number of actuation cycles. The difference between the number of required cycles before customer acceptance is orders of magnitude, with expendable being the lowest (less than 10,000 cycles for a contemporary second-stage actuator, for example) to millions of cycles for commercial passenger aircraft, with the Shuttle Orbiter's main engine TVC actuator requiring an operating life on the order of 400,000 actuation cycles.[28] In terms of up-front cost, most of the cost to conduct these tests is likely to be a fixed cost involved with setting up the test program, with the variable cost per order of magnitude of testing being of lesser contribution than traditionally perceived.

[28] E.T. Raymond and C.C. Chenoweth, "Aircraft Flight Control Actuation System Design," Society of Automotive Engineers, Inc., Warrendale, PA, (SAE Publication R-123) 1993, p. 265.

Table 20. Shuttle Vehicle Servicing Fluids

SERVICING FLUID Chemical and Physical Characteristics Controlled by Reference Document SE-S-0073 Rev. G	FLUID USE & CLEANLINESS					SHUTTLE SUBSYSTEM IDENTIFICATION									
	Ref. Table SE-S-0073	Operational Fluids	Assembly Fluids	Cleaning Fluids	Maximum Sub-system Cleanliness Level Req'd	Main Propulsion (Orbiter, ET and SSME)	Aft Propulsion Orbit Maneuvering & Reaction Control	Reaction Control	Auxiliary Power Unit - Orbiter (APU)	Auxiliary Power Unit - SRB (APU)	SSME and Orbiter Hydraulic	SRB Hydraulic	Fuel Cells/Cryo	Environmental and Thermal Control / Life Support	Purge, Vent, and Drain
Propellant Pressurizing Agent, Helium	6.3-1	X		X	100 A	100 A	200	200	100	NA	NA	100	200 A	300 A	
Liquid Oxygen	6.3-2	X			800 A	800 A									
Gaseous Nitrogen	6.3-3	X	X	X	100 A	100 A	200	200	100	100	200	200			
Gaseous and Liquid Oxygen	6.3-4	X			200 A								200 A	200 A	
Gaseous and Liquid Nitrogen	6.3-5	X		X	200 A								200 A	200 A	Class 5,000
Gaseous and Liquid Hydrogen	6.3-6	X			200	400							200		
Hydraulic Fluid	6.3-7	X			190						190	190			
Water (Grade A or B)	6.3-8	X		X	100/200 A	1000	200 A	200 A	100	100	200	200		200 A	
Propellant, Monomethylhydrazine	6.3-9	X			200		200	200							
Propellant, Nitrogen Tetroxide (MON-3)	6.3-10	X			200 A		200 A	200 A							
Propellant, Hydrazine	6.3-11	X		X	100				100	100					
Argon	6.3-12	X	X		100 A	100 A	200	200	100	NA			200 A	100 A	Class 5,000
Propellant, Mixed Oxides of Nitrogen (Deleted)	6.3-13	*					*	*							
Lubricating Oil	6.3-14	X			300				300	300					
Conditioned Air - Purge, Vent, and Drain	6.3-15	X		X	Class 5,000	1000									Class 5,000
Potable Water	6.3-16	X			300									300	
Carbon Dioxide	6.3-17	X		X	100	NA								100	
Ammonia	6.3-18	X			300A									300 A	
Biocide Flush Fluid	6.3-19	X			300									300	
Refrigerant 21 (New and Recycled)	6.3-20	X		X	300									300	
Avionics Fire Extinguishing Fluid	6.3-21	X												NA	
Conditioned Air - ECS	6.3-22	X			Class 100,000										Class 100,000
Fuel Cell Cooling Fluid	6.3-23	X			300								300		
Breathing Oxygen	6.3-24			X	200 A									200 A	
Isopropyl Alcohol	6.3-25			X	100/200 A		200	200	100	100	100	100		200 A	200
Trichlorotrifluoroethane	6.3-26			X	200 A	400 A	200 A	200 A					200 A	200 A	200 A
Denatured Ethyl Alcohol	6.3-27			X	300									300	
Trichloroethylene	6.3-28			X	800 A	800 A									
Breathing Air Mixture	6.3-29			X	200 A									200 A	200 A
Airlock LCG Cooling Water	6.3-30	X			300									300	
Ferry Flight WCL Fluid	6.3-31	X			300									300	
Heat Transport Water	6.3-32	X		X	200/300 A				200	NA				300 A	
Shock Strut Hydraulic Fluid - Orbiter	6.3-33	X			190						190	NA			
Missile Grade Air (Deleted)	6.3-34			*		*									
Refrigerant 114	6.3-35	X			300									300	
Breathing Air (EMU Ground Test Only)	6.3-36			X	200 A									200 A	
Nitric Oxide (Deleted)	6.3-37	*					*	*							
EMU Gaseous Oxygen	6.3-38	X			100 A		*	*						100 A	
Propellant, Nitrogen Tetroxide (MON-10)	6.3-39	*					*	*							
RTG Heat Transport Fluid	6.3-40	X			300									300	
APU/WSB Pre-Flush Lubricating Oil	6.3-41			X	300				300	NA					
WMS Flush Fluid	6.3-42			X	300									300	
Waste Tank Cleaning Fluid	6.3-43			X	300									300	
Isopropyl Alcohol	6.3-44			X	400 A	400 A									
WMS Servicing Fluid	6.3-45			X	300									300	
HCFC-225	6.3-46		X	X	200 A	400 A	200 A	200 A					200 A	200 A	
ZBA25	6.3-47			X	400	400									
Refrigerant 124 (HCFC-124)	6.3-48	X		X	300									300	
Perfluorohexane (PF-5060) (Lear Romec Quick Disconnects Only)	6.3-49			X	300									300	
Vertrel MCA (Decafluoropentane [62%] and Trans-Dichloroethylene [38%]	6.3-50			X	200 A	400 A		200 A					200 A	200 A	200 A
HFE-7100 (Methoxy-nonafluorobutane)	6.3-51			X	200 A	400 A		200					200 A	200 A	200 A
Vertrel XF (1,1,1,2,3,4,4,5,5,5-Decafluoropentane)	6.3-52			X	400 A	400 A									
HCFC-225 G	6.3-53			X	200 A	400 A		200 A					200 A	200 A	200 A
PGME/Water Azeotrope	6.3-54	X			200						200	NA			

NOTES
(1) Numbered values represent acceptable particulate contamination level
(2) "*" Reference source does not address applications in subsystems control.
(3) "A" indicates requirement to verify minimum threshold of non-volitile residue (NVR)

Traditional Design Approach to Flight Reliability

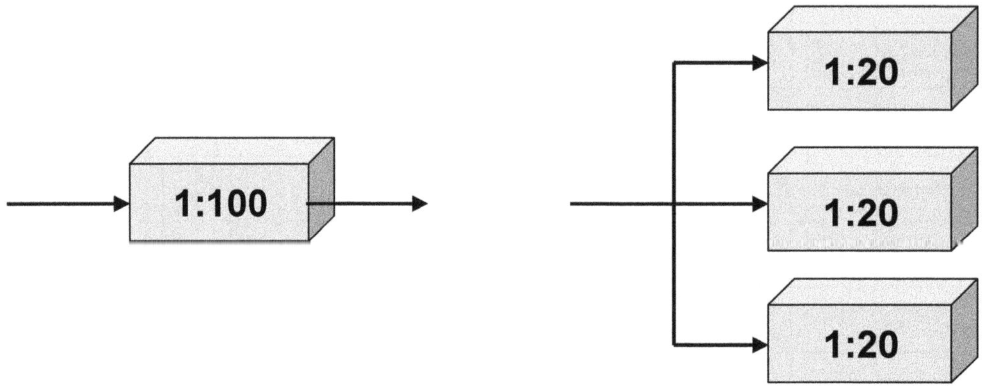

- Greater flight reliability, but increased unplanned troubleshooting and repair--in space, or on the ground
- Decreased ground personnel safety if devices are hazardous or toxic in nature

Alternative Design Approach to Flight Reliability

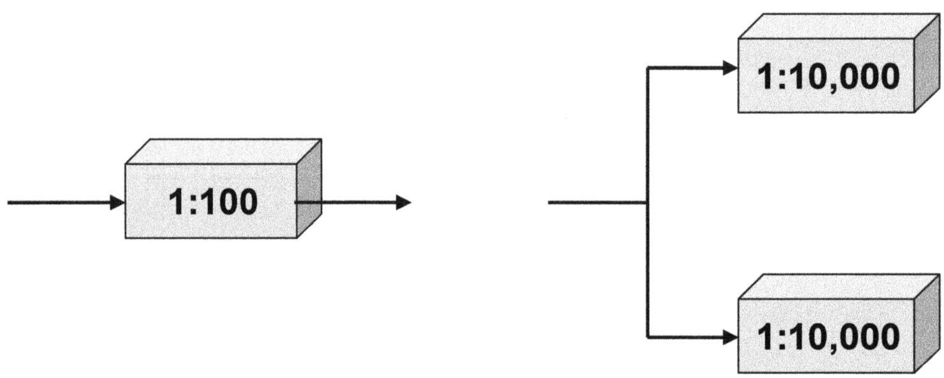

- Greater flight reliability
- Decreased unplanned troubleshooting and repair on two counts (fewer parts in the system exposed to failure and more robust parts)
- Increased ground personnel safety, even if devices are hazardous when compared to traditional design approach

Figure 51. Design Redundancy Pros and Cons

7.4 Built-In Flight System Integrity Between Routine Flights

Careful planning, design, and operational readiness validation are needed to maintain a reusable flight element's integrity between flights. This means maintaining structural integrity between flights by attaining a design that is tolerant of corrosion issues (similar to modern passenger and military aircraft) without detailed inspections of delicate designs. It also means that no maintenance, planned or unplanned, is likely to occur between routine flights, and therefore, no requirement to destroy previously certified structural integrity by opening up internal compartments, access doors with sensitive thermal/pressure seals, and so forth. The foregoing discussion on dependability will be useful in pursuing this area.

7.5 Need for Operationally-Relevant Technology Integration Environments

Critical to pursuing the foregoing operator needs is the ability to represent, on the ground, system concepts operating in environments relevant to the maintainer. Until there is greater investment in such infrastructure, it will be very difficult for the designer and the operator to communicate highly complex and often subtle and latent operability and supportability issues.

7.6 Summary

A detailed characterization of the operability, dependability, supportability, and recurring affordability has been completed of the world's first attempt at reusable space flight hardware: America's Space Shuttle. Many design causes and quantified operational effects were documented, along with an accounting of specific operator needs, requirements, and possible future design and technology maturation strategies. The itemization of the seemingly overwhelming amount of work involved in turning around Shuttle flight hardware should not cause us to retreat from designing highly reusable space transportation. The identification of problem areas and modern engineering design methods, along with a new generation of designers and operators, can make respectable improvements and succeed in overcoming the identified shortfalls. They must be overcome because the future of humankind in space depends on improving space transportation operations and infrastructure.

This page intentionally left blank.

APPENDIX A. GENERIC FUNCTIONAL BREAKDOWN STRUCTURE (FBS)

1.0	LANDING/RECOVERY		
	1.01	Facility and Equipment Preps for Vehicle Landing / Recovery	
		1.01.01	Functional verification of ground recovery systems prior to flight element arrival comitment
		1.01.02	Servicing of Ground Recovery systems for arrival
		1.01.03	Landing Facility Contamination control
	1.02	Element Landing and Recovery Operations	
		1.02.01	Activate ground landing systems and position resources for element arrival
		1.02.02	Position recovery equipment and provide services to vehicle element upon return
		1.02.03	Perform active system shutdown, safing and check out for return to spaceport
	1.03	Payload Removal Operations	
		1.03.01	Internal payload item removal
		1.03.02	Payload handling and transport operations
		1.03.03	Attached payload item removal
		1.03.04	Crew and passenger egress
	1.04	Postflight element ground handling and transport	
		1.04.01	Postflight element handling operations
		1.04.02	Element ground transport from recovery point
	1.05	Landing/Recovery Facility and Equipment Periodic Maintenance	
		1.05.01	Interval Maintenance of Landing Recovery Ground Systems
		1.05.02	Recovery Ground System and Facility Modifications and Process Changes
	1.06	Remote Landing Site Recovery and Ferry Return Operations	
		1.06.01	Flight element system preparations for ferry
		1.06.02	Ferry equipment and facility preps
		1.06.03	Flight element mate to ferry systems
		1.06.04	Element Ferry operations support
2.0	FLIGHT ELEMENT TURNAROUND		
	2.01	Facility Preps for Vehicle Turnaround	
		2.01.01	Functional Verification of ground systems prior to Vehicle arrival
		2.01.02	Servicing and staging of ground systems prior to vehicle arrival
		2.01.03	Contamination control
	2.02	Positioning Vehicle, Connection to Services, Gaining Access, and Protection	
		2.02.01	Vehicle Transport and Alignment
		2.02.02	Ground access equipment positioning
		2.02.03	Connection to facility services
		2.02.04	Establish protective enclosures and install/remove flight vehicle protective covers
		2.02.05	Removal of flight equipment covers/panels to gain access
		2.02.06	Opening and closing of vehicle hatches and hinged/articulated doors
	2.03	Vehicle Element Safing	
		2.03.01	Hazardous Fluid Safing
		2.03.02	Pyro/Ordnance Safing
		2.03.03	Vehicle Purges for Personnel/Orbiter Safety
	2.04	Vehicle Element Systems Servicing	
		2.04.01	Landing Gear Mechanical Servicing
		2.04.02	Ordnance Handling and Installation

		2.04.03	Maintenance of Fuselage/PLB Vent Filters & Window Cavity Purge
		2.04.04	Flight Crew Systems Servicing
		2.04.05	Acreage Tile & TPS Blanket Servicing
		2.04.06	TPS Moldline Penetration, Leading Edge/Nose & Aerosurface Hingeline Seal Servicing
		2.04.07	Fluid Drain and Deservicing
		2.04.08	Fluid Servicing and Fill
		2.04.09	Routine Vehicle Cleaning & Filter Inspections
		2.04.10	Flight Pressurizations
		2.04.11	Navigation & Instrumentation Component Servicing
		2.04.12	Routine Replacement of Expendable and Limited Life Items
		2.04.13	Vehicle electrical power application, and data processing monitoring & control operations
	2.05	Inspection & Checkout	
		2.05.01	Component/Subsystem Functional Verification
		2.05.02	Structural integrity inspections and mechanism functional verifications
		2.05.03	Propellant/fluid/gas system integrity verifications & inspections
		2.05.04	Vehicle power up/down, switch-lists, functional checks and on-board s/w config
	2.06	Vehicle Payload Accommodations Turnaround	
		2.06.01	Reconfigure unique payload thermal protection/insulation
		2.06.02	Repositioning vehicle-fixed mechanisms
		2.06.03	Remove/Install mission-unique mechanisms
		2.06.04	Reconfigure vehicle supplied fluid services
		2.06.05	Reconfigure vehicle supplied electrical and information services
		2.06.06	Access to Payload accommodations
		2.06.07	Payload accommodations cleaning and closeout
		2.06.08	Payload accommodations turnaround unplanned troubleshooting and repair
		2.06.09	Payload Accommodations Inspections & Testing
	2.07	Processing Support Systems & Functions	
		2.07.01	Purge Operations & Monitoring
		2.07.02	Hazard Monitoring Systems Operation
		2.07.03	Vehicle System Reconfigurations
	2.08	Turnaround Unplanned Troubleshooting and Repair	
		2.08.01	Troubleshooting, Replacement, Disposition of Failed/Suspect LRUs
		2.08.02	Troubleshooting and Repair of Leaks
		2.08.03	Fluid/Pneumatic System Decontamination/Cleaning
		2.08.04	Troubleshooting, Retest and Repair of Electrical cables, connectors
		2.08.05	Repair or Replacement of Thermal Protection Systems (TPS) Hardware
		2.08.06	Structural Repair/Refurbishment
		2.08.07	Repair of Damaged Ducts, Tubes & Hoses
		2.08.08	Troubleshooting & Repair of Mechanisms and Thermal/Pressure Seals
		2.08.09	Troubleshooting and Repair of Ground Equipment
	2.09	Modifications and Special Tests	
		2.09.01	Flight Element Modifications and Process Changes
		2.09.02	Special Tests and Fleet System Checks
		2.09.03	Payload-Driven Flight Element Weight/CG Alterations
	2.10	Removal of Access/Umbilicals and Closeout	
		2.10.01	Removal of Umbilicals and Access equipment
		2.10.02	Flight Element System Closeouts for Vehicle and Ground Launch System Integration
		2.10.03	Preps for Vehicle Element Transfer

	2.11	Turnaround Facility and Equipment Periodic Maintenance	
		2.11.01	Interval Maintenance of Ground Systems
		2.11.02	Ground System and Facility Modifications and Process Changes
3.0	FLIGHT ELEMENT ASSEMBLY		
	3.01	Element Transport, Lift and Handling for Assembly	
		3.01.01	Facility Preps for Element Assembly Operations
		3.01.02	Ground Transport of Major Element Segments
		3.01.03	Lift, Positioning and Connection to Assembly Facility Support Fixtures
		3.01.04	Activate Element Environmental and Safety Controls
	3.02	Element Structural Assembly	
		3.02.01	Flight Segment Handling Operations
		3.02.02	Flight Segment Structural Connections
		3.02.03	Structural Assembly Verification and Closeout
	3.03	Subsystem / Component Installations	
		3.03.01	Subsystem / Component Handling Ops
		3.03.02	Facility Preps for Installations
		3.03.03	Flight Component Installation onto Flight Element
		3.03.04	Non-Flight Component Installations
		3.03.05	Flight Component Installation Closeouts
	3.04	Flight Element Assembly Functional Verifications	
		3.04.01	Passive System Verifications
		3.04.02	Active System Verifications
		3.04.03	Inspections and Alignment Verifications
	3.05	Assembly Unplanned Troubleshooting & Repair	
		3.05.01	Unplanned Element Disassembly/Reassembly
		3.05.02	Flight Element Troubleshooting and Repair
		3.05.03	Flight Element Modifications and Process Changes
	3.06	Postassembly Operations	
		3.06.01	Reconfigure/Relocate Facility and Ground Equipment
		3.06.02	Flight Element Preparations for Integration
	3.07	Assembly Facility and Equipment Periodic Maintenance	
		3.07.01	Interval Maintenance of Assembly Ground Systems
		3.07.02	Assembly Ground System and Facility Modifications and Process Changes
	3.08	Routine Reusable Flight Element Module/Component Disassembly	
		3.08.01	Disassembly Preps and Attach/closeout hardware removal
		3.08.02	Flight Element Component disassembly, handling, and transport
		3.08.03	Post-Disassembly
4.0	FLIGHT VEHICLE INTEGRATION		
	4.01	Element Transportation, Handling, Lift and Mate	
		4.01.01	Facility Preps for Flight Element Integration Operations
		4.01.02	Ground Transport of Flight Elements
		4.01.03	Element Lift, Position and Mate
		4.01.04	Postmate Servicing and Interface Closeouts
	4.02	Postmating Functional Verification	
		4.02.01	Element to Ground System Interfaces

		4.02.02	Element to Element Interfaces
	4.03	Integration Unplanned Troubleshooting & Repair	
		4.03.01	Disconnection of Flight Elements
		4.03.02	Interface Troubleshooting and Repair
		4.03.03	Postintegration Flight Element Troubleshooting and Repair
		4.03.04	Postintegration Ground Element/Equipment Troubleshooting and Repair
	4.04	Perform integration of payload / cargo	
		4.04.01	Facility Preparations for Payload Installation
		4.04.02	Payload Installation into Vehicle
		4.04.03	Postinstallation operations
		4.04.04	Payload handling and transport operations
		4.04.05	Payload Inspections and Testing
	4.05	Configure systems for Vehicle Transfer to Launch Point	
		4.05.01	Disconnect/Stow integration and mating ground h/w
		4.05.02	Configure Vehicle systems for launch functions
		4.05.03	Preps for Integrated Vehicle Transfer/Rotation to launch position
	4.06	Integration Facility and Equipment Periodic Maintenance	
		4.06.01	Interval Maintenance of Integration Ground Systems
		4.06.02	Integration Ground System and Facility Modifications and Process Changes
5.0	FLIGHT VEHICLE LAUNCH/DEPARTURE		
	5.01	Facility/GSE Refurbishment, Reservicing & Preps for Vehicle Arrival	
		5.01.01	Launch Equipment Refurbishment/Reservicing
		5.01.02	Contamination Control
		5.01.03	Functional verifications of ground systems prior to flight vehicle mating
		5.01.04	Servicing of ground systems prior to flight vehicle mating
	5.02	Position Vehicle at Launch Point, Mate to Ground Facility and Services & Provide Access	
		5.02.01	Vehicle Transport and Alignment
		5.02.02	Mate Umbilicals, Environmental Protection and Access positioning
	5.03	Functional Verification of Flight/Ground Systems for Launch	
		5.03.01	Periodic Maintenance / Calibrations
		5.03.02	System Integrity Verification
		5.03.03	System Functionality Verification
		5.03.04	Human systems verification and training
	5.04	Launch Systems Unplanned Troubleshooting/Repair	
		5.04.01	Component Troubleshooting, Removal and Replacements
		5.04.02	Ground Launch Structural Systems Repair
	5.05	Crew Ingress, Late Payload Stowage & Functional Verification for Flight	
		5.05.01	Flight Crew Ingress/Egress
		5.05.02	Late Payload Stowage
		5.05.03	Payload Functional Verification for Flight
		5.05.04	Unplanned Payload Troubleshooting and Repair
	5.06	Vehicle System Servicing at Launch Point	
		5.06.01	Local Servicing with facility/GSE by remote control room
		5.06.02	Local Manual Servicing at the Launch Facility
		5.06.03	Ordnance Hookups

	5.07	Access/Umbilical Removal & Closeout for Flight	
		5.07.01	Payload Closeout tasks
		5.07.02	Access Removal tasks
		5.07.03	Umbilical Removal tasks
		5.07.04	Environmental Cover Removal tasks
	5.08	Remote Automated Servicing and Propellant Loading	
		5.08.01	Cryogenic Fluids transfer
		5.08.02	Thermal Management Fluid Servicing
		5.08.03	Tank Gaseous Pressurization
		5.08.04	Flight Vehicle Press
		5.08.05	Environmental Ctl purging
		5.08.06	Engine Conditioning Purges
		5.08.07	Remote Fuel Cell Gas Supply
	5.09	Launch the Vehicle	
		5.09.01	Coordinate the network of local and offsite support functions
		5.09.02	Activate Flight Vehicle and Ground Systems for Launch
		5.09.03	Obtain Weather and Range Clearances for Launch
		5.09.04	Configure Flight Vehicle for Terminal (Final) Launch Operations
		5.09.05	Final Launch Sequencing
		5.09.06	Configure systems for launch abort/recycle, as needed
	5.10	Postlaunch Securing	
		5.10.01	Postlaunch System Safing Operations
		5.10.02	Postlaunch Inspection of Facilities and Equipment
	5.11	Launch Facility and Equipment Periodic Maintenance	
		5.11.01	Interval Maintenance of Launch Ground Systems
		5.11.02	Launch Ground System and Facility Modifications and Process Changes
6.0	FLIGHT ELEMENT SHIPPING, RECEIVING & ACCEPTANCE		
	6.01	Element Receiving and Handling Ops	
		6.01.01	Receiving Facility and Equipment Preps and Functional Verification
		6.01.02	Element Transport Receiving and Positioning Ops
		6.01.03	Element Offload from Transport system
	6.02	Element Ground Transfer and Storage Operations at the Spaceport	
		6.02.01	Element Ground Handling equipment and fixtures
		6.02.02	Element Environmental Shelter and Conditioning
		6.02.03	Remove and Stow Transportation Hardware
	6.03	Element Inspection and Receiving Acceptance	
		6.03.01	Receiving Inspection
		6.03.02	Hardware Inventory
		6.03.03	Softgoods Inventory
	6.04	Conditioning if required (purging, temperature and humidity control)	
		6.04.01	Facility/GSE preps for Element Storage
		6.04.02	Activate/Monitor Element conditioning systems
	6.05	Perform design modifications (deferred work)	
		6.05.01	Element/Subsystem modifications
		6.05.02	Fleet-wide tests & inspections

	6.06	Receiving Inspection discrepancy resolution & repairs	
		6.06.01	Element Repair & Rework
		6.06.02	Inventory Discrepancy Resolutions
	6.07	Element Shipment from Spaceport	
		6.07.01	Facility and Equipment Preps for Flight Element Hardware Shipment
		6.07.02	Shipment Kit installation on the element
		6.07.03	Element Handling and Connection with Transport System
		6.07.04	Softgoods preparation for shipment
7.0	FLIGHT ELEMENT, SUBSYSTEM & COMPONENT DEPOT MAINTENANCE		
	7.01	Vehicle overhaul, inspection/verification, and modification (structural, flight controls, etc.)	
	7.02	Modular element overhaul, remanufacture and inspection/verification	
	7.03	Hot test propulsion hardware	
	7.04	Spaceport software upgrades (non-flight)	
8.0	TRAFFIC/FLIGHT CONTROL		
	8.01	Ground/flight vehicle inter-communications systems management and control	
	8.02	Weather advisory for launch, landing, and ground operations	
	8.03	Vehicle-related launch/flight/landing/ground operations control and monitoring	
	8.04	Ascent/Reentry flight safety monitor and control	
	8.05	Audio/visual monitor of ground operations	
9.0	OFFLINE PAYLOAD/CARGO PROCESSING		
	9.01	Receipt	
	9.02	Packaging	
	9.03	Assembly	
	9.04	Test & Checkout	
	9.05	Encapsulation	
	9.06	Payload Commodity Servicing	
	9.07	Transport to Vehicle Integration location	
	9.08	Pre-Flight support	
	9.09	Passenger pre-flight activities	
	9.10	Postflight payload processing	
10.0	SPACEPORT SUPPORT INFRASTRUCTURE		
	10.01	Shops and Labs	
	10.02	Photographic Services	
	10.03	Fire Protection, Medical	
	10.04	Security	
	10.05	Library (technical documents)	
	10.06	Utilities	
	10.07	Roads and Grounds	
	10.08	Foods Services	
	10.09	Heavy Equipment	
	10.10	Communication/Information Services	
	10.11	Ground Transportation Services	
	10.12	Environmental Compatibility Management	
	10.13	Pyrotechnic Storage and Handling	
	10.14	Personal Environmental Protection Equipment	

	10.15	Facility Maintenance Services and Shops
11.0	FLIGHT SYSTEM-UNIQUE LOGISTICS	
	11.01	Propellants (acquisition, storage, distribution, conditioning verification)
	11.02	Other fluids and gases and unique consumables
	11.03	LRU replacement hardware (flight and ground systems)
12.0	TRANSPORTATION SYSTEMS OPS PLANNING & MANAGEMENT	
	12.01	Customer Relations
	12.02	Vehicle Manifesting and Scheduling
	12.03	Ground Systems Scheduling and Management
	12.04	Software Production
	12.05	Personnel Management
	12.06	Sustaining Operations Engineering Work Control
	12.07	Public Affairs
	12.08	Economic Development
	12.09	Business Management
	12.10	Advanced Planning
	12.11	Safety, Reliability & Quality Assurance (SR&QA)
13.0	COMMUNITY SUPPORT SERVICES INFRASTRUCTURE	
	13.01	Shelter
	13.02	Connecting Utility Infrastructure
	13.03	Transportation Support
	13.04	Educational Support
	13.05	Community Police/Fire Protection
	13.06	Community Resources Infrastructure and Services
	13.07	Consumer Retail Support
	13.08	Community Medical Support
	13.09	Financial Institutions

Table A2. Generic Functional Breakdown Structure (FBS)

FLIGHT VEHICLE OPERATIONS			
A1	RECEIVE AND ACCEPT FLIGHT ELEMENTS		
	A1.01	Element Receiving and Handling Ops	
		A1.01.01	Receiving facility and equipment preps and functional verification
		A1.01.02	Element transport receiving and positioning ops
		A1.01.03	Element offload from transport system
	A1.02	Element Ground Transfer and Storage Operations at the Spaceport	
		A1.02.01	Element ground handling equipment and fixtures
		A1.02.02	Element environmental shelter and conditioning
		A1.02.03	Remove and stow transportation hardware
	A1.03	Element Inspection and Receiving Acceptance	
		A1.03.01	Receiving inspection
		A1.03.02	Hardware inventory
		A1.03.03	Soft goods inventory
	A1.04	Conditioning if Required (Purging, Temperature, and Humidity Control)	
		A1.04.01	Facility/GSE preps for element storage
		A1.04.02	Activate/monitor element conditioning systems
	A1.05	Perform Design Modifications (Deferred Work)	
		A1.05.01	Element/subsystem modifications
		A1.05.02	Fleetwide tests and inspections
	A1.06	Receiving Inspection Discrepancy Resolution and Repairs	
		A1.06.01	Element repair and rework
		A1.06.02	Inventory discrepancy resolutions
B1	ASSEMBLY FLIGHT ELEMENTS		
	B1.01	Element Transport, Lift, and Handling for Assembly	
		B1.01.01	Facility preps for element assembly operations
		B1.01.02	Ground transport of major element segments
		B1.01.03	Lift, positioning, and connection to assembly facility support fixtures
		B1.01.04	Activate element environmental and safety controls
	B1.02	Element Structural Assembly	
		B1.02.01	Flight segment handling operations
		B1.02.02	Flight segment structural connections
		B1.02.03	Structural assembly verification and closeout
	B1.03	Subsystem/Component Installations	
		B1.03.01	Subsystem/component handling ops
		B1.03.02	Facility preps for installations
		B1.03.03	Flight component installation onto flight element
		B1.03.04	Nonflight component installations
		B1.03.05	Flight component installation closeouts
	B1.04	Flight Element Assembly Functional Verifications	
		B1.04.01	Passive system verifications
		B1.04.02	Active system verifications
		B1.04.03	Inspections and alignment verifications

	B1.05	Assembly Unplanned Troubleshooting and Repair	
		B1.05.01	Unplanned element disassembly/reassembly
		B1.05.02	Flight element troubleshooting and repair
		B1.05.03	Flight element modifications and process changes
	B1.06	Postassembly Operations	
		B1.06.01	Reconfigure/relocate facility and ground equipment
		B1.06.02	Flight element preparations for integration
C1	SERVICE AND CHECKOUT FLIGHT ELEMENTS		
	C1.01	Vehicle Element Systems Servicing	
		C1.01.01	Landing-gear mechanical servicing
		C1.01.02	Ordnance handling and installation
		C1.01.03	Maintenance of fuselage/PLB vent filters and window cavity purge
		C1.01.04	Flight crew systems servicing
		C1.01.05	Acreage tile and TPS blanket servicing
		C1.01.06	TPS mold line penetration, leading edge/nose, and aerosurface hingeline seal servicing
		C1.01.07	Fluid drain and deservicing
		C1.01.08	Fluid servicing and fill
		C1.01.09	Routine vehicle cleaning and filter inspections
		C1.01.10	Flight pressurizations
		C1.01.11	Navigation and instrumentation component servicing
		C1.01.12	Routine replacement of expendable and limited-life items
		C1.01.13	Vehicle electrical power application; and data processing, monitoring, and control operations
	C1.02	Inspection and Checkout	
		C1.02.01	Component/subsystem functional verification
		C1.02.02	Structural integrity inspections and mechanism functional verifications
		C1.02.03	Propellant/fluid/gas system integrity verifications and inspections
		C1.02.04	Vehicle power-up/down, switch lists, functional checks, and onboard s/w configuration
	C1.03	Removal of Access/Umbilicals and Closeout	
		C1.03.01	Removal of umbilicals and access equipment
		C1.03.02	Flight element system closeouts for vehicle and ground launch system integration
		C1.03.03	Preps for vehicle element transfer
D	INTEGRATE FLIGHT SYSTEM AND PAYLOAD		
	D.01	Element Transportation, Handling, Lift, and Mate	
		D.01.01	Facility preps for flight element integration operations
		D.01.02	Ground transport of flight elements
		D.01.03	Element lift, position, and mate
		D.01.04	Postmate servicing and interface closeouts
	D.02	Postmating Functional Verification	
		D.02.01	Element-to-ground-system interfaces
		D.02.02	Element-to-element interfaces
	D.03	Integration Unplanned Troubleshooting and Repair	
		D.03.01	Disconnection of flight elements
		D.03.02	Interface troubleshooting and repair
		D.03.03	Postintegration flight element troubleshooting and repair
		D.03.04	Postintegration ground element/equipment troubleshooting and repair

	D.04	Perform Integration of Payload/Cargo	
		D.04.01	Facility preparations for payload installation
		D.04.02	Payload installation into vehicle
		D.04.03	Postinstallation operations
		D.04.04	Payload handling and transport operations
		D.04.05	Payload inspections and testing
	D.05	Configure Systems for Vehicle Transfer to Launch Point	
		D.05.01	Disconnect/stow integration and mating ground h/w
		D.05.02	Configure vehicle systems for launch functions
		D.05.03	Preps for integrated vehicle transfer/rotation to launch position
	D.06	Position Vehicle at Launch Point, Mate to Ground Facility and Services, and Provide Access	
		D.06.01	Vehicle transport and alignment
		D.06.02	Mate umbilicals, environmental protection, and access positioning
E	EXECUTE DEPARTURE OPERATIONS		
	E.01	Functional Verification of Flight/Ground Systems for Launch	
		E.01.01	Periodic maintenance/calibrations
		E.01.02	System integrity verification
		E.01.03	System functionality verification
		E.01.04	Human systems verification and training
	E.02	Launch Systems Unplanned Troubleshooting/Repair	
		E.02.01	Component troubleshooting, removal, and replacements
		E.02.02	Ground launch structural systems repair
	E.03	Crew Ingress, Late Payload Stowage, and Functional Verification for Flight	
		E.03.01	Flight crew ingress/egress
		E.03.02	Late payload stowage
		E.03.03	Payload functional verification for flight
		E.03.04	Unplanned payload troubleshooting and repair
	E.04	Vehicle System Servicing at Launch Point	
		E.04.01	Local servicing with facility/GSE by remote control room
		E.04.02	Local manual servicing at the launch facility
		E.04.03	Ordnance hookups
	E.05	Access/Umbilical Removal and Closeout for Flight	
		E.05.01	Payload closeout tasks
		E.05.02	Access removal tasks
		E.05.03	Umbilical removal tasks
		E.05.04	Environmental cover removal tasks
	E.06	Remote Automated Servicing and Propellant Loading	
		E.06.01	Cryogenic fluids transfer
		E.06.02	Thermal management fluid servicing
		E.06.03	Tank Gaseous pressurization
		E.06.04	Flight vehicle press
		E.06.05	Environmental control purging
		E.06.06	Engine conditioning purges
		E.06.07	Remote fuel cell gas supply
	E.07	Launch the Vehicle	
		E.07.01	Coordinate the network of local and offsite support functions
		E.07.02	Activate flight vehicle and ground systems for launch

		E.07.03	Obtain weather and range clearances for launch
		E.07.04	Configure flight vehicle for terminal (final) launch operations
		E.07.05	Final launch sequencing
		E.07.06	Configure systems for launch abort/recycle, as needed
F	MONITOR AND MANAGE THE FLIGHT		
	F.01	Ascent/Reentry Flight Safety Monitor and Control	
G	LAND/RECOVER FLIGHT ELEMENTS AND PAYLOAD		
	G.01	Facility and Equipment Preps for Vehicle Landing/Recovery	
		G.01.01	Functional verification of ground recovery systems prior to flight element arrival
		G.01.02	Servicing of ground recovery systems for arrival
		G.01.03	Landing facility contamination control
	G.02	Element Landing and Recovery Operations	
		G.02.01	Activate ground landing systems and position resources for element arrival
		G.02.02	Position recovery equipment and provide services to vehicle element upon return
		G.02.03	Perform active system shutdown, safing, and checkout for return to spaceport
	G.03	Payload Removal Operations	
		G.03.01	Internal payload item removal
		G.03.02	Payload handling and transport operations
		G.03.03	Attached payload item removal
		G.03.04	Crew and passenger egress
	G.04	Postflight Element Ground Handling and Transport	
		G.04.01	Postflight element handling operations
		G.04.02	Element ground transport from recovery point
	G.05	Remote Landing Site Recovery and Ferry Return Operations	
		G.05.01	Flight element system preparations for ferry
		G.05.02	Ferry equipment and facility preps
		G.05.03	Flight element mate to ferry systems
		G.05.04	Element ferry operations support
H1	PREPARE FLIGHT ELEMENTS FOR TURNAROUND		
	H1.01	Facility Preps for Vehicle Turnaround	
		H1.01.01	Functional verification of ground systems prior to vehicle arrival
		H1.01.02	Servicing and staging of ground systems prior to vehicle arrival
		H1.01.03	Contamination control
	H1.02	Positioning Vehicle, Connection to Services, Gaining Access, and Protection	
		H1.02.01	Vehicle transport and alignment
		H1.02.02	Ground access equipment positioning
		H1.02.03	Connection to facility services
		H1.02.04	Establish protective enclosures and install/remove flight vehicle protective covers
		H1.02.05	Removal of flight equipment covers/panels to gain access
		H1.02.06	Opening and closing of vehicle hatches and hinged/articulated doors
	H1.03	Vehicle Element Safing	
		H1.03.01	Hazardous fluid safing
		H1.03.02	Pyro/ordnance safing
		H1.03.03	Vehicle purges for personnel/Orbiter safety
	H1.04	Processing Support Systems and Functions	
		H1.04.01	Purge operations and monitoring

		H1.04.02	Hazard monitoring systems operation
		H1.04.03	Vehicle system reconfigurations
	H1.05	Routine Reusable Flight Element Module/Component Disassembly	
		H1.05.01	Disassembly preps and attach/closeout hardware removal
		H1.05.02	Flight element component disassembly, handling, and transport
		H1.05.03	Post-disassembly
I1	RESTORE FLIGHT ELEMENTS		
	I1.01	Troubleshooting, replacement, disposition of failed/suspect LRUs	
	I1.02	Troubleshooting and repair of leaks	
	I1.03	Fluid/pneumatic system decontamination/cleaning	
	I1.04	Troubleshooting, retest, and repair of electrical cables, connectors	
	I1.05	Repair or replacement of TPS hardware	
	I1.06	Structural repair/refurbishment	
	I1.07	Repair of damaged ducts, tubes, and hoses	
	I1.08	Troubleshooting and repair of mechanisms and thermal/pressure seals	
	I1.09	Troubleshooting and repair of ground equipment	
J	RESTORE GROUND SYSTEMS FOR REUSE		
	J.01	Postlaunch Securing	
		J.01.01	Postlaunch system safing operations
		J.01.02	Postlaunch inspection of facilities and equipment
	J.02	Facility/GSE Refurbishment, Reservicing, and Preps for Vehicle Arrival	
		J.02.01	Launch equipment refurbishment/reservicing
		J.02.02	Contamination control
		J.02.03	Functional verifications of ground systems prior to flight vehicle mating
		J.02.04	Servicing of ground systems prior to flight vehicle mating
PAYLOAD ELEMENT OPERATIONS			
A2	RECEIVE AND ACCEPT PAYLOAD ELEMENTS		
	A2.01	Payload Element Receipt	
B2	ASSEMBLY PAYLOAD ELEMENTS		
	B2.01	Payload Element Packaging	
	B2.02	Payload Element Assembly	
C2	SERVICE AND CHECKOUT PAYLOAD ELEMENTS		
	C2.01	Test and Checkout	
	C2.02	Encapsulation	
	C2.03	Payload Commodity Servicing	
	C2.04	Transport to Vehicle Integration Location	
	C2.05	Preflight Support	
	C2.06	Passenger Preflight Activities	
H2	PREPARE PAYLOAD ELEMENTS FOR TURNAROUND OR DISPOSAL		
	H2.01	Vehicle Payload Accommodations Turnaround	
		H2.01.01	Reconfigure unique payload thermal protection/insulation
		H2.01.02	Repositioning vehicle-fixed mechanisms
		H2.01.03	Remove/install mission-unique mechanisms
		H2.01.04	Reconfigure vehicle-supplied fluid services

		H2.01.05	Reconfigure vehicle-supplied electrical and information services
		H2.01.06	Access to PAYLOAD accommodations
		H2.01.07	Payload accommodations cleaning and closeout
		H2.01.08	Payload accommodations turnaround unplanned troubleshooting and repair
		H2.01.09	Payload accommodations inspections and testing
	H2.02	Postflight payload processing	
I2	RESTORE PAYLOAD ELEMENTS		
FLIGHT & GROUND TRAFFIC CONTROL AND SAFETY OPERATIONS			
K	TRAFFIC CONTROL AND SAFETY		
	K.01	Ground/flight Vehicle Intercommunications Systems Management and Control	
	K.02	Weather Advisory for Launch, Landing, and Ground Operations	
	K.03	Vehicle-Related Launch/Flight/Landing/Ground Operations Control and Monitoring	
	K.04	Audio/Visual Monitor of Ground Operations	
ENABLING OPERATIONS			
L	SPACEPORT LOGISTICS		
M1	MANAGEMENT OF FLIGHT SYSTEM OPERATIONS		
M2	MANAGEMENT OF THE SPACEPORT		
N	SPACEPORT-PROVIDED SUPPORT SERVICES		
O	OFFLINE MAINTENANCE, REPAIR, AND OVERHAUL		
	O.01	Landing/Recovery Facility and Equipment Periodic Maintenance	
		O.01.01	Interval maintenance of landing recovery ground systems
		O.01.02	Recovery ground system; facility modifications; and process changes
	O.02	Mods and Special Tests	
		O.02.01	Flight element modifications and process changes
		O.02.02	Special Tests and fleet system checks
	O.03	Turnaround Facility and Equipment Periodic Maintenance	
		O.03.01	Interval maintenance of ground systems
		O.03.02	Ground system and facility modifications, and process changes
	O.04	Assembly Facility and Equipment Periodic Maintenance	
		O.04.01	Interval maintenance of assembly ground systems
		O.04.02	Assembly ground system and facility modifications, and process changes
	O.05	Integration Facility and Equipment Periodic Maintenance	
		O.05.01	Interval maintenance of integration ground systems
		O.05.02	Integration ground system and facility modifications, and process changes
	O.06	Launch Facility and Equipment Periodic Maintenance	
		O.06.01	Interval maintenance of launch ground systems
		O.06.02	Launch ground system and facility modifications, and process changes
	O.07	Vehicle Overhaul, Inspection/Verification, and Modification (Structural, Flight Controls, etc.)	
	O.08	Modular Element Overhaul, Remanufacture and Inspection/Verification	
	O.09	Hot-Test Propulsion Hardware	
	O.10	Spaceport Software Upgrades (Nonflight)	

P	PUBLIC AND COMMUNITY SUPPORT SERVICES	
	P.01	Shelter
	P.02	Connecting Utility Infrastructure
	P.03	Transportation Support
	P.04	Educational Support
	P.05	Community Police/Fire Protection
	P.06	Community Resources Infrastructure and Services
	P.07	Consumer Retail Support
	P.08	Community Medical Support
	P.09	Financial Institutions

APPENDIX B. STS PROGRAM COST BURDEN TABLE

1994 STS WBS	Functional Breakdown Structure (FBS)	8 Flt/Year Baseline Annual Ops Cost $FY94M	8 Flt/Year Baseline Materials & ODC $FY94M	8 Flt/Year Baseline Labor $FY94M	8 Flts/Year Baseline Headcount	Calculated Fixed Cost-to-Variable Cost Ratio @ 8 Flts/Year*
Shuttle Operations		**$3,363.40**	**$920.70**	**$2,442.70**	**28,305**	
TOTAL EXTERNAL TANK		**$372.40**	**$122.50**	**$249.90**	**2,376**	
Mission Analysis		**$26.80**	**$0.90**	**$25.90**	**209**	
Launch Support Services	**12**	$6.10	$0.00	$6.10	49	1
Flight Support	**12**	$15.60	$0.00	$15.60	128	1
Technical Directives	**12**	$5.10	$0.90	$4.20	32	1
Production		**$300.90**	**$82.90**	**$218.00**	**2,041**	
Build and Support	**11**	$261.20	$72.70	$188.50	1,710	0.78
Facilities Self-Sustaining	**10**	$39.70	$10.20	$29.50	331	1
Project Support		**$44.70**	**$38.70**	**$6.00**	**126**	
Plant Operations		$25.70	$25.70	$0.00	0	
Replacement Equipment	**10**	$13.60	$13.60	$0.00	0	1
Utilities	**10**	$10.80	$10.80	$0.00	0	1
Rehab Equipment	**10**	$1.00	$1.00	$0.00	0	1
Special Studies	**12**	$0.30	$0.30	$0.00	0	1
Logistics		$3.40	$3.40	$0.00	0	
Refurbishment	**11**	$0.50	$0.50	$0.00	0	0.82
ET Transportation	**6**	$0.90	$0.90	$0.00	0	0.82
Government Bills of Lading	**6**	$0.10	$0.10	$0.00	0	0.82
Pressurants	**11**	$1.90	$1.90	$0.00	0	0.82
MAF Communications		$2.70	$2.00	$0.70	14	
Labor	**10**	$0.70	$0.00	$0.70	14	1
GSA FTS	**10**	$0.80	$0.80	$0.00	0	1
Maintenance	**10**	$0.70	$0.70	$0.00	0	1
Equipment/Supplies/Materials	**10**	$0.40	$0.40	$0.00	0	1
Local Phone Service	**10**	$0.10	$0.10	$0.00	0	1
Slidell Computer Complex		$9.00	$3.70	$5.30	106	
ADPE Purchases	**10**	$0.40	$0.40	$0.00	0	1
Labor	**10**	$5.30	$0.00	$5.30	106	1
ADPE Lease/Maintenance	**10**	$3.30	$3.30	$0.00	0	1
Technical Evaluation and Analysis		$3.90	$3.90	$0.00	6	
Science and Engineering	**12**	$2.60	$2.60	$0.00	0	1
Rockwell Support	**12**	$1.00	$1.00	$0.00	0	1
Computer Labor Support	**12**	$0.30	$0.00	$0.30	6	1
TOTAL SOLID ROCKET MOTOR (SRM)		**$404.20**	**$162.90**	**$241.30**	**2,727**	
Sustaining Engineering	**12**	$67.20	$4.00	$63.20	632	0.73
Touch & Support for Manufacturing & Refurbishment Labor	**11**	$193.60	$15.50	$178.10	2,095	0.73
SRM Propellant	**11**	$102.00	$102.00	$0.00	0	0.26
Expendable/Reusable Hardware	**11**	$13.60	$13.60	$0.00	0	0.73
Tooling Maintenance & Computer Support	**11**	$20.00	$20.00	$0.00	0	0.73
Freight	**11**	$7.80	$7.80	$0.00	0	0.26
Institutional Support	**12**	$0.00	$0.00	$0.00	0	0.73
TOTAL SOLID ROCKET BOOSTER (SRB)		**$152.00**	**$68.30**	**$83.70**	**985**	
Touch & Support Labor	**7**	$39.00	$1.60	$37.40	440	0.6
Expendable/Reusable Hardware	**11**	$27.50	$27.50	$0.00	0	0.6
Sustaining Engineering & Management	**12**	$46.20	$4.60	$41.60	489	0.6
Vendor Refurbishment of Reusable H/W	**11**	$18.70	$18.70	$0.00	0	0.6

1994 STS WBS	Functional Breakdown Structure (FBS)	8 Flt/Year Baseline Annual Ops Cost $FY94M	8 Flt/Year Baseline Materials & ODC $FY94M	8 Flt/Year Baseline Labor $FY94M	8 Flts/Year Baseline Headcount	Calculated Fixed Cost-to-Variable Cost Ratio @ 8 Flts/Year*
Travel, Computer & ODC	**7**	$14.70	$14.70	$0.00	0	0.6
KSC Support, Comm. & Sys Analysis	**12**	$5.90	$1.20	$4.70	56	0.6
TOTAL ENGINE (Sustaining Engineering)		**$125.30**	**$41.40**	**$83.90**	**599**	
Flight Support	**12**	$29.00	$2.90	$26.10	186	0.76
Anomaly Resolution	**12**	$22.30	$2.20	$20.10	143	1
Inventory Management & Warehousing	**7**	$6.80	$0.70	$6.10	44	1
Hardware Rerfurbishment	**7**	$21.10	$7.40	$13.70	98	0.37
New Hardware Spares	**7**	$44.80	$26.90	$17.90	128	0.72
Transportation	**7**	$1.30	$1.30	$0.00	0	0.7
TOTAL ORBITER & GFE (JSC)		**$177.30**	**$36.80**	**$140.50**	**1,173**	
Sustaining Engineering & Launch Spt	**7**	**$108.30**	**$21.70**	**$86.60**	**693**	**0.8**
Orbiter Support		**$50.90**	**$6.10**	**$44.80**	**407**	
PICS	**11**	$0.30	$0.00	$0.30	2	0.72
NASA Std Initiators (NSI)	**11**	$0.40	$0.00	$0.40	3	0.72
Pyros, Standard Operations	**11**	$1.60	$0.20	$1.40	13	0.72
RMS-Ops & Support	**9**	$3.20	$0.40	$2.80	26	0.7
RMS-Sustaining Engineering	**12**	$4.70	$0.60	$4.10	38	0.72
RMS-Program Management	**12**	$1.70	$0.20	$1.50	14	0.72
FCE Operations Management	**12**	$0.50	$0.10	$0.40	4	0.72
EMU/EVA Field Support/O&R	**9**	$1.20	$0.10	$1.10	10	0.95
EMU Logistics	**11**	$1.30	$0.20	$1.10	10	0.72
FEPC Tasks	**9**	$35.40	$4.20	$31.20	283	0.6
SSA Provisions (FEPC)	**9**	$0.40	$0.00	$0.40	3	1
Parachute Maintenance	**11**	$0.20	$0.00	$0.20	2	0.72
Flight Data Support	**12**	**$10.40**	**$5.20**	**$5.20**	**42**	**0.83**
Orbiter /ET Disconnects	**12**	**$7.70**	**$3.90**	**$3.90**	**31**	**0.45**
TOTAL ORBITER LOGISTICS & GSE (KSC)		**$174.00**	**$40.50**	**$133.50**	**1,110**	
Spares	**11**	$44.30	$17.70	$26.60	222	0.48
Overhaul & Repair	**11**	$64.60	$12.90	$51.70	431	0.82
Manpower to Support Logistics, Procurement, Engineering	**11**	$33.10	$0.00	$33.10	276	0.72
Tile Spares & Maintenance	**11**	$26.30	$7.90	$18.40	153	0.32
GSE Sustaining Engineering	**11**	$5.70	$2.00	$3.70	29	0.6
TOTAL PROPELLANT (KSC Launch Ops)		**$16.50**	**$16.50**	**$0.00**	**0**	
Propellant	**11**	$16.50	$16.50	$0.00	0	0.44
TOTAL LAUNCH OPERATIONS (KSC)		**$619.40**	**$115.90**	**$503.50**	**7,547**	
Shuttle Processing		**$206.10**	**$18.70**	**$187.30**	**2,864**	
Orbiter Operations		$132.40	$17.20	$115.20	1,797	
Orbiter Maintenance	**2**	$62.00	$8.20	$53.80	807	0.4
Orbiter Shop Operations	**2**	$8.90	$1.60	$7.20	117	0.4
Orbiter Modifications	**2**	$6.50	$0.60	$6.00	89	0.4
Orbiter Landing Operations	**1**	$12.60	$4.80	$7.80	107	0.4
Orbiter Processing Support	**12**	$24.80	$2.00	$22.80	398	0.4
Orbiter Tile Operations	**2**	$17.70	$0.00	$17.70	279	0.4
SRB Operations		$16.70	$0.00	$16.70	250	
SRB Processing Operations	**3,4**	$4.80	$0.00	$4.80	75	0.4
SRB Stacking	**3,4**	$4.70	$0.00	$4.70	74	0.4
SRB Retrieval & Disassembly Operations	**1**	$4.00	$0.00	$4.00	51	0.4
SRB Shop Operations	**10**	$1.50	$0.00	$1.40	25	0.4
SRB Modifications	**12**	$0.10	$0.00	$0.10	1	0.4

1994 STS WBS	Functional Breakdown Structure (FBS)	8 Flt/Year Baseline Annual Ops Cost $FY94M	8 Flt/Year Baseline Materials & ODC $FY94M	8 Flt/Year Baseline Labor $FY94M	8 Flts/Year Baseline Headcount	Calculated Fixed Cost-to-Variable Cost Ratio @ 8 Flts/Year*
SRB Processing Support	12	$1.70	$0.00	$1.70	25	0.4
ET Operations		$4.30	$0.00	$4.30	67	
ET Processing Operations	3,4	$2.90	$0.00	$2.90	45	0.4
ET Shop Operations	10	$0.30	$0.00	$0.30	5	0.4
ET Modifications	12	$0.10	$0.00	$0.10	2	0.4
ET Processing Support	12	$1.00	$0.00	$1.00	15	0.4
Launch Operations		$41.20	$0.30	$40.90	601	
Integrated Vehicle Servicing	5	$12.40	$0.00	$12.40	181	0.4
Integrated Vehicle Test & Launch Ops	5	$18.50	$0.00	$18.50	259	0.4
Launch Operations Support	12	$10.20	$0.20	$10.00	161	0.4
Payload Operations		$11.40	$1.20	$10.20	148	
Payload Integration and Support Services	2	$10.30	$0.10	$10.20	148	0.4
Payload Operations Support	12	$1.10	$1.10	$0.00	0	0.4
Systems Engineering/Support		**$11.30**	**$0.11**	**$11.10**	**171**	
Engineering Services	12	$3.50	$0.10	$3.40	62	0.5
Systems Engineeering	12	$7.80	$0.10	$7.80	109	0.5
Facility Operations & Maintenance		**$113.60**	**$28.70**	**$84.90**	**1,298**	
Facility O&M Support Operations	10	$17.80	$3.20	$14.60	235	0.52
Facility Maintenance		$49.60	$3.40	$46.20	684	
OPF Maintenance	2	$5.00	$0.10	$4.90	70	0.52
HMF Maintenance	7	$1.40	$0.00	$1.40	21	0.52
VAB Maintenance	3,4	$4.30	$0.00	$4.30	62	0.52
LCC Maintenance	8	$0.60	$0.00	$0.60	8	0.52
MLP Maintenance	3,4	$6.60	$0.10	$6.50	95	0.52
Transporter Maintenance	3,4	$1.60	$0.00	$1.60	26	0.52
PAD A Maintenance	5	$9.10	$0.30	$8.80	135	0.52
PAD B Maintenance	5	$10.40	$0.30	$10.10	147	0.52
SLS Maintenance	1	$0.40	$0.00	$0.40	7	0.52
CLS Maintenance	1	$0.70	$0.60	$0.10	1	0.52
Logistics Facilities Maintenance	11	$0.70	$0.10	$0.60	10	0.52
RPSF Maintenance	3,4	$0.70	$0.00	$0.70	10	0.52
SRB Retrieval Vessel Maintenance	1	$1.80	$0.60	$1.20	16	0.52
Miscellaneous Facility Maintenance	10	$5.10	$0.70	$4.40	66	0.52
Dredging Operations	6	$0.40	$0.40	$0.00	0	0.52
Processing Control Center Maintenance	8	$0.40	$0.00	$0.40	6	0.52
OSB Maintenance	10	$0.50	$0.20	$0.30	4	0.52
Launch Equipment Shops (LES)		$18.90	$11.50	$7.40	108	
Launch Equipment Shops (LES)	10	$5.20	$0.30	$4.90	76	0.52
Decontamination/Cleaning/Refurb/Shops	10	$3.80	$3.80	$0.10	2	0.52
Janitorial Services	10	$3.40	$3.40	$0.00	1	0.52
Corrosion Control	10	$6.40	$4.00	$2.30	30	0.52
Facility Systems	10	$5.00	$1.60	$3.40	56	0.52
Maintenance Service Contracts	10	$3.80	$3.80	$0.00	0	0.52
Inventory Spares and Repair	10	$2.10	$1.60	$0.50	8	0.52
System Equipment		$16.40	$3.60	$12.80	209	
SE Maintenance	10	$13.50	$0.70	$12.80	209	0.52
SE Acquisition	10	$0.00	$0.00	$0.00	0	0.52
Capital Equipment Procurements	10	$2.90	$2.90	$0.00	0	0.52
LPS/Instrumentation & Calibration (I&C)		**$50.60**	**$5.90**	**$44.60**	**697**	
LPS Engineering and Software		$12.10	$1.40	$10.70	158	
LPS Engineering	12	$3.70	$1.00	$2.70	40	0.53
LPS S/W Development & Maintenance	12	$5.10	$0.40	$4.80	69	0.53
LPS Software Production	12	$3.20	$0.00	$3.20	49	0.53
LPS O&M		$25.50	$0.30	$25.20	398	
Checkout, Control & Monitor Subsystem	8	$11.00	$0.10	$10.90	168	0.53

1994 STS WBS	Functional Breakdown Structure (FBS)	8 Flt/Year Baseline Annual Ops Cost $FY94M	8 Flt/Year Baseline Materials & ODC $FY94M	8 Flt/Year Baseline Labor $FY94M	8 Flts/Year Baseline Headcount	Calculated Fixed Cost-to-Variable Cost Ratio @ 8 Flts/Year*
CDS Operations	**8**	$4.20	$0.00	$4.20	66	0.53
Record & Playback System O&M	**8**	$3.10	$0.00	$3.10	48	0.53
LPS Maintenance/Support Engineering	**12**	$7.20	$0.20	$7.00	115	0.53
Instrumentation & Calibration		$13.00	$4.30	$8.70	141	
Instrumentation	**10**	$6.40	$0.10	$6.30	101	0.53
Calibration	**10**	$6.60	$4.20	$2.40	40	0.53
Modifications		**$13.90**	**$3.00**	**$10.90**	**156**	
OPF Modifications	**2**	$1.90	$0.50	$1.30	19	0.69
HMF Modifications	**7**	$0.10	$0.00	$0.10	2	0.69
VAB Modifications	**3,4**	$0.70	$0.20	$0.40	6	0.69
LCC Modifications	**8**	$0.20	$0.10	$0.10	1	0.69
MLP Modifications	**3,4**	$0.70	$0.50	$0.30	4	0.69
Transporter Modifications	**3,4**	$0.00	$0.00	$0.00	0	0.69
PAD A Modifications	**5**	$0.50	$0.20	$0.30	5	0.69
PAD B Modifications	**5**	$0.60	$0.40	$0.30	4	0.69
SLS Modifications	**1**	$0.00	$0.00	$0.00	0	0.69
CLS Modifications	**1**	$0.10	$0.10	$0.00	0	0.69
RPSF Modifications	**3,4**	$0.10	$0.00	$0.10	1	0.69
Miscellaneous Facility Modifications	**10**	$1.00	$0.30	$0.70	10	0.69
SE Modifications	**10**	$0.60	$0.00	$0.50	6	0.69
LPS Hardware Modifications	**8**	$7.00	$0.40	$6.80	99	0.69
Istrumentation & Calibration Modifications	**10**	$0.00	$0.00	$0.00	0	0.69
Communication Modifications	**8**	$0.30	$0.30	$0.00	0	0.69
PAD B Block Modification	**5**	$0.00	$0.00	$0.00	0	0.69
Technical Operations Support		**$113.10**	**$45.80**	**$67.30**	**1,017**	
Safety, Reliability, Maintainability & Quality		$24.50	$5.30	$19.10	282	
Safety	**12**	$12.30	$4.90	$7.40	108	0.38
Reliability	**12**	$1.90	$0.00	$1.90	32	0.38
Quality Assurance	**12**	$10.30	$0.40	$9.90	142	0.38
Logistics		$41.00	$28.80	$12.20	219	
Logistics Engineering	**10**	$3.20	$0.00	$3.20	48	0.38
Systems & Audit	**10**	$0.90	$0.00	$0.80	13	0.38
Receiving Service Center	**10**	$1.00	$1.00	$0.00	0	0.38
Supply	**10**	$28.20	$22.00	$6.10	117	0.38
Transportation	**10**	$4.60	$2.50	$2.10	40	0.38
Procurement Service Center	**10**	$3.30	$3.30	$0.00	0	0.38
Facility/SE Engineering		$19.60	$3.20	$16.50	233	
Systems Integration/Design Engineering	**12**	$12.50	$0.60	$11.90	165	0.38
Special Engineering Projects	**12**	$2.60	$0.10	$2.50	35	0.38
Ground Systems Change Control	**12**	$2.10	$0.00	$2.10	33	0.38
Technical Data/Documentations Service	**12**	$2.40	$2.40	$0.00	0	0.38
Operations Management		$6.60	$0.80	$5.90	89	
Manifest Planning	**12**	$3.80	$0.70	$3.10	46	0.38
Flt Element/Mission-Related Change Ctl	**12**	$1.70	$0.10	$1.60	25	0.38
Configuration Management Office	**12**	$1.10	$0.00	$1.10	18	0.38
Non-IWCS H/W, S/W and Maintenance	**12**	$1.60	$1.30	$0.40	6	0.38
Launch Team Training System (LTTS) Pgm	**12**	$1.40	$0.00	$1.40	22	0.38
Integ Work Ctl System (IWCS) Development		$18.30	$6.50	$11.90	168	
IWCS Shop Floor Control Project	**12**	$1.80	$0.00	$1.80	26	0.38
IWCS Work Preparation Support System	**12**	$1.50	$0.30	$1.20	17	0.38
IWCS Automated Reqments Management	**12**	$0.70	$0.00	$0.70	11	0.38
IWCS Computer Aided Schedule & Planning	**12**	$1.30	$0.00	$1.30	19	0.38
IWCS Project Integration	**12**	$6.80	$6.10	$0.80	10	0.38
IWCS Operations, Management & Support	**12**	$6.20	$0.20	$6.00	86	0.38

1994 STS WBS	Functional Breakdown Structure (FBS)	8 Flt/Year Baseline Annual Ops Cost $FY94M	8 Flt/Year Baseline Materials & ODC $FY94M	8 Flt/Year Baseline Labor $FY94M	8 Flts/Year Baseline Headcount	Calculated Fixed Cost-to-Variable Cost Ratio @ 8 Flts/Year*
Program Operations Support		**$32.50**	**$6.50**	**$26.00**	**431**	
Program Administration		$10.90	$0.60	$10.40	159	
Contract/Financial Management	12	$5.20	$0.20	$5.00	69	0.49
Management Planning & Procedures	12	$0.90	$0.00	$0.90	14	0.49
Team Member Manage-ment/Administration	12	$4.90	$0.40	$4.50	75	0.49
Training	12	$14.10	$1.50	$12.60	204	0.49
Human Resources		$7.50	$4.40	$3.00	68	
Security	12	$3.00	$0.10	$2.90	67	0.49
Human Resources Service Center	12	$4.50	$4.40	$0.20	1	0.49
Communications		**$19.70**	**$1.20**	**$18.50**	**326**	
Voice Communications O&M	8	$7.10	$0.20	$6.90	120	0.58
Wideband Transmission & Navaids O&M	8	$6.00	$0.50	$5.40	97	0.58
Cable and Wire O&M	8	$2.60	$0.20	$2.50	45	0.58
Communications Support	8	$2.90	$0.20	$2.70	49	0.58
OIS-D Implementation	8	$1.10	$0.10	$1.00	16	0.58
Base Operations Contract (BOC)	10	**$20.80**	**$2.10**	**$18.70**	**208**	**0.77**
Launch Support Services	12	**$35.00**	**$3.50**	**$31.50**	**350**	**0.82**
Weather Support	8	**$2.90**	**$0.30**	**$2.60**	**29**	**0.7**
TOTAL PAYLOAD OPERATIONS (KSC)		**$35.70**	**$8.00**	**$27.70**	**378**	
P/L Transportation & Interface Verification	9	**$29.70**	**$7.40**	**$22.30**	**318**	0.36
P/L Processing GSE Sustaining Engrg	9	**$6.00**	**$0.60**	**$5.40**	**60**	0.36
TOTAL MISSION OPERATIONS (JSC)		**$292.60**	**$58.80**	**$233.90**	**3,118**	
Mission Operations Facilities		**$148.30**	**$32.40**	**$116.00**	**1,546**	
Control Center Operations	8	$53.50	$3.50	$50.00	667	0.78
Integrated Training Facility Operations	10	$22.70	$1.30	$21.40	285	0.78
Integrated Planning System Operations	10	$6.30	$1.00	$5.30	71	0.78
Shuttle Avionics Integration Lab (SAIL)	10	$18.30	$1.20	$17.10	228	0.78
Flight Operations Trainer	10	$3.40	$0.30	$3.20	42	0.78
Software Production/Software Dev. Facility	10	$16.60	$1.00	$15.60	208	0.78
Mockup & Integration Lab	10	$1.00	$0.10	$0.90	12	0.78
Control Center Systems Division	10	$2.10	$0.50	$1.60	21	0.78
Integrated Planning System Office	12	$0.70	$0.10	$0.60	8	0.78
Simulator and Traininbg Systems Division	12	$0.40	$0.10	$0.30	4	0.78
STSOC Material	10	$23.30	$23.30	$0.00	0	0.78
Mission Planning & Operations		**$74.90**	**$5.30**	**$69.60**	**928**	
Systems Division	12	$15.30	$1.50	$13.80	184	0.78
Ops Division	8	$10.00	$0.20	$9.80	131	0.78
Training Divivion	12	$10.00	$0.60	$9.40	125	0.78
Flight Design Division	12	$34.40	$2.60	$31.80	424	0.78
Recon Division	12	$5.20	$0.40	$4.80	64	0.78
Program & Doc. Support/Management		**$69.40**	**$21.10**	**$48.30**	**644**	
STSOC Support	12	$52.40	$10.90	$41.60	554	0.78
Flight Software Support	12	$2.80	$0.50	$2.30	31	0.78
Shuttle Data Support	8	$11.30	$9.10	$2.20	29	0.78
MOD Directorate Office	8	$2.90	$0.70	$2.30	30	0.78
TOTAL CREW OPERATIONS (JSC)		**$50.70**	**$28.80**	**$21.90**	**327**	
Aircraft Maintenance & Ops		**$46.40**	**$28.80**	**$17.60**	**279**	
T-38 Training Aircraft	12	$25.40	$16.40	$9.00	159	0.7
Shuttle Training Aircraft	12	$16.90	$8.80	$8.10	111	0.76
Shuttle Carrier Aircraft	12	$2.90	$2.40	$0.50	9	0.82
Heavy Aircraft Training	12	$0.40	$0.40	$0.00	0	0.67

1994 STS WBS	Functional Breakdown Structure (FBS)	8 Flt/Year Baseline Annual Ops Cost $FY94M	8 Flt/Year Baseline Materials & ODC $FY94M	8 Flt/Year Baseline Labor $FY94M	8 Flts/Year Baseline Headcount	Calculated Fixed Cost-to-Variable Cost Ratio @ 8 Flts/Year*
Astronaut Support	12	$0.80	$0.80	$0.00	0	1
STSOC Flt Crew Ops Directorate Support	**12**	**$4.30**	**$0.00**	**$4.30**	**48**	**0.51**
TOTAL CREW TRAINING & MEDICAL OPS (JSC)	**12**	**$21.10**	**$2.10**	**$19.10**	**191**	**0.75**
TOTAL PROGRAM OF-FICE/HEADQUARTERS		**$180.30**	**$47.20**	**$133.10**	**1,046**	
Program Office		**$157.00**	**$27.90**	**$129.10**	**1,012**	
Management, SE&I, Flight Analysis	12	$81.40	$12.20	$69.20	494	0.83
Payload Integration	12	$41.10	$10.30	$30.80	257	0.5
STSOC Mission Integration Support	12	$7.40	$0.70	$6.70	56	0.62
Other Support	12	$1.50	$0.20	$1.40	11	0.5
Landing Site Support	12	$3.70	$3.00	$0.70	5	0.78
Config Mgmt, Mission Verif, & PRCB	12	$7.90	$0.40	$7.50	54	0.78
ADP Facility & Ops, MIC Support, Publica-tions	12	$11.10	$0.00	$11.10	123	0.78
ADP Equipment	12	$1.00	$1.00	$0.00	0	0.78
Program Office Support	12	$1.90	$0.20	$1.70	12	0.78
Headquarters		**$23.30**	**$19.30**	**$4.10**	**34**	
Systems Engineering & Integration Support	12	$4.50	$0.50	$4.10	34	0.83
Auditing Services Tax	12	$17.90	$17.90	$0.00	0	0.83
EEE Parts Program	11	$0.90	$0.90	$0.00	0	0.83
TOTAL INSTITUTION		**$477.60**	**$51.40**	**$426.20**	**5,328**	
Institution JSC		**$146.90**	**$13.90**	**$133.00**	**1,662**	
CS Direct Labor & Travel	12	$67.20	$3.40	$63.80	798	0.83
CS Indirect Labor & Travel	12	$14.00	$0.70	$13.30	166	0.83
Operation of Installation	12	$65.70	$9.90	$55.80	698	0.83
Institution MSFC		**$67.70**	**$7.80**	**$59.90**	**749**	
CS Direct Labor & Travel	12	$20.40	$1.00	$19.40	242	0.83
CS Indirect Labor & Travel	12	$3.10	$0.20	$2.90	37	0.83
Operation of Installation	12	$44.20	$6.60	$37.60	470	0.83
Institution KSC		**$195.20**	**$19.50**	**$175.70**	**2,196**	
CS Direct Labor & Travel	12	$82.00	$4.10	$77.90	974	0.83
CS Indirect Labor & Travel	12	$15.80	$0.80	$15.00	188	0.83
Operation of Installation	12	$97.40	$14.60	$82.80	1,035	0.83
Institution Headquarters		**$57.90**	**$8.70**	**$49.20**	**615**	
Operation of Installation	12	$57.90	$8.70	$49.20	615	0.83
Institution SSC		**$9.90**	**$1.50**	**$8.40**	**105**	
Operation of Installation	12	$9.90	$1.50	$8.40	105	0.83
TOTAL PMS		**$75.80**	**$37.90**	**$37.90**	**379**	
MSFC	12	**$20.00**	**$10.00**	**$10.00**	**100**	0.83
JSC	12	**$33.00**	**$16.50**	**$16.50**	**165**	0.83
KSC	12	**$19.90**	**$10.00**	**$10.00**	**100**	0.83
SSC	12	**$2.90**	**$1.50**	**$1.50**	**15**	0.83
TOTAL NETWORK SUPPORT Tracking, Telemetry, Comm. & Data Proc-essing	**8**	**$72.30**	**$72.30**	**$0.00**	**0**	**0.6**
TOTAL SYSTEMS ENGINEERING		**$116.12**	**$9.50**	**$106.72**	**1,021**	
MSFC Propulsion Systems Engineering		**$22.42**	**$1.40**	**$21.12**	**250**	
Institutional Program Support		$8.00	$0.20	$7.80	98	
Computer/SPO	12	$2.85	$0.10	$2.70	27	0.83

1994 STS WBS	Functional Breakdown Structure (FBS)	8 Flt/Year Baseline Annual Ops Cost $FY94M	8 Flt/Year Baseline Materials & ODC $FY94M	8 Flt/Year Baseline Labor $FY94M	8 Flts/Year Baseline Headcount	Calculated Fixed Cost-to-Variable Cost Ratio @ 8 Flts/Year*
Data Reduction	**12**	$2.75	$0.00	$2.80	40	0.83
Information Services/HOSC	**12**	$1.80	$0.00	$1.80	24	0.83
Information Services Direct	**12**	$0.40	$0.00	$0.40	5	0.83
Facilities	**10**	$0.20	$0.10	$0.10	1	0.83
Science & Engineering		$4.22	$0.00	$4.22	60	
Technical Tasks	**12**	$0.56	$0.00	$0.56	7	0.83
Mission Operations (EO) HOSC	**12**	$3.66	$0.00	$3.66	52	0.83
Weather Support	**12**	$0.70	$0.40	$0.40	4	0.6
General Shuttle Support (Integ. Contractor)		$9.50	$0.80	$8.70	88	
Rockwell Prime	**12**	$7.40	$0.00	$7.40	68	0.83
Administrative Operations Support	**12**	$0.50	$0.00	$0.50	9	0.83
Small Business (Facility & HOSC Equip)	**12**	$1.60	$0.80	$0.80	11	0.83
JSC Engineering Directorate		**$71.70**	**$6.30**	**$65.40**	**545**	
Engineering Analysis	**12**	$18.10	$0.90	$17.20	143	0.84
Flight Software Support	**12**	$53.60	$5.40	$48.20	402	0.84
White Sands Test Facility	**12**	**$9.70**	**$0.00**	**$9.70**	**108**	**0.83**
JSC Center Ops	**12**	**$6.30**	**$0.90**	**$5.40**	**67**	**0.83**
Ames	**12**	**$6.00**	**$0.90**	**$5.10**	**51**	**0.83**

* Data Derived from "Shuttle Operations Zero-Base Cost Study," presentation to Dr. Lenoir, July 2, 1991.

This page intentionally left blank.

APPENDIX C. STS DESIGN ROOT CAUSE ANALYSIS KNOWLEDGE BASE

RCA TEAM MEMBER	EXPERTISE	ORGANIZATION
RCA Steering Team		
R. Johnson, Steering Group Lead	Systems Engineering Management	NASA KSC/YA-D4
R. Harrison	Shuttle Program Management	NASA KSC/PH
F. Izquierdo	Shuttle Engineering Management	NASA KSC/PH-A
C. McCleskey	Systems Engineering & Analysis	NASA KSC/YA-D4
R. Rhodes	Systems Engineering Analysis	NASA KSC/YA-D4
C. Stevenson	Shuttle Systems Engineering	NASA KSC/PH-A
RCA Study Team		
C. McCleskey, Study Team Lead	Systems Engineering & Analysis	NASA KSC/YA-D4
N. Berger	Electrical Systems Engineering	NASA KSC/PK-F
T. Hoffmann	Payload Integration, Shuttle Project Engineering	NASA KSC/PK-M
K. Ingoldsby	Systems Engineering & Analysis, Database Design	Lockheed Martin
F. Jones	Thermal Protection	NASA KSC/PK-H
D. Johnson	Systems Analysis	KSC/ASRC
A. Menendez	Guidance & Control, Avionics	NASA KSC/PK-K
J. Najarro	Payload Integration	NASA KSC/PK-H
D. Stambolian	Shuttle Processing	NASA KSC/PK-M
J. Tatum	Airframe, Structures and Mechanical Systems	NASA KSC/PK-H
M. Wilhoit	Propulsion and Fluid Systems Management	NASA KSC/PK-G

This page intentionally left blank.

APPENDIX D. EXAMPLE WORK CONTENT ANALYSIS DATA (STS-86)

Table 21 presents the most important data fields used in the analysis of Shuttle direct, on-vehicle operations. While presenting only one processing flow as an example, eight such tables (representing eight Shuttle processing flows) were used in the analysis. Further explanation of the data fields can be found in 4.2.3.

The records are grouped alphabetically according to the "Data Set" field, then by date. To understand the relationships between the Data Sets and the sequence of events, refer to Figure 52.

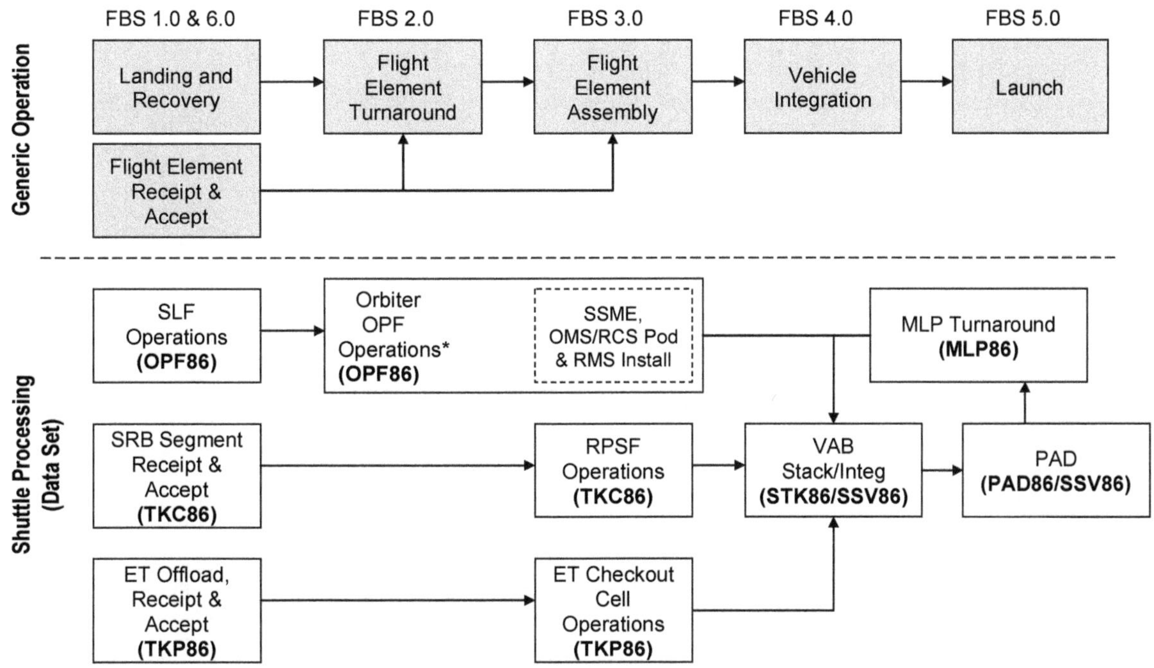

Figure 52. STS-86 Data Set Relationships to Overall Processing Flow and FBS

Table 21. Example Direct Work Content Data (STS–86)

No.	DATA SET	START TIME DD:MMM-YYYY:H	PROCEDURE NAME	DESCRIPTION	DURATION (Hours)	SYSTEM	DESIGN DISCIPLINE	GENERIC OPS FBS NUMBER (Appx A1)
1	MLP86	15-MAY-1997:5		STS-84 LAUNCH @ PAD-A	0		Structures , Mechanisms, Veh Handling	5.09.05
2	MLP86	15-MAY-1997:5	Y2007	PMS REFURBISHMENT	76	332	Ground Systems & Facilities	5.10.02
3	MLP86	15-MAY-1997:8	S2005	POST LAUNCH SECURING	8	VOP	Structures , Mechanisms, Veh Handling	5.10.02
4	MLP86	15-MAY-1997:15	S2005	POST LAUNCH SECURING	3	VOP	Structures , Mechanisms, Veh Handling	5.10.02
5	MLP86	15-MAY-1997:17	S2005	POST LAUNCH SECURING	24	VOP	Structures , Mechanisms, Veh Handling	5.10.02
6	MLP86	15-MAY-1997:17	V2270/B2036L	INSTALL ESP'S & LIGHTS(AS REQ'D)	0		Propulsion	5.10.02
7	MLP86	16-MAY-1997:9	WAC:0259-0567	PAINT 0 LEVEL(EXCEPT SRB HOLES)	10		Ground Systems & Facilities	5.10.02
8	MLP86	19-MAY-1997:4	S2005	POST LAUNCH SECURING	2	VOP	Structures , Mechanisms, Veh Handling	5.10.02
9	MLP86	19-MAY-1997:6	V2270/B2036	INSTALL ESP'S & LIGHTS(AS REQ'D)	0		Propulsion	5.10.02
10	MLP86	19-MAY-1997:6	S2005	POST LAUNCH SECURING	1	VOP	Structures , Mechanisms, Veh Handling	5.10.02
11	MLP86	19-MAY-1997:9	WAC:	PAINT 0 LEVEL(EXCEPT SRB HOLES)	10		Ground Systems & Facilities	5.10.02
12	MLP86	19-MAY-1997:9		FIRE INSPECTOR:ISSUE BURN PERMITS(AS REQ'D)	0		Structures , Mechanisms, Veh Handling	5.10.02
13	MLP86	19-MAY-1997:10		HIGH CREW/STRUCTURES:PROVIDE NECESSARY ACCESS	0		Structures , Mechanisms, Veh Handling	5.10.02
14	MLP86	19-MAY-1997:16	B5150E	SUPPORT EQUIPMENT INSTLN/REMOVAL	0		Ground Systems & Facilities	5.10.02
15	MLP86	19-MAY-1997:16		REMOVE S.S. SPOOL PIECES/ACCESS SRB HOLES	0		Structures , Mechanisms, Veh Handling	5.10.02
16	MLP86	19-MAY-1997:17	B5032	SRB POST LAUNCH REFURB OF HDP'S	8	SRM	Structures , Mechanisms, Veh Handling	5.10.02
17	MLP86	19-MAY-1997:17	WAC:	MLP GIRDER INSPECTION(INTERNAL)	8		Ground Systems & Facilities	5.10.02
18	MLP86	20-MAY-1997:1	B5032	SRB POST LAUNCH REFURB OF HDP'S	4	SRM	Structures , Mechanisms, Veh Handling	5.10.02
19	MLP86	20-MAY-1997:4	G6105G/G6205J	DISC FROM FACILITY PRESSURE	1		Propulsion	5.10.02
20	MLP86	20-MAY-1997:9	A3204.002D	MLP-2 TO PARKSITE(CTS:0600)	9		Ground Systems & Facilities	5.01.01
21	MLP86	20-MAY-1997:18		MLP LOCAL SYSTEMS CHECKS/MAINTENANCE	264		Structures , Mechanisms, Veh Handling	5.01.01
22	MLP86	20-MAY-1997:18		BLOWDOWN SYSTEM/SAMPLE	6		Structures , Mechanisms, Veh Handling	5.01.01
23	MLP86	20-MAY-1997:24	G6105/G6205	CONNECT TO FACILITY PRESSURE	2		Propulsion	5.01.01
24	MLP86	20-MAY-1997:24	Y2007	PMS REFURBISHMENT	384	332	Ground Systems & Facilities	5.01.01
25	MLP86	21-MAY-1997:1	B5032	SRB POST LAUNCH REFURB OF HDP'S	17	SRM	Structures , Mechanisms, Veh Handling	5.01.01
26	MLP86	21-MAY-1997:9	WAC:0259-0567	GEN LAUNCH DAMAGE REPRS(WLDNG,STRCTRL REPRS,CLEANING)	258		Ground Systems & Facilities	5.01.01
27	MLP86	21-MAY-1997:9	WAC:	INSPECT/INSTALL SACRIFICIAL PLATES	64		Ground Systems & Facilities	5.01.01
28	MLP86	21-MAY-1997:9	IPR:0524	REMOVE SEIZED COUPLERS (0258-2150)LES(LCL CLR)	104		Cmd, Ctl & Health Mngmt	5.04.01

No.	DATA SET	START TIME DD:MMM-YYYY:H	PROCEDURE NAME	DESCRIPTION	DURATION (Hours)	SYSTEM	DESIGN DISCIPLINE	GENERIC OPS FBS NUMBER (Appx A1)
29	MLP86	21-MAY-1997:10		INSTL FLOATS,PIC BOARDS,SFTY NETS & LIGHTS(AS REQ'D)	0		Structures , Mechanisms, Veh Handling	5.01.01
30	MLP86	21-MAY-1997:17	WAC:	MLP GIRDER INSPECTION(INTERNAL)	8		Ground Systems & Facilities	5.01.01
31	MLP86	21-MAY-1997:17	B5146J	SRB UMB T-0 REFURB/PREPS(L/H)	2		Structures , Mechanisms, Veh Handling	5.01.01
32	MLP86	21-MAY-1997:19	B5146	01-028 UMB TUNNEL PURGE & CABLE REMOVAL	4	SRM	Structures , Mechanisms, Veh Handling	5.01.01
33	MLP86	21-MAY-1997:23	B5146J	SRB UMB T-0 REFURB/PREPS(R/H)	1		Structures , Mechanisms, Veh Handling	5.01.01
34	MLP86	21-MAY-1997:23	B5146	01-032 GCA ELEC/MECH DISCONNECTION	8	SRM	Structures , Mechanisms, Veh Handling	5.01.01
35	MLP86	21-MAY-1997:24	B5146	01-028 UMB TUNNEL PURGE & CABLE REMOVAL	2	SRM	Structures , Mechanisms, Veh Handling	5.01.01
36	MLP86	22-MAY-1997:1	B5032	REFURB HDP ACCESSORIES	24	SRM	Structures , Mechanisms, Veh Handling	5.01.01
37	MLP86	22-MAY-1997:2	B5146	01-032 GCA ELEC/MECH DISCONNECTION	4	SRM	Structures , Mechanisms, Veh Handling	5.01.01
38	MLP86	22-MAY-1997:9	WAC:	NDE PHOTO OPTICAL CRITICAL WELDS	8		Ground Systems & Facilities	5.01.01
39	MLP86	22-MAY-1997:9	B5146	01-051 GCA REMOVAL(10 FT CLR/CRANE)	2	SRM	Structures , Mechanisms, Veh Handling	5.01.01
40	MLP86	22-MAY-1997:9	A77-1336-0010/0259-5239	MODIFY 1336 PLTFRMS(INTERFERENCE)ESP LWR & RAISE FIT CHECK ONLY	24		Ground Systems & Facilities	5.11.02
41	MLP86	22-MAY-1997:9	TPS:S72-0812-00-002-044	REPLACE SIGNAL CONDITIONER FOR TSM DEF. MOD	40		Cmd, Ctl & Health Mngmt	5.11.02
42	MLP86	22-MAY-1997:10		MOBILE CRANE:REM GCA	0		Structures , Mechanisms, Veh Handling	5.01.01
43	MLP86	22-MAY-1997:10	B5146	01-051 GCA REMOVAL(10 FT CLR/CRANE)	2	SRM	Structures , Mechanisms, Veh Handling	5.10.01
44	MLP86	22-MAY-1997:11		MOBILE CRANE:REM GCA	0		Structures , Mechanisms, Veh Handling	5.01.01
45	MLP86	22-MAY-1997:12	B5146	01-070 TUNNEL COMPONENT REMOVAL OPS	2	SRM	Structures , Mechanisms, Veh Handling	5.10.01
46	MLP86	22-MAY-1997:14	B5146	01-077 GCA FLEXHOSE POST LAUNCH PRG(10 FT CLR/GN2)	2	SRM	Structures , Mechanisms, Veh Handling	5.10.01
47	MLP86	23-MAY-1997:9		PCN:W202455 PHASE ONE HALON MOD(COMPTS.9A,7A,21A)	104		Structures , Mechanisms, Veh Handling	5.11.02
48	MLP86	23-MAY-1997:10	B5032	SHIP TO LOGISTICS	0	SRM	Structures , Mechanisms, Veh Handling	5.01.01
49	MLP86	27-MAY-1997:9	WAC:	INSPECT 4 NORTH BELLY BANDS(CONTINGENT UPON LAUNCH DAMAGE)	4		Ground Systems & Facilities	5.01.01
50	MLP86	27-MAY-1997:9	G2125K	LOX SYS FLANGE/BONNET TORQUE TEST(NON-LPS)	72		Propulsion	5.01.03
51	MLP86	27-MAY-1997:9	G2225I	LH2 SYS FLANGE/TORQUE TEST(NON-LPS)	40		Propulsion	5.01.03
52	MLP86	27-MAY-1997:9	0254-6803	P/L POWER SUPPLY UPGRADE/RECEPTACLE(FPC:FIE)	56		Power Management	5.11.02
53	MLP86	27-MAY-1997:9	TPS:C72-0810-135/136	ITCU REWIRE PNLS (5 TOT)	96		Environmental Ctl & Life Spt	5.11.02
54	MLP86	27-MAY-1997:9	0254-8049	REPLACE ECLSS COOLING UNITS W/FIXED SYS.(FIE)	72		Environmental Ctl & Life Spt	5.11.02
55	MLP86	27-MAY-1997:17	B5115J	REFURB MLP HDP PIC CABLES	152		Cmd, Ctl & Health Mngmt	5.01.01
56	MLP86	28-MAY-1997:10	IPR:0524	FPC J. HERBERT CO. ON SITE	0		Structures , Mechanisms, Veh Handling	5.11.02
57	MLP86	29-MAY-1997:9		QUAL/LES A&R INSPECT HDP'S	16		Structures , Mechanisms, Veh Handling	5.01.01
58	MLP86	29-MAY-1997:9	WAC:	MLP GIRDER INSPECTION(INTERNAL)	40		Ground Systems & Facilities	5.01.01

No.	DATA SET	START TIME DD:MMM-YYYY:H	PROCEDURE NAME	DESCRIPTION	DURATION (Hours)	SYSTEM	DESIGN DISCIPLINE	GENERIC OPS FBS NUMBER (Appx A1)
59	MLP86	30-MAY-1997:2	WAC:	MLP GIRDER INSPECTION(EXTERNAL S-2/4)	24		Ground Systems & Facilities	5.01.01
60	MLP86	30-MAY-1997:9	WAC:0259-0565	TIE BACK BOLT INSPECTION	24		Ground Systems & Facilities	5.01.01
61	MLP86	30-MAY-1997:9		HIGH CREW/STRUCTURES:PROVIDE NECESSARY ACCESS	0		Structures , Mechanisms, Veh Handling	5.01.01
62	MLP86	30-MAY-1997:9	WAC:	HEAT SHIELD INSPECTION/PIN REPAIR	120		Ground Systems & Facilities	5.01.01
63	MLP86	31-MAY-1997:7	WAC:0259-0567	GEN LAUNCH DAMAGE REPRS(WLDNG,STRCTRL REPRS,CLEANING)	9		Ground Systems & Facilities	5.01.01
64	MLP86	31-MAY-1997:9	A77-1336-00100259-5239	MODIFY 1336 PLTMS (INTERFERENCE)ESP LWR & RAISE FIT CHECK ONLY	8		Ground Systems & Facilities	5.11.02
65	MLP86	02-JUN-1997:9	PR:A77-1336-0010	REMOVE AND MODIFY 1336 PLTFRMS(INTERFERENCE)ESP LWR & RAISE REQ'D	16		Ground Systems & Facilities	5.11.02
66	MLP86	02-JUN-1997:9	IPR:0524	REMOVE SEIZED COUPLERS (0258-2150)LES	6		Structures , Mechanisms, Veh Handling	5.11.02
67	MLP86	03-JUN-1997:9	WAC:S72-0814-00-002	CUT PORT AND REWELD(0259-0291)SIDE 2	24		Ground Systems & Facilities	5.04.02
68	MLP86	03-JUN-1997:13		ENGINEERING EVALUATION	4		Structures , Mechanisms, Veh Handling	5.01.01
69	MLP86	03-JUN-1997:13		EG & G REPORT AVAILABLE	0		Structures , Mechanisms, Veh Handling	5.01.01
70	MLP86	04-JUN-1997:9	S060787	TSM DEFLECTION MOD	24		Ground Systems & Facilities	5.11.02
71	MLP86	04-JUN-1997:14		DEFINE NECESSARY REPAIR LIST	0		Structures , Mechanisms, Veh Handling	5.01.01
72	MLP86	06-JUN-1997:9	0254-6803	WALKDOWN FOR FINAL SYSTEM BUY OFF	0		Ground Systems & Facilities	5.11.02
73	MLP86	07-JUN-1997:6	WAC:0259-0567	GEN LAUNCH DAMAGE REPRS(WLDNG,STRCTRL REPRS,CLEANING)	8		Ground Systems & Facilities	5.01.01
74	MLP86	09-JUN-1997:9	V9051	HGDS HUMS VALIDATION(NIB)	32	HWS	Safety Management & Control	5.01.03
75	MLP86	09-JUN-1997:9	V9043	B/U HGDS DAILY OPS & MAINT(NIB)	32	HWS	Safety Management & Control	5.01.04
76	MLP86	09-JUN-1997:9	WAC:PV6-324638	POST LAUNCH GIRDER REPAIRS IDENTIFIED DURING INSPEC-TIONS	16		Structures , Mechanisms, Veh Handling	5.04.02
77	MLP86	11-JUN-1997:9	TPS S70-0871-002-017	SYSTEM LEAK CHECKS	4		Propulsion	5.11.02
78	MLP86	11-JUN-1997:13	0254-8049	FINAL BUY OFF WALKDOWN	4		Ground Systems & Facilities	5.11.02
79	MLP86	12-JUN-1997:9		VERIFY NOZZLE ACCESS PLATFORMS INSTALLED	0		Structures , Mechanisms, Veh Handling	5.01.01
80	MLP86	12-JUN-1997:9	WAC:0259-0567	PAINT SRB HOLES/HDP'S/GCA TUNNELS	2		Ground Systems & Facilities	5.01.01
81	MLP86	13-JUN-1997:3	G6105/G6205	DISC FROM FACILITY PRESSURE	1		Propulsion	4.01.01
82	MLP86	13-JUN-1997:7	WAC:	PAINT SRB HOLES/HDP'S/GCA TUNNELS	4		Ground Systems & Facilities	5.01.01
83	MLP86	13-JUN-1997:10	A3206.001G	MLP-2 TO VAB/HB-1(CTS:0800)	10		Ground Systems & Facilities	4.01.01
84	MLP86	16-JUN-1997:9	S060787	TSM DEFLECTION MOD	72		Ground Systems & Facilities	5.11.02
85	MLP86	16-JUN-1997:10		SPHERICAL BEARINGS AVAILABLE	0		Structures , Mechanisms, Veh Handling	4.01.01
86	MLP86	16-JUN-1997:12		BLOWDOWN SYSTEM/SAMPLE	4		Structures , Mechanisms, Veh Handling	4.01.01
87	MLP86	16-JUN-1997:12	Y2007	PMS REFURBISHMENT	157	332	Ground Systems & Facilities	4.01.01

No.	DATA SET	START TIME DD:MMM-YYYY:H	PROCEDURE NAME	DESCRIPTION	DURATION (Hours)	SYSTEM	DESIGN DISCIPLINE	GENERIC OPS FBS NUMBER (Appx A1)
88	MLP86	16-JUN-1997:12		PCN:W202455 PHASE ONE HALON MOD(COMPTS.9A,7A,21A)	56		Structures , Mechanisms, Veh Handling	5.11.02
89	MLP86	16-JUN-1997:13		MLP SYSTEM TESTS/VALIDATIONS	621		Structures , Mechanisms, Veh Handling	4.01.01
90	MLP86	16-JUN-1997:14		INSTL FLOATS,PIC BOARDS & SAFETY NETS(AS REQ'D)	0		Structures , Mechanisms, Veh Handling	4.01.01
91	MLP86	16-JUN-1997:16	G6105G/G6205J	CONNECT TO FACILITY PRESSURE	2		Propulsion	4.01.01
92	MLP86	16-JUN-1997:16		FACILITY PNEU ACTIVATE GN2 PURGE	0		Structures , Mechanisms, Veh Handling	4.01.01
93	MLP86	17-JUN-1997:9		TRUSS/GIRDER DEFLECTION SURVEY	8		Structures , Mechanisms, Veh Handling	4.01.01
94	MLP86	18-JUN-1997:1	B5115J	REFURB MLP HDP PIC CABLES	65		Cmd, Ctl & Health Mngmt	4.01.01
95	MLP86	20-JUN-1997:1	B5146	01-070 TUNNEL COMPONENT REMOVAL OPERATIONS	2	SRM	Structures , Mechanisms, Veh Handling	4.01.01
96	MLP86	20-JUN-1997:3	B5146	03-000 SRB JOINT HEATER UMB GCA INST TUNNEL OPS	1	SRM	Structures , Mechanisms, Veh Handling	4.01.01
97	MLP86	20-JUN-1997:4	B5146	02-000 REPLACE CONNECTOR SAVERS/ELEC CHECKOUT	9	SRM	Structures , Mechanisms, Veh Handling	4.01.01
98	MLP86	20-JUN-1997:4	B5146	01-077 GCA GN2 FLEXHOSE POST LAUNCH PRG(10 FT CLR/GN2)	2	SRM	Structures , Mechanisms, Veh Handling	4.01.01
99	MLP86	20-JUN-1997:6	B5146	02-000 REPLACE CONNECTOR SAVERS/ELEC CHECKOUT	7	SRM	Structures , Mechanisms, Veh Handling	4.01.01
100	MLP86	20-JUN-1997:12	B5146	03-007 GCA PRE-STACK OPERATIONS	2	SRM	Structures , Mechanisms, Veh Handling	4.01.01
101	MLP86	20-JUN-1997:13	B5146	03-000 SRB JOINT HEATER UMB GCA INST TUNNEL OPS	2	SRM	Structures , Mechanisms, Veh Handling	4.01.01
102	MLP86	20-JUN-1997:14	B5146	03-014 GCA STACKING OPS(10 FT CLR/CRANE)	1	SRM	Structures , Mechanisms, Veh Handling	4.01.01
103	MLP86	23-JUN-1997:3	B5150E	INSTALL PIC BDS. NORTH HDP'S	0		Ground Systems & Facilities	4.01.01
104	MLP86	23-JUN-1997:7	B5146	GCA STACKING OPS(10 FT CLR/CRANE)	1		Structures , Mechanisms, Veh Handling	4.01.01
105	MLP86	23-JUN-1997:7		CRANE:INST GCA	0		Structures , Mechanisms, Veh Handling	4.01.01
106	MLP86	23-JUN-1997:8	B5146	03-007 GCA PRE-STACK OPERATIONS	2	SRM	Structures , Mechanisms, Veh Handling	4.01.01
107	MLP86	23-JUN-1997:8	B5146	03-032 GCA ASSEMBLY-MECHANICAL CONNECTION	3	SRM	Structures , Mechanisms, Veh Handling	4.01.01
108	MLP86	23-JUN-1997:9	B5146	03-014 GCA STACKING OPS(10 FT CLR/CRANE)	2	SRM	Structures , Mechanisms, Veh Handling	4.01.01
109	MLP86	23-JUN-1997:11	B5146	03-051 GCA GN2 PURGE SYS VERIF(10 FT CLR/GN2)	1	SRM	Structures , Mechanisms, Veh Handling	4.01.01
110	MLP86	23-JUN-1997:11	B5146	GCA STACKING OPS(10 FT CLR/CRANE)	1		Structures , Mechanisms, Veh Handling	4.01.01
111	MLP86	23-JUN-1997:11		CRANE:INST GCA	0		Structures , Mechanisms, Veh Handling	4.01.01
112	MLP86	23-JUN-1997:12	B5146	GCA GN2 PURGE SYS VERIF(10 FT CLR/GN2)	1		Structures , Mechanisms, Veh Handling	4.01.01
113	MLP86	23-JUN-1997:12	B5146	03-032 GCA ASSEMBLY-MECHANICAL CONNECTION	4	SRM	Structures , Mechanisms, Veh Handling	4.01.01
114	MLP86	23-JUN-1997:13	B5146	03-063 CARRIER MATING TOOL PREPS FOR STACK	3	SRM	Structures , Mechanisms, Veh Handling	4.01.01
115	MLP86	23-JUN-1997:16	B5146	03-051 GCA GN2 PURGE SYS VERIF(10 FT CLR/GN2)	1	SRM	Structures , Mechanisms, Veh Handling	4.01.01
116	MLP86	23-JUN-1997:17	B5303_(2)	INSTALL SPHERICAL BEARINGS/BUSHINGS	4		Structures , Mechanisms, Veh Handling	4.01.01
117	MLP86	23-JUN-1997:17		HB-1&2 DOORS CLOSED(CONFIGURED FOR OPTICS)	56		Structures , Mechanisms, Veh Handling	4.01.01
118	MLP86	23-JUN-1997:17	B5146	GCA GN2 PURGE SYS VERIF(10 FT CLR/GN2)	1		Structures , Mechanisms, Veh Handling	4.01.01

No.	DATA SET	START TIME DD:MMM-YYYY:H	PROCEDURE NAME	DESCRIPTION	DURATION (Hours)	SYSTEM	DESIGN DISCIPLINE	GENERIC OPS FBS NUMBER (Appx A1)
119	MLP86	23-JUN-1997:18	B5146	03-063 CARRIER MATING TOOL PREPS FOR STACK	3	SRM	Structures , Mechanisms, Veh Handling	4.01.01
120	MLP86	24-JUN-1997:10		FACILITY PNEU SECURE GN2 PURGE	0		Structures , Mechanisms, Veh Handling	4.01.01
121	MLP86	24-JUN-1997:17	B5303_(2)	INSTALL PIC BDS. SIDE 2	0		Structures , Mechanisms, Veh Handling	4.01.01
122	MLP86	24-JUN-1997:17	B5303_(2)	OPTICS PRE-OPS SETUPS	8		Structures , Mechanisms, Veh Handling	4.01.01
123	MLP86	25-JUN-1997:1	B5303_(2)	OPTICS	24		Structures , Mechanisms, Veh Handling	4.01.01
124	MLP86	26-JUN-1997:1	B5303_(2)	CALCULATE SHIM PACK DATA	2		Structures , Mechanisms, Veh Handling	4.01.01
125	MLP86	26-JUN-1997:1		STS-86 SRB STACK	88		Structures , Mechanisms, Veh Handling	4.01.03
126	MLP86	26-JUN-1997:3	B5303_(2)	POSITION TVC PLATFORMS	4		Structures , Mechanisms, Veh Handling	4.01.01
127	MLP86	26-JUN-1997:3	B5303_(2)	INSTALL SHIM PACKS	4		Structures , Mechanisms, Veh Handling	4.01.01
128	MLP86	26-JUN-1997:6		CRANE:POSI TVC'S	0		Structures , Mechanisms, Veh Handling	4.01.01
129	MLP86	26-JUN-1997:8		CRANE PM'S/BEAM PREPS	1		Structures , Mechanisms, Veh Handling	4.01.01
130	MLP86	30-JUN-1997:1		STS-86 SRB STACK	75		Structures , Mechanisms, Veh Handling	4.01.03
131	MLP86	03-JUL-1997:11	B2157J	SRB JOINT HTR/GEI SYS VAL(C-9/10)	9		Power Management	4.01.01
132	MLP86	07-JUL-1997:9	V2456	LO2 TSM REFURB	150	LAA	Structures , Mechanisms, Veh Handling	4.01.01
133	MLP86	07-JUL-1997:9	V9051	HGDS HUMS VALIDATION(NIB)	40	HWS	Safety Management & Control	4.01.01
134	MLP86	07-JUL-1997:9	V2397	LH2 TSM REFURB	151	LAA	Structures , Mechanisms, Veh Handling	4.01.01
135	MLP86	07-JUL-1997:10	TPS:TPD	HGDS PAMS VALIDATION(NIB)	0		Safety Management & Control	4.01.01
136	MLP86	08-JUL-1997:9	V2263	NICAD CELL SYS OPS & MAINT	24	GSP	Power Management	4.01.01
137	MLP86	08-JUL-1997:9	V9017	HGDS END TO END VALIDATION(NIB)	100		Safety Management & Control	4.01.01
138	MLP86	08-JUL-1997:9	V9043	B/U HGDS DAILY OPS & MAINT(NIB)	88	HWS	Safety Management & Control	4.01.01
139	MLP86	08-JUL-1997:9	B2157	SRB JOINT HTR/GEI SYS VAL(C-9/10)	24		Power Management	4.01.01
140	MLP86	08-JUL-1997:23		STS-86 SRB STACK	196		Structures , Mechanisms, Veh Handling	4.01.03
141	MLP86	10-JUL-1997:17	V2285	B/U BATTERY SWITH OVER VAL & CALIBRATION	8		Power Management	4.01.01
142	MLP86	11-JUL-1997:9	V2401.001	PIC SYSTEM VALIDATION(C-10)	8	GSP	Power Management	4.01.01
143	MLP86	11-JUL-1997:9	G2265F	FILL LINE AREA COMPNT C/O(NON-LPS)	8		Propulsion	4.01.01
144	MLP86	11-JUL-1997:9	G2145H	LOX MLP PURGE SYS LEAK TEST(NON-LPS)	8		Propulsion	4.01.01
145	MLP86	14-JUL-1997:3	S060787	REINSTALL LOX 8" F/H	6		Propulsion	4.01.01
146	MLP86	14-JUL-1997:9	V2401.001	PIC SYSTEM VALIDATION(C-10)	24	GSP	Power Management	4.01.01
147	MLP86	21-JUL-1997:9	B2023.001C	SHUTTLE RANGE SAFETY-GSE VERIFICATION(C-5)	4		Cmd, Ctl & Health Mngmt	4.01.01
148	MLP86	22-JUL-1997:9	G2247H	PNEU LEAK & FUNCTIONAL, GLAND & BONNET TORQUING(C-4)	12		Propulsion	4.01.01
149	MLP86	23-JUL-1997:9	V2342	OVPU VALIDATION	8	GSP	Power Management	4.01.01

No.	DATA SET	START TIME DD:MMM-YYYY:H	PROCEDURE NAME	DESCRIPTION	DURATION (Hours)	SYSTEM	DESIGN DISCIPLINE	GENERIC OPS FBS NUMBER (Appx A1)
150	MLP86	24-JUL-1997:9		STS-86 ET MATE	40		Structures , Mechanisms, Veh Handling	4.01.03
151	MLP86	25-JUL-1997:9	V2521	AC/DC POWER VALIDATION(C-10)	24	GSP	Power Management	4.01.01
152	MLP86	25-JUL-1997:9	B2040P	SRB/TVC APU LUBE OIL SERVICING	16		Power Management	4.01.01
153	MLP86	28-JUL-1997:9	G2169F	LOX MLP PNUE SYS C/O(C-3)	24		Propulsion	4.01.01
154	MLP86	28-JUL-1997:9	B1009AA	SRB TVC/GSE CONN VALIDATION(C-8)	112		Power Management	4.01.01
155	MLP86	29-JUL-1997:9	V9002.04	HYD PWR UP/DWN	144	HYD	Power Management	4.01.01
156	MLP86	29-JUL-1997:9	V2521	AC/DC POWER VALIDATION(C-10)	56	GSP	Power Management	4.01.01
157	MLP86	29-JUL-1997:9	G2266F	XFER LINE AREA COMPNT C/O(C-4)	24		Propulsion	4.01.01
158	MLP86	30-JUL-1997:9	V2481.001	P/L GSE DATA I/F SYS FAC VAL	16		Payload Accommodations	4.01.01
159	MLP86	30-JUL-1997:9	G2165K	LOX MLP CRYO SYS MAINT & LEAK TEST(C-3)	8		Propulsion	4.01.01
160	MLP86	30-JUL-1997:17	G2267G	VENT LINE AREA COMPNT C/O(NON-LPS)	16		Propulsion	4.01.01
161	MLP86	31-JUL-1997:9	V2481.002	P/L GSE DATA I/F SYS FAC VAL(RM 10A)	16		Payload Accommodations	4.01.01
162	MLP86	31-JUL-1997:17	G2274C	S72-0685-04 VALIDATION(C-4)	16		Propulsion	4.01.01
163	MLP86	31-JUL-1997:17	T2308J	LO2 TK PRESS & PRG PNL/SYS LEAK & FUNCT(C-3)	16		Propulsion	4.01.01
164	MLP86	01-AUG-1997:15		PPU (S/N:003) TO MLP '0' LEVEL	0		Structures , Mechanisms, Veh Handling	4.01.01
165	MLP86	01-AUG-1997:17	G2343E	LH2 RPLNSH VLV/VCA C/O(C-4)	8		Propulsion	4.01.01
166	MLP86	02-AUG-1997:9	T2303	LH2 0685-5 PNL VAL(C-4)	8	LH2	Propulsion	4.01.01
167	MLP86	04-AUG-1997:9	T2305L	S72-0685-05 PNL/SYS LK,FNCTNL & VAL(C-3/4)	8		Propulsion	4.01.01
168	MLP86	04-AUG-1997:9	V3594.002	ECLSS CFC REDUCTION/ELIMINATION(MLP)	8		Environmental Ctl & Life Spt	4.01.01
169	MLP86	04-AUG-1997:9	V2273	ECS SYSTEM VALIDATION	4		Thermal Management	4.01.01
170	MLP86	04-AUG-1997:13	V2273	ECS SYSTEM VALIDATION	8		Thermal Management	4.01.01
171	MLP86	05-AUG-1997:9	G2240J	LH2 HDWR SAFING/AUTO CYCLING(C-4)	4		Propulsion	4.01.01
172	MLP86	06-AUG-1997:9	B1009	SRB TVC/GSE(C-8)	48		Power Management	4.01.01
173	MLP86	11-AUG-1997:1		OV-104/ORBITER MATE	48		Structures , Mechanisms, Veh Handling	4.01.01
174	MLP86	11-AUG-1997:2		ORBITER ROLL TO VAB	0		Structures , Mechanisms, Veh Handling	4.01.01
175	MLP86	11-AUG-1997:9	V2093/V2094	T-0 UMB CARRIER PLT MATE	32		Propulsion	4.02.01
176	MLP86	11-AUG-1997:12	V9021.002	SAMPLE TRICKLE PURGE PANELS(S70-1228)	0	ORP	Propulsion	4.01.01
177	MLP86	12-AUG-1997:9	Z2400	PRELIM/FINAL LIVIS CHKS	8	332	Ground Systems & Facilities	4.01.01
178	MLP86	13-AUG-1997:2	V1149	T-0 UMBIL I/F CKS	40	MPS	Propulsion	4.02.01
179	MLP86	15-AUG-1997:9	Z2400	PRELIM/FINAL LIVIS CHKS	8	332	Ground Systems & Facilities	4.01.01
180	MLP86	29-AUG-1997:1	G6105/G6205	DISC FROM FACILITY PRESSURE	2		Propulsion	4.01.01

No.	DATA SET	START TIME DD:MMM-YYYY:H	PROCEDURE NAME	DESCRIPTION	DURATION (Hours)	SYSTEM	DESIGN DISCIPLINE	GENERIC OPS FBS NUMBER (Appx A1)
181	OPF86	20-MAY-1997:9	V5032.001	POSU 1 MERL- OFFLINE PYRO INTERRUPT BOX C/O	8	PYR	Structures , Mechanisms, Veh Handling	2.01.02
182	OPF86	20-MAY-1997:9	V1091	OPF INTEGRATED ROLL-IN PREPS (2 DAYS PRIOR)	2	FCP	Power Management	2.01.02
183	OPF86	20-MAY-1997:9	V1158VL1	POSU'S 2B,3B,4-7,20,25,27 GSE VLV'S/ SCRUBBER CONFIG.	32	OMS	Propulsion	2.03.01
184	OPF86	20-MAY-1997:11	V1091	OPF INTEGRATED ROLL-IN PREPS (2 DAYS PRIOR)	6	FCP	Power Management	2.01.02
185	OPF86	21-MAY-1997:9	V1091	OPF INTEGRATED ROLL-IN PREPS (2 DAYS PRIOR) FROM: 86V351A2 - Note 1	8	FCP	Power Management	2.01.02
186	OPF86	21-MAY-1997:9	V5032.001	SEQ. 02-04- OFFLINE PYRO INTERRUPT BOX C/O	8	PYR	Structures , Mechanisms, Veh Handling	2.01.02
187	OPF86	21-MAY-1997:9	V1196	GSE POSU'S 3,9,10,12,14-16,32,51 (PRIOR TO ROLLIN)	16	APU	Power Management	2.01.02
188	OPF86	22-MAY-1997:9	V1091	OPF INTEGRATED ROLL-IN PREPS (2 DAYS PRIOR) FROM: 86V351A2 - Note 1	32	FCP	Power Management	2.01.02
189	OPF86	22-MAY-1997:17	V9002.04	OPS 170 GSE PUMP CIRC/SAMPLING	8	HYD	Propulsion	3.08.01
190	OPF86	23-MAY-1997:13	MPS-5-12-097	PANEL 679-6 LEAK CHECK(GSE-LPS)	4	MPS	Propulsion	2.05.03
191	OPF86	24-MAY-1997:10		KSC LANDING	4	OSO	Ground Systems & Facilities	1.02.02
192	OPF86	24-MAY-1997:10	S0028	KSC LANDING/CONVOY OPS/TOW TO OPF(REF MIN XX ON SLF KICS)	9	OSO	Structures , Mechanisms, Veh Handling	1.04.02
193	OPF86	24-MAY-1997:10	V5126	TEMPILABEL INSTL/RMVL (BOOK RELEASE ONLY)		TPS	Thermal Management	2.04.05
194	OPF86	24-MAY-1997:10	V5155	SPACEHAB ACCESS PLATFORMS INSTLN (BOOK RELEASE)		PLO	Structures , Mechanisms, Veh Handling	2.06.06
195	OPF86	24-MAY-1997:18		ORBITER AT OPF DOOR		OSO	Structures , Mechanisms, Veh Handling	2.02.01
196	OPF86	24-MAY-1997:19	S0028	SPOT ORB IN OPF	1	OSO	Structures , Mechanisms, Veh Handling	1.04.02
197	OPF86	24-MAY-1997:20	V1091	OPF INTEGRATED ROLL-IN OPS	2	FCP	Power Management	2.03.01
198	OPF86	24-MAY-1997:22	V1091	OPF INTEGRATED ROLL-IN OPS	18	FCP	Power Management	2.03.01
199	OPF86	24-MAY-1997:22	V9001VL1	ORBITER POWER UP		OTC	Power Management	2.04.13
200	OPF86	24-MAY-1997:23		POWER UP(SAT & SUN)	18	OTC	Power Management	2.04.13
201	OPF86	25-MAY-1997:1		POWER UP(THIRD SHIFT)	8	OTC	Power Management	2.04.13
202	OPF86	25-MAY-1997:7	V3555	FWD PURGE CIRCUIT # 1 DOWN (FOR STATIC AIR CKS)		PVD	Safety Management & Control	2.07.01
203	OPF86	25-MAY-1997:8	V6054.002	POST FLT INSPECT - LRCS PRIMARY THRUSTER INSPECTIONS	16	OMS	Propulsion	2.05.03
204	OPF86	25-MAY-1997:9	STR-A0267//V35-00001/2/3/	INSTALL AFT ACCESS PLTFMS/MPS BACKSHELL PROT CVRS/AV BAYS 4,5,6 BLNKT RMVLS	48	STR	Structures , Mechanisms, Veh Handling	2.02.02
205	OPF86	25-MAY-1997:9	V1158 VL1	POSU 8&9 STATIC AIR PANEL RMVL	32	OMS	Propulsion	2.03.01
206	OPF86	25-MAY-1997:9	V80-06026//V9002.08	SEQ 3 INSTL ELEVON HOIST PTS	8	HYD	Power Management	2.04.06
207	OPF86	25-MAY-1997:9	V5006	POSU 2 PLBD STRONGBACK INSTL	16	MEQ	Structures , Mechanisms, Veh Handling	2.05.02
208	OPF86	25-MAY-1997:9	V31-14142	MIDBODY 02/N2 VENT INSP	16	QC	Structures , Mechanisms, Veh Handling	2.05.02
209	OPF86	25-MAY-1997:16	V1091	OPF INTEGRATED ROLLIN OPS	2	FCP	Power Management	2.03.01
210	OPF86	25-MAY-1997:18	V1091	OPF INTEGRATED ROLL-IN OPS	2	FCP	Power Management	2.03.01

No.	DATA SET	START TIME DD:MMM-YYYY:H	PROCEDURE NAME	DESCRIPTION	DURATION (Hours)	SYSTEM	DESIGN DISCIPLINE	GENERIC OPS FBS NUMBER (Appx A1)
211	OPF86	25-MAY-1997:20	V1091	OPF INTEGRATED ROLL-IN OPS	6	FCP	Power Management	2.03.01
212	OPF86	25-MAY-1997:20	V1091	CRYO OFFLOAD COMPLETE		FCP	Power Management	2.03.01
213	OPF86	25-MAY-1997:20		CRYO OFFLOAD COMPLETE		FCP	Power Management	2.03.01
214	OPF86	27-MAY-1997:1	V1275.001B	PLBD HINGE LINE WATERPROOFING	2	TPS	Thermal Management	2.04.05
215	OPF86	27-MAY-1997:1	V3575	BRIDGE BUCKET CHECKOUT	344	GSE	Structures , Mechanisms, Veh Handling	2.06.06
216	OPF86	27-MAY-1997:1	V3576.003	BRIDGE BUCKET CLEANING	344	GSE	Structures , Mechanisms, Veh Handling	2.11.01
217	OPF86	27-MAY-1997:9	V5032.002	SEQ. 03 PYRO INTERRUPT BOX INSTL	4	PYR	Structures , Mechanisms, Veh Handling	2.01.02
218	OPF86	27-MAY-1997:9	V1196	POSU 1,4-8,23,24,25,28,29	40	OMS	Power Management	2.01.02
219	OPF86	27-MAY-1997:9	V66-50006	C/M PRIMARY ACCESS LADDER INSTALLATION(NOT REQ'D)		STR	Structures , Mechanisms, Veh Handling	2.02.02
220	OPF86	27-MAY-1997:9	V1091	OPF INTEGRATED ROLL-IN OPS POI'S	20	FCP	Power Management	2.03.01
221	OPF86	27-MAY-1997:9	V5012.013	SEQ. 03 FIREX DISCONNECTS	3	PYR	Structures , Mechanisms, Veh Handling	2.04.02
222	OPF86	27-MAY-1997:9	V5017.002AV//FCS-549	CREW SYSTEMS DESTOWAGE/EPICS DESTOW	24	FCS	Cockpit & Crew Cabin	2.04.04
223	OPF86	27-MAY-1997:9	V9046.006	VOLUME 'A' COVER/SECURITY LOCK INSTLN	4	FCS	Cockpit & Crew Cabin	2.04.04
224	OPF86	27-MAY-1997:9		(CHIT J4177) DETERMINE WATERPROOFING VS. UNWATER-PROOFING	16	TPS	Structures , Mechanisms, Veh Handling	2.04.05
225	OPF86	27-MAY-1997:9	V6049.003	LESS C/P INSP/REWORK	48	TPS	Thermal Management	2.04.05
226	OPF86	27-MAY-1997:9	V1263	POSU 4 MPS GSE INSTALLATION FOR HE TANK FILL	20	MPS	Propulsion	2.04.08
227	OPF86	27-MAY-1997:9	V1239G//V1240F//V1230I	F/M/A CLEANING	72	GSE	Structures , Mechanisms, Veh Handling	2.04.09
228	OPF86	27-MAY-1997:9	V6029.001	RCS THRUSTER DESICCANT INSPECTION (OMS)	72	GSE	Propulsion	2.04.09
229	OPF86	27-MAY-1997:9	V3576.001	PLBD WALKDOWN CLEANLINESS INSP.	2	PLO	Payload Accommodations	2.05.02
230	OPF86	27-MAY-1997:9	V30-14253/254/681/682//V3	LMLG/RMLG & WHEEL WELL INSP	8	QC	Structures , Mechanisms, Veh Handling	2.05.02
231	OPF86	27-MAY-1997:9	V31-15109	RH/LH ET UMB/DR AREA INSP	4	QC	Structures , Mechanisms, Veh Handling	2.05.02
232	OPF86	27-MAY-1997:9	V6028.003	WING LEADING EDGE RCC INSPECTIONS/CVR INSTLS	4	TPS	Thermal Management	2.05.02
233	OPF86	27-MAY-1997:9	V30-13201/204//V31-13102/	NLG/WHEEL WELL INSP	8	QC	Structures , Mechanisms, Veh Handling	2.05.02
234	OPF86	27-MAY-1997:9	V80-04400//V63-50006//V30	DOOR 44/SURROUNDING AREA INSP (PRIOR TO LADDER INSTL)	8	QC	Structures , Mechanisms, Veh Handling	2.05.02
235	OPF86	27-MAY-1997:9	V31-15104/106	MIDDECK INSP	24	QC	Cockpit & Crew Cabin	2.05.04
236	OPF86	27-MAY-1997:9		PAYLOAD DOWNLOAD/PLB & AFD DECONFIGURE(SUMMARY BAR)	152	PLO	Structures , Mechanisms, Veh Handling	2.06.05
237	OPF86	27-MAY-1997:9	FCS-546	MISSION UNIQUE DESTOW	24	FCS	Payload Accommodations	2.06.07
238	OPF86	27-MAY-1997:9	FCS-545	TEHM DESTOW	8	FCS	Payload Accommodations	2.06.07
239	OPF86	27-MAY-1997:9	V6049.009A/V6028.003	PLBD HINGE INSPECTION	8	TPS	Thermal Management	2.08.08

No.	DATA SET	START TIME DD:MMM-YYYY:H	PROCEDURE NAME	DESCRIPTION	DURATION (Hours)	SYSTEM	DESIGN DISCIPLINE	GENERIC OPS FBS NUMBER (Appx A1)
240	OPF86	27-MAY-1997:9	V6051	ET DOOR INSP/T/B	80	TPS	Thermal Management	2.08.08
241	OPF86	27-MAY-1997:9	V9002.04	OPS 170 GSE PUMP CIRC/SAMPLING FROM: 86V053D - Note 1	10	HYD	Propulsion	3.08.01
242	OPF86	27-MAY-1997:12	V5006	POSU 2 PLBD STRONGBACK INSTL	17	MEQ	Structures , Mechanisms, Veh Handling	2.05.02
243	OPF86	27-MAY-1997:12	V6028.003	TPS POST FLT INSPECTION	253	TPS	Thermal Management	2.05.02
244	OPF86	27-MAY-1997:12		TILE PROCESSING	140	TPS	Thermal Management	2.08.05
245	OPF86	27-MAY-1997:13	V31-15109	RH/LH ET UMB/DR AREA INSP	4	QC	Structures , Mechanisms, Veh Handling	2.05.02
246	OPF86	27-MAY-1997:13	V6054.001E1//V6029.001F1	POST FLT INSPECT - FRCS VERNIERS/PRIMARY THRUSTER INSP/UTPA INSTL	28	OMS	Propulsion	2.05.03
247	OPF86	27-MAY-1997:15	V9021.001	OP 20 RUN 2 OMS/RCS TRICKLE PURGE ACTIVATION (FLT CAPS INSTL)	2	OMS	Propulsion	2.04.08
248	OPF86	27-MAY-1997:17	V1009.003O1/.004T1/.005T2	JC'S MPS/SSME POST FLT ROLL-IN OPS	24	MPS	Propulsion	2.02.03
249	OPF86	27-MAY-1997:17	V9021.001	OMS/RCS TRICKLE PURGE ACTIVATION (96 HR OMRS)	4	OMS	Propulsion	2.04.08
250	OPF86	27-MAY-1997:17		POWER UP	2	OTC	Power Management	2.04.13
251	OPF86	27-MAY-1997:17	V9002.10 SEQ 20,100,210//	INSTL HYD TRAY & FLEXHOSES/DUMP HYD BOOT STRAP	16	HYD	Propulsion	2.07.03
252	OPF86	27-MAY-1997:21	V9021.001	OP 20 RUN 2 OMS/RCS TRICKLE PURGE ACTIVATION (96 HR OMRS)	4	OMS	Propulsion	2.04.08
253	OPF86	28-MAY-1997:9	V1314	OP 10 OPEN ALL VENT DOORS	4	PVD	Safety Management & Control	2.02.06
254	OPF86	28-MAY-1997:9	V1158VL1	POSU 8/9 OMS/FRCS STATIC AIR CKS		OMS	Propulsion	2.03.01
255	OPF86	28-MAY-1997:9	V1026.001	POSU'S WCS REMOVAL	8	FCP	Environmental Ctl & Life Spt	2.04.04
256	OPF86	28-MAY-1997:9	V1200K2//V75-50001	DAM UNIT INSTL(POST LNDG)	4	INS	Cmd, Ctl & Health Mngmt	2.04.11
257	OPF86	28-MAY-1997:9		POWER UP FROM: 86V000A - Note 1	8	OTC	Power Management	2.04.13
258	OPF86	28-MAY-1997:9	V31-15101	AFT PENTHOUSE INSP	16	QC	Structures , Mechanisms, Veh Handling	2.05.02
259	OPF86	28-MAY-1997:9	V6029.001	RUN 1 FRCS UTPA INSTALLATION	16	OMS	Propulsion	2.05.03
260	OPF86	28-MAY-1997:9	V6054.003/.002F//V6029.00	POST FLT INSPECT LRCS/RRCS VERNIER THRUSTER INSP/UTPA INSTL	8	NDE	Propulsion	2.05.03
261	OPF86	28-MAY-1997:9	V9021.001	OME TRICKLE PURGE MONITORING (OMS)	48	GSE	Propulsion	2.07.01
262	OPF86	28-MAY-1997:9	V3555	MONITOR ORBITER PURGE AIR (PVD)	1432	GSE	Safety Management & Control	2.07.01
263	OPF86	28-MAY-1997:9	V9048	ET DOOR TO 90	2	MEQ	Structures , Mechanisms, Veh Handling	2.07.03
264	OPF86	28-MAY-1997:13	V1263	MPS/SSME POST FLIGHT OPF ROLL-IN OPS	12	MPS	Propulsion	2.04.08
265	OPF86	28-MAY-1997:13	V1200	ORB FLT RECORDER DUMP TO GSE (POST LNDG)	4	INS	Cmd, Ctl & Health Mngmt	2.04.13
266	OPF86	28-MAY-1997:13	V5006	PLBD LATCH FUNCTIONAL/POSIT LH TO 145 & RH TO 160	4	MEQ	Structures , Mechanisms, Veh Handling	2.05.02
267	OPF86	28-MAY-1997:16	V1276.001	POSU 2,3,4 C&T STAT.SET-UP/RTS DISPLAY/REC SET-UP AND CTS	1	COM	Communications	2.05.01

No.	DATA SET	START TIME DD:MMM-YYYY:H	PROCEDURE NAME	DESCRIPTION	DURATION (Hours)	SYSTEM	DESIGN DISCIPLINE	GENERIC OPS FBS NUMBER (Appx A1)
268	OPF86	28-MAY-1997:17	V5006	PLBD'S OPEN		MEQ	Structures , Mechanisms, Veh Handling	2.02.06
269	OPF86	28-MAY-1997:17		PLBD'S OPEN COMPLETE		MEQ	Structures , Mechanisms, Veh Handling	2.02.06
270	OPF86	28-MAY-1997:17	V1158VL1	POSU'S 2B,3B,4-7,20,25,27 GSE VLV'S/ SCRUBBER CONFIG. FROM: 86V204A1 - Note 1	16	OMS	Propulsion	2.03.01
271	OPF86	28-MAY-1997:17	V1026.001	WCS REMOVAL	4	FCP	Environmental Ctl & Life Spt	2.04.04
272	OPF86	28-MAY-1997:17	V5006	POSITION DOOR TO 145 DEGREES	4	MEQ	Structures , Mechanisms, Veh Handling	2.05.02
273	OPF86	28-MAY-1997:17	V1276.001	SEQ 4 & 5 ADA CHKS/DEPLOY KU BAND ANTENNA	1	COM	Communications	2.05.02
274	OPF86	28-MAY-1997:17	V6012	ORBITER HYDRAULIC SYSTEM INSPECTION	8	HYD	Power Management	2.05.03
275	OPF86	28-MAY-1997:17	V9002.01	HYDRAULIC POWER UP POSU'S	12	HYD	Power Management	2.07.03
276	OPF86	28-MAY-1997:17	V9002.06	SEQ 4 DEMATE ENGINE I/F QDS	4	SME	Propulsion	3.08.01
277	OPF86	28-MAY-1997:17	V80-05907/05933/05935	REMOVE CARRIER PANELS ENG 1,2,3	8	STR	Structures , Mechanisms, Veh Handling	3.08.01
278	OPF86	28-MAY-1997:19	V1158VL1	POSU 8/9 OMS/FRCS STATIC AIR CKS		OMS	Propulsion	2.03.01
279	OPF86	28-MAY-1997:21	V5057	SEQ 15,17 INSTL ENG 2,3,1 BELLOWS CVRS	4	SME	Propulsion	2.02.04
280	OPF86	28-MAY-1997:21	V1026.001	WCS REMOVAL POI'S	4	FCP	Environmental Ctl & Life Spt	2.04.04
281	OPF86	28-MAY-1997:21	V5006	PLBD & LATCH FUNCT POI'S	4	MEQ	Structures , Mechanisms, Veh Handling	2.05.02
282	OPF86	28-MAY-1997:21	APU-133	POSU 46,47 APU TOXIC VAPOR CKS	4	APU	Power Management	2.08.01
283	OPF86	28-MAY-1997:21	V1264	GSE CABLE CONNECTION	1	COM	Communications	2.10.02
284	OPF86	28-MAY-1997:22	V1264	SEQ 29 DA COUNTERWEIGHT INSTLN.	1	COM	Communications	2.10.02
285	OPF86	29-MAY-1997:9	CE-192	SPACEHAB TO ORB ELECTRICAL DEMATES	4	PLO	Payload Accommodations	1.03.03
286	OPF86	29-MAY-1997:9	V34-00018/0007/0008//V63-	OPEN BAYS/INSTALL ACCESS	8	STR	Structures , Mechanisms, Veh Handling	2.02.05
287	OPF86	29-MAY-1997:9	V1158VL1	POSU 8/9 OMS/FRCS STATIC AIR CKS		OMS	Propulsion	2.03.01
288	OPF86	29-MAY-1997:9	V5032.003	POSU 1-2-1.3 MERL-AFT DISASSY	8	PYR	Structures , Mechanisms, Veh Handling	2.03.02
289	OPF86	29-MAY-1997:9	VITT-061	CREW MODULE DECONFIG	4	FCS	Cockpit & Crew Cabin	2.04.04
290	OPF86	29-MAY-1997:9	V6049.001	NLGD THERMAL BARRIER INSP/REWORK	40	TPS	Thermal Management	2.04.06
291	OPF86	29-MAY-1997:9	V6049.002	NLGD PREFLIGHT EVALUATION	8	TPS	Thermal Management	2.04.06
292	OPF86	29-MAY-1997:9	V6059.001A//.004	ELEVON COVE REWORK/C/P	40	TPS	Thermal Management	2.04.06
293	OPF86	29-MAY-1997:9		POWER UP FROM: 86V000A - Note 1	8	OTC	Power Management	2.04.13
294	OPF86	29-MAY-1997:9	V1200	MADS RECORDER DUMP	10	INS	Cmd, Ctl & Health Mngmt	2.04.13
295	OPF86	29-MAY-1997:9	V1276.001	SEQ 8 FREEDOM/BALANCE/ENCODER CHECKS	1	COM	Communications	2.05.01
296	OPF86	29-MAY-1997:9	V6044.001	MSBLS DESSICANT CHECKS (RUN 1)	2	COM	Communications	2.05.01
297	OPF86	29-MAY-1997:9	V00-10102	CHIN PANEL EXTERIOR INSPECTION	32	TPS	Thermal Management	2.05.02

No.	DATA SET	START TIME DD:MMM-YYYY:H	PROCEDURE NAME	DESCRIPTION	DURATION (Hours)	SYSTEM	DESIGN DISCIPLINE	GENERIC OPS FBS NUMBER (Appx A1)
298	OPF86	29-MAY-1997:9	V31-14152	KU-BAND COM ANTENNA INSPECTION	12	QC	Communications	2.05.02
299	OPF86	29-MAY-1997:9	V33-10001	POST FLIGHT WINDOW INSPECT	8	NDE	Structures , Mechanisms, Veh Handling	2.05.02
300	OPF86	29-MAY-1997:9	V30-14324/396/389//V31-14	VENT DOOR INSP (FULL OPEN)	16	QC	Structures , Mechanisms, Veh Handling	2.05.02
301	OPF86	29-MAY-1997:9	V1011.02	SSME INSP (RCKTDYNE)	24	SME	Propulsion	2.05.03
302	OPF86	29-MAY-1997:9	E0984	S/H H2O DEMATES	2	PLO	Payload Accommodations	2.06.04
303	OPF86	29-MAY-1997:9	E5006	S/H DESTOW	4	PLO	Payload Accommodations	2.06.04
304	OPF86	29-MAY-1997:9	APU-133	POSU 46,47 APU TOXIC VAPOR CKS	8	APU	Power Management	2.08.01
305	OPF86	29-MAY-1997:9	STR-A0270/A0271	(WAS STR-4564/4565)INSPECT/RMV TEMPORARY RESTRAINTS	8	STR	Structures , Mechanisms, Veh Handling	2.08.06
306	OPF86	29-MAY-1997:9	APU-A002	(CHIT J5031)TEMPILABEL INSTL	8	APU	Power Management	2.09.01
307	OPF86	29-MAY-1997:9	G41-20017	SSME DAVIT CRANE POSITIONING		SME	Propulsion	3.08.02
308	OPF86	29-MAY-1997:9	V5058	INTERFACE TEST PLATE INSTL	16	SME	Propulsion	3.08.02
309	OPF86	29-MAY-1997:11	V3555	FWD PURGE CIRCUIT # 1 UP (AFTER STATIC AIR CKS)		PVD	Safety Management & Control	2.07.01
310	OPF86	29-MAY-1997:13	FI-025-V3//V3575	REMOVE SPACEHAB CATERPILLAR BLANKET (SPIKE REQ'D)(MDAC)	4	PLO	Payload Accommodations	2.06.01
311	OPF86	29-MAY-1997:13	PT-040	EPS COVER INSTALLATION (CUSTOMER REQ'D)	1	PLO	Payload Accommodations	2.06.03
312	OPF86	29-MAY-1997:13	V9046.003	DEPIN EVA SLIDEWIRES	1	PLO	Payload Accommodations	2.06.03
313	OPF86	29-MAY-1997:13	FI-025-V3	SPACEHAB FLEX SECTION DISCONNECT(MDAC)	4	PLO	Payload Accommodations	2.06.03
314	OPF86	29-MAY-1997:13	V9002.05K1//V5057	SEQ 12 POSIT SSME TO NULL/INSTL MID STROKE LKS/ELEV FULL UP/B/F DWN/RSB CLOSED	4	HYD	Power Management	2.07.03
315	OPF86	29-MAY-1997:14	FI-XXX	INSTL HAB ACCESS(WORKED ON LANDING)		PLO	Payload Accommodations	2.06.06
316	OPF86	29-MAY-1997:15	V9001	SSME CONTROLLER INITIAL POWER UP(UA-A0047 MEC T/S)	4	SME	Propulsion	3.03.03
317	OPF86	29-MAY-1997:17	CE-192	SPACEHAB TO ORB ELECTRICAL DEMATES FROM: 86V435 - Note 1	4	PLO	Payload Accommodations	1.03.03
318	OPF86	29-MAY-1997:17	V5032.003	SEQ 03-AFT PYRO DISASSY SETUPS	2	PYR	Structures , Mechanisms, Veh Handling	2.01.02
319	OPF86	29-MAY-1997:17	V1196	POSU 1,4-8,23,24,25,28,29	8	OMS	Power Management	2.01.02
320	OPF86	29-MAY-1997:17	STR-A0267//V35-00001/2/3/	INSTALL AFT ACCESS PLTFMS/MPS BACKSHELL PROT CVRS/AV BAYS 4,5,6 BLNKT RMVLS FROM: 86V219 - Note 1	16	STR	Structures , Mechanisms, Veh Handling	2.02.02
321	OPF86	29-MAY-1997:17	V6029.001	RUN 2 LRCS/RRCS UTPA INSTALLATION	8	OMS	Propulsion	2.05.03
322	OPF86	29-MAY-1997:17	V6054.001E1//V6029.001F1	RUN 1 FRCS UTPA INSTALLATION FROM: 86V204AF - Note 1	12	OMS	Propulsion	2.05.03
323	OPF86	29-MAY-1997:17	V31-15104/106	MIDDECK INSP FROM: 86V272C - Note 1	8	QC	Cockpit & Crew Cabin	2.05.04
324	OPF86	29-MAY-1997:17	V80-05907/05933/05935	REMOVE CARRIER PANELS ENG 1,2,3 FROM: 86V214 - Note 1	16	STR	Structures , Mechanisms, Veh Handling	3.08.01
325	OPF86	29-MAY-1997:17	V1011.03	SSME TURBO PUMP TORQUE CHECKS (LOW PRESS)(POST FLT)	4	SME	Propulsion	3.08.01
326	OPF86	29-MAY-1997:17	V5058	POSU'S SSME REMOVE	8	SME	Propulsion	3.08.02

No.	DATA SET	START TIME DD:MMM-YYYY:H	PROCEDURE NAME	DESCRIPTION	DURATION (Hours)	SYSTEM	DESIGN DISCIPLINE	GENERIC OPS FBS NUMBER (Appx A1)
327	OPF86	29-MAY-1997:19	V5032.003	SEQ 04 AFT EXP. PYRO VERIF.	1	PYR	Structures , Mechanisms, Veh Handling	2.01.02
328	OPF86	29-MAY-1997:20	V5032.003	SEQ 05-07 AFT EXPENDED PYRO DISCONN. & REMOVAL	2	PYR	Structures , Mechanisms, Veh Handling	2.01.02
329	OPF86	29-MAY-1997:21	V5032.002	SEQ. 04 NLG STRUT THRUSTER DISC.	2	PYR	Structures , Mechanisms, Veh Handling	2.01.02
330	OPF86	29-MAY-1997:23	V5032.002	SEQ. 05 FWD SEP BOLT ELEC DISC./RMVL	2	PYR	Structures , Mechanisms, Veh Handling	2.01.02
331	OPF86	30-MAY-1997:9	V1009.003O1/.004T1/.005T2	JC'S MPS/SSME POST FLT ROLL-IN OPS FROM: 86V215A - Note 1	16	MPS	Propulsion	2.02.03
332	OPF86	30-MAY-1997:9	V34-00018/0007/0008//V63-	OPEN BAYS/INSTALL ACCESS FROM: 86V080C - Note 1	8	STR	Structures , Mechanisms, Veh Handling	2.02.05
333	OPF86	30-MAY-1997:9	V1158VL1	POSU 10 1307 BULKHEAD XDUCER SNIFF CKS	8	OMS	Propulsion	2.03.01
334	OPF86	30-MAY-1997:9	V1158VL1	POSU'S 2B,3B,4-7,20,25,27 GSE VLV'S/ SCRUBBER CONFIG. FROM: 86V204A1 - Note 1	8	OMS	Propulsion	2.03.01
335	OPF86	30-MAY-1997:9	V33-10001	POST FLIGHT WINDOW INSPECT FROM: 86V279A - Note 1	8	NDE	Structures , Mechanisms, Veh Handling	2.05.02
336	OPF86	30-MAY-1997:9	V1011.01	POI'S SSME BEARING DRYING	8	SME	Propulsion	2.05.03
337	OPF86	30-MAY-1997:9	UA-A0046	(CHIT J5050)GPS/INS REMOVAL	8	COM	Communications	2.08.01
338	OPF86	30-MAY-1997:9	STR-A0270/A0271	(WAS STR-4564/4565)INSPECT/RMV TEMPORARY RESTRAINTS FROM: 86V34 - Note 1	4	STR	Structures , Mechanisms, Veh Handling	2.08.06
339	OPF86	30-MAY-1997:13	S0065.002A//N5284	POSU'S SPACEHAB DOWNLOAD	12	PLO	Payload Accommodations	1.03.02
340	OPF86	30-MAY-1997:13	V1165.001	MLG WHEEL/TIRE ASSEMBLY REMOVAL/BRAKE INSP	8	MEQ	Structures , Mechanisms, Veh Handling	2.04.01
341	OPF86	30-MAY-1997:13	V1165.003	NLG WHEEL/TIRE ASSEMBLY REMOVAL	4	MEQ	Cmd, Ctl & Health Mngmt	2.04.01
342	OPF86	30-MAY-1997:13	V9021.001	DEACTIVATE OMS TRICKLE PURGE	4	OMS	Propulsion	2.04.08
343	OPF86	30-MAY-1997:13	V1165.011	MLG SEP HARNESS REMOVAL	4	INS	Cmd, Ctl & Health Mngmt	2.04.12
344	OPF86	30-MAY-1997:13		POWER UP FROM: 86V000A - Note 1	20	OTC	Power Management	2.04.13
345	OPF86	30-MAY-1997:13	V30-15430	DETAILED B/F DRAIN INSP(BOROSCOPE W/VIDEO)	8	QC	Structures , Mechanisms, Veh Handling	2.05.02
346	OPF86	30-MAY-1997:13	V35-50002	SALAD BOWL REMOVAL	4	STR	Structures , Mechanisms, Veh Handling	2.10.02
347	OPF86	30-MAY-1997:13	V9001	SSME CONTROLLER INITIAL POWER UP(UA-A0047 MEC T/S) FROM: 86V079A - Note 1	4	SME	Propulsion	3.03.03
348	OPF86	30-MAY-1997:13	V41-40021/40022/40023/400	ENG 1,2,3 SSME DOME HEAT SHIELDS REMOVAL	12	OHE	Propulsion	3.08.01
349	OPF86	30-MAY-1997:17	V1158VL2	FUEL AND OXID QD MATES/LEAK CHECKS	8	OMS	Propulsion	2.03.01
350	OPF86	30-MAY-1997:17	V1165.013	NLG SEP HARNESS REMOVAL	4	INS	Cmd, Ctl & Health Mngmt	2.04.12
351	OPF86	30-MAY-1997:17	V30-15305	AFT INTERNAL DETAIL INSP	16	NDE	Structures , Mechanisms, Veh Handling	2.05.02
352	OPF86	30-MAY-1997:17	MPS-1281	PD3 CONTAMINATION INVEST	4	MPS	Propulsion	2.08.01
353	OPF86	30-MAY-1997:17	V1011.03	SSME TURBO PUMP TORQUE CHECKS (LOW PRESS)(POST FLT) FROM: 86V176 - Note 1	4	SME	Propulsion	3.08.01
354	OPF86	31-MAY-1997:7	V1158VL1	HI BAY WALK DOWN	2	OMS	Propulsion	2.03.01

No.	DATA SET	START TIME DD:MMM-YYYY:H	PROCEDURE NAME	DESCRIPTION	DURATION (Hours)	SYSTEM	DESIGN DISCIPLINE	GENERIC OPS FBS NUMBER (Appx A1)
355	OPF86	31-MAY-1997:7		POWER UP(THIRD SHIFT) FROM: 86V000C - Note 1	2	OTC	Power Management	2.04.13
356	OPF86	31-MAY-1997:9	V1158VL2	OXID BALL VALVE DRAIN/QD DEMATE	3	OMS	Propulsion	2.03.01
357	OPF86	31-MAY-1997:9		POWER UP(SAT & SUN) FROM: 86V000B - Note 1	5	OTC	Power Management	2.04.13
358	OPF86	31-MAY-1997:11	V1158VL1	HI BAY WALK DOWN	1	OMS	Propulsion	2.03.01
359	OPF86	31-MAY-1997:12	V1158VL2	OMS FUEL BALL VALVE DRAIN/QD DEMATE	3	OMS	Propulsion	2.03.01
360	OPF86	31-MAY-1997:13	V80-05859	C/P DOOR RMVL	2	STR	Structures , Mechanisms, Veh Handling	2.02.04
361	OPF86	31-MAY-1997:13	V41-40021/40022/40023/400	ENG 1,2,3 SSME DOME HEAT SHIELDS REMOVAL FROM: 86V178A - Note 1	12	OHE	Propulsion	3.08.01
362	OPF86	31-MAY-1997:15	V1158VL1	OPEN HI BAY/RECONFIGURE ORB FROM REMOTE	2	OMS	Propulsion	2.03.01
363	OPF86	31-MAY-1997:15	V1158VL2	HYPER OFFLOAD COMPLETE		OMS	Propulsion	2.03.01
364	OPF86	31-MAY-1997:15		HYPER OFFLOAD COMPLETE		OMS	Propulsion	2.04.07
365	OPF86	31-MAY-1997:15	V5067	C/M HATCH CLOSE	2	MEQ	Structures , Mechanisms, Veh Handling	2.07.03
366	OPF86	31-MAY-1997:17	V1158VL2	SEQ 15/17 GSE EDUCTION	8	OMS	Propulsion	2.03.01
367	OPF86	31-MAY-1997:21	V80-05917/05928/05901/059	59-1,59-2,59-17,59-28 DOOR RMVLS	2	STR	Structures , Mechanisms, Veh Handling	2.02.05
368	OPF86	01-JUN-1997:1	V1275.003D	CREW HATCH/UPPER SURFACE WATERPROOF-ING(OPPORTUNITY BASIC)	12	TPS	Thermal Management	2.04.05
369	OPF86	01-JUN-1997:9		COM,ECL,EPD,INS,MEQ,OEL PROCESSING/HISTORICAL PWR DOWN PROCESSING(SAT. & SUN.) FROM: 86V350B - Note 1 FIXEDFLAG 2	4		Structures , Mechanisms, Veh Handling	2.08.04
370	OPF86	02-JUN-1997:4	S0065.002A//N5284	SPACEHAB DOWNLOAD	9	PLO	Payload Accommodations	1.03.02
371	OPF86	02-JUN-1997:7	V1003	POSU'S ORB POWER SYSTEM VALIDATIONS	2	EPD	Power Management	2.05.01
372	OPF86	02-JUN-1997:8	V5067	C/M HATCH OPEN	2	MEQ	Structures , Mechanisms, Veh Handling	2.07.03
373	OPF86	02-JUN-1997:9	V1026.002	POSU'S WMS POST FLIGHT SERVICING	32	FCP	Environmental Ctl & Life Spt	2.04.04
374	OPF86	02-JUN-1997:9	V00-10101	BODY FLAP/COVE INSP & REWORK(DELETED NOT REQ'D OMDP ONLY PER J. ROSENBAUER)		TPS	Thermal Management	2.04.06
375	OPF86	02-JUN-1997:9	V1053.003	REMOVE CO2 SENSOR	4	ECL	Environmental Ctl & Life Spt	2.04.12
376	OPF86	02-JUN-1997:9	V1003	ORB POWER SYSTEM VALIDATIONS	8	EPD	Power Management	2.05.01
377	OPF86	02-JUN-1997:9	V00-10101	B/F CAVITY,END PLATES & DOOR BEAM FRSI INSP(DELETED NOT REQ'D)		TPS	Thermal Management	2.05.02
378	OPF86	02-JUN-1997:9	V43-70001	7 DAY DECAY CHECK		OMS	Propulsion	2.05.03
379	OPF86	02-JUN-1997:9	V1291.002	SEQ 2 FES INTERNAL CORROSION INSPECT (BORESCOPE)(NDE)	4	ECL	Environmental Ctl & Life Spt	2.05.03
380	OPF86	02-JUN-1997:9	V1181	FRCS ACCESS PANEL REMOVAL	4	OMS	Propulsion	2.05.03
381	OPF86	02-JUN-1997:9	V9002.08	SEQ-4 LOWER ELEVONS	12	HYD	Power Management	2.05.04
382	OPF86	02-JUN-1997:9	ECL-401	FES PURGE	16	ECL	Environmental Ctl & Life Spt	2.07.01

No.	DATA SET	START TIME DD:MMM-YYYY:H	PROCEDURE NAME	DESCRIPTION	DURATION (Hours)	SYSTEM	DESIGN DISCIPLINE	GENERIC OPS FBS NUMBER (Appx A1)
383	OPF86	02-JUN-1997:9	V9002.01/.05	BF UP/RSB OPEN/RETRACT NLG STRUT(DELETED AS DUP 86VDPF)		HYD	Propulsion	2.07.03
384	OPF86	02-JUN-1997:9	V9002.011/.014	DEPIN MLG STRUT(DELETED NOT REQ'D)		MEQ	Propulsion	2.07.03
385	OPF86	02-JUN-1997:9	V1009.004T1/.005T2/.006E/	MPS CONFIGURE FOR SYSTEM TESTING	16	MPS	Propulsion	2.07.03
386	OPF86	02-JUN-1997:9	V9002	INSTALL ELEVON LOCKS(FOR DPF TEST)(DELETED DUP. OF 86V9002C)		MEQ	Propulsion	2.07.03
387	OPF86	02-JUN-1997:9		LESS C/P INSP/REWORK(DELETED AS DUP OF 86V223C)		TPS	Thermal Management	2.08.05
388	OPF86	02-JUN-1997:9	V3599	ORBITER ALARM TEST	8	GSE	Ground Systems & Facilities	2.11.01
389	OPF86	02-JUN-1997:9	V41-40021/40022/40023/400	ENG 1,2,3 SSME DOME HEAT SHIELDS REMOVAL	32	OHE	Propulsion	3.08.01
390	OPF86	02-JUN-1997:9	V1011.03	SSME TURBO PUMP TORQUE CHECKS (LOW PRESS)(POST FLT) FROM: 86V176 - Note 1	4	SME	Propulsion	3.08.01
391	OPF86	02-JUN-1997:9	V5058	POSU'S SSME REMOVE FROM: 86V139A - Note 1	16	SME	Propulsion	3.08.02
392	OPF86	02-JUN-1997:13	V80-06001,002,3,4,9,10,11	OPEN FLIPPER DRS 1-4,12	20	STR	Structures , Mechanisms, Veh Handling	2.02.06
393	OPF86	02-JUN-1997:13	V9002.08	SEQ-5,6,7,8 INSTL FERRY LOCKS	12	HYD	Power Management	2.04.01
394	OPF86	02-JUN-1997:13	V6012	ORBITER HYDRAULIC SYSTEM INSPECTION FROM: 86V053 - Note 1	76	HYD	Power Management	2.05.03
395	OPF86	02-JUN-1997:13	V1181	POSU 15 FRCS SPILL PROTECTION	4	OMS	Propulsion	2.05.03
396	OPF86	02-JUN-1997:13	V9048	ET DOOR TO CENTERLINE	2	MEQ	Structures , Mechanisms, Veh Handling	2.07.03
397	OPF86	02-JUN-1997:13	UA-A0047	DBIA CONNECTOR REMOVAL	4	EPD	Power Management	2.08.04
398	OPF86	02-JUN-1997:13	V5058	SSME DEFOAM FOR ENGINE REMOVAL	12	SME	Propulsion	3.08.02
399	OPF86	02-JUN-1997:17	V34-00018/0007/0008//V63-	OPEN BAYS/INSTALL ACCESS FROM: 86V080C - Note 1	8	STR	Structures , Mechanisms, Veh Handling	2.02.05
400	OPF86	02-JUN-1997:17	V5139	FWD RCC INSTALLATION (REMOVE G/F & REWORK F/B)	8	TPS	Thermal Management	2.04.06
401	OPF86	02-JUN-1997:17	V35-40004/7	REMOVE LO2/LH2 ET UMB MONOBALL	8	STR	Structures , Mechanisms, Veh Handling	2.04.12
402	OPF86	02-JUN-1997:17	V33-40001	LIOH CANNISTER RMVL	4	ECL	Environmental Ctl & Life Spt	2.04.12
403	OPF86	02-JUN-1997:17	V1181	FRCS FLT CAP REMOVAL	4	OMS	Propulsion	2.05.03
404	OPF86	02-JUN-1997:19	V63-06303/06304	OPEN WINGS	6	STR	Thermal Management	2.10.02
405	OPF86	02-JUN-1997:21	S0065.002	SPACEHAB REMOVAL COMP		PLO	Payload Accommodations	1.03.03
406	OPF86	02-JUN-1997:21		PAYLOAD REMOVAL COMP		PLO	Structures , Mechanisms, Veh Handling	1.03.03
407	OPF86	02-JUN-1997:21	V1181	FRCS SCUPPER INSTLN	4	OMS	Propulsion	2.05.03
408	OPF86	03-JUN-1997:9	V1165.001	MLG WHEEL/TIRE ASSEMBLY REMOVAL/BRAKE INSP	8	MEQ	Structures , Mechanisms, Veh Handling	2.04.01
409	OPF86	03-JUN-1997:9		POWER UP FROM: 86V000A - Note 1	44	OTC	Power Management	2.04.13
410	OPF86	03-JUN-1997:9	V1076	POSU'S WCCS SERVICING	16	PVD	Safety Management & Control	2.05.01

No.	DATA SET	START TIME DD:MMM-YYYY:H	PROCEDURE NAME	DESCRIPTION	DURATION (Hours)	SYSTEM	DESIGN DISCIPLINE	GENERIC OPS FBS NUMBER (Appx A1)
411	OPF86	03-JUN-1997:9	V1003	ORB POWER SYSTEM VALIDATIONS FROM: 86V087 - Note 1	4	EPD	Power Management	2.05.01
412	OPF86	03-JUN-1997:9	V30-14105/111/115/118/121	PLBDS INSP	24	QC	Structures , Mechanisms, Veh Handling	2.05.02
413	OPF86	03-JUN-1997:9	V31-13109/112/114/115/117	FWD FUSELAGE/CM INSP	40	QC	Structures , Mechanisms, Veh Handling	2.05.02
414	OPF86	03-JUN-1997:9	V30-13207//V33-10001	(NDE)FWD WINDOW FRAME INSP	8	QC	Structures , Mechanisms, Veh Handling	2.05.02
415	OPF86	03-JUN-1997:9	V1181	FRCS FUNCTIONAL & C/O	24	OMS	Propulsion	2.05.03
416	OPF86	03-JUN-1997:9	V9002.08	SEQ-4 LOWER ELEVONS FROM: 86V9002B - Note 1	2	HYD	Power Management	2.05.04
417	OPF86	03-JUN-1997:9	ECL-401	FES PURGE	688	ECL	Environmental Ctl & Life Spt	2.07.01
418	OPF86	03-JUN-1997:9	V6049.003	LESS C/P REWORK	272	TPS	Thermal Management	2.08.08
419	OPF86	03-JUN-1997:13	V9002.08	ORB HYD CONN FROM: 86V9002C - Note 1	4	HYD	Power Management	2.04.01
420	OPF86	03-JUN-1997:13	V1008.002	POSU'S TACAN ACTIVATION	4	COM	Communications	2.05.01
421	OPF86	03-JUN-1997:13	V6054.002	LH OME INSPECT	4	NDE	Propulsion	2.05.03
422	OPF86	03-JUN-1997:13	V5058	POSU'S SSME REMOVE FROM: 86V139A - Note 1	36	SME	Propulsion	3.08.02
423	OPF86	03-JUN-1997:17	V35-40004/7	REMOVE LO2/LH2 ET UMB MONOBALL FROM: 86V270G - Note 1	8	STR	Structures , Mechanisms, Veh Handling	2.04.12
424	OPF86	03-JUN-1997:24		HEAT SHIELD REMOVAL COMPLETE		SME	Propulsion	3.08.01
425	OPF86	04-JUN-1997:9	V34-00018/0007/0008//V63-	OPEN BAYS/INSTALL ACCESS FROM: 86V080C - Note 1	12	STR	Structures , Mechanisms, Veh Handling	2.02.05
426	OPF86	04-JUN-1997:9	V9028.003	SEQ 20 DISC NLG DR ROLLER	4	MEQ	Power Management	2.04.01
427	OPF86	04-JUN-1997:9	V1026.002	WMS POST FLIGHT SERVICING	8	FCP	Environmental Ctl & Life Spt	2.04.04
428	OPF86	04-JUN-1997:9	V1271	REPLACE ICEPHOBIC COAT ON H2O/URINE DUMP THRML BARRI-ERS	4	TPS	Thermal Management	2.04.06
429	OPF86	04-JUN-1997:9	V1239G//V1240F//V1230I	F/M/A CLEANING FROM: 86V1239 - Note 1	520	GSE	Structures , Mechanisms, Veh Handling	2.04.09
430	OPF86	04-JUN-1997:9	V6029.001	RCS THRUSTER DESICCANT INSPECTION (OMS) FROM: 86V6029 - Note 1	520	GSE	Propulsion	2.04.09
431	OPF86	04-JUN-1997:9	V1008.002	TACAN ACTIVATION & SELF TEST	6	COM	Communications	2.05.01
432	OPF86	04-JUN-1997:9	LAF-0393	LINE CLAMP NUT	8	SME	Structures , Mechanisms, Veh Handling	2.05.02
433	OPF86	04-JUN-1997:9	V30-13416	SEP BOLT INSPECTION	8	QC	Structures , Mechanisms, Veh Handling	2.05.02
434	OPF86	04-JUN-1997:9	V9048	ET DOOR TO 90	2	MEQ	Structures , Mechanisms, Veh Handling	2.07.03
435	OPF86	04-JUN-1997:9	MPS-1283	LH2 RECIRC MANIFOLD DENT INSP	5	MPS	Propulsion	2.08.01
436	OPF86	04-JUN-1997:9	STR-A0266	(MCR 11620)FWD ET INSP/REWORK	8	STR	Structures , Mechanisms, Veh Handling	2.08.06
437	OPF86	04-JUN-1997:9	IPR-0006	EPS ELECTRICAL T/S	2	EPD	Structures , Mechanisms, Veh Handling	2.08.08
438	OPF86	04-JUN-1997:12	V1088	INT/EXT LIGHTING FUNCTIONAL	2	EPD	Power Management	2.05.04
439	OPF86	04-JUN-1997:14	MPS-1283	LH2 RECIRC MANIFOLD DENT INSP(X-RAY)	2	MPS	Propulsion	2.08.01

No.	DATA SET	START TIME DD:MMM-YYYY:H	PROCEDURE NAME	DESCRIPTION	DURATION (Hours)	SYSTEM	DESIGN DISCIPLINE	GENERIC OPS FBS NUMBER (Appx A1)
440	OPF86	04-JUN-1997:15		SSME HYSTER AVAIL		SME	Propulsion	3.08.02
441	OPF86	04-JUN-1997:17	V9028.002	POSU'S NLG STRUT HYD LEVEL CK & GN2 LK CKS	8	MEQ	Structures , Mechanisms, Veh Handling	2.04.01
442	OPF86	04-JUN-1997:17	V41-10015/10016/10017	DOME H/S INSPECT ENG 1,2,3	72	QC	Propulsion	2.05.02
443	OPF86	04-JUN-1997:17	V1181	FRCS FUNCTIONAL & C/O	8	OMS	Propulsion	2.05.03
444	OPF86	04-JUN-1997:17	CE-194	EPS DEMATES(SPIKE)	4	PLO	Payload Accommodations	2.06.08
445	OPF86	04-JUN-1997:21	V1262.009	EMERG CREW EGRESS VENT VERIF (FWD)	4	PVD	Safety Management & Control	2.05.02
446	OPF86	04-JUN-1997:23	V3581	OPEN L/R XSLATOR PLATFORM FOR SSME REMOVAL	2	GSE	Propulsion	3.08.01
447	OPF86	05-JUN-1997:1	V5058	REMOVE SSME 2/3/1	28	SME	Propulsion	3.08.02
448	OPF86	05-JUN-1997:9	PYR-115	POSU'S ODS PYRO DEMATES & SAFING	4	PYR	Structures , Mechanisms, Veh Handling	2.03.02
449	OPF86	05-JUN-1997:9	V5012.002	POSU'S POST FLT DRAG CHUTE OPERATIONS	8	PYR	Structures , Mechanisms, Veh Handling	2.03.02
450	OPF86	05-JUN-1997:9	V1165.008	NLG STRUT HYD LVL CHECKS	16	MEQ	Structures , Mechanisms, Veh Handling	2.04.01
451	OPF86	05-JUN-1997:9	V1026.002	POSU'S WMS POST FLIGHT SERVICING FROM: 86V011B - Note 1	36	FCP	Environmental Ctl & Life Spt	2.04.04
452	OPF86	05-JUN-1997:9	V1295.001	COMMANDER & PILOT SEAT FUNCTIONAL	4	FCS	Cockpit & Crew Cabin	2.04.04
453	OPF86	05-JUN-1997:9	V1271	REPLACE ICEPHOBIC COAT ON H2O/URINE DUMP THRML BARRI-ERS FROM: 86V028 - Note 1	8	TPS	Thermal Management	2.04.06
454	OPF86	05-JUN-1997:9	FWD-3817/18/21/22/26/27	NLGD TPS TILE PROCESS/THERMAL BARRIERS	8	TPS	Thermal Management	2.04.06
455	OPF86	05-JUN-1997:9	V9021.001	OP 70 FRCS T/P ACTIVATION	2	OMS	Propulsion	2.04.08
456	OPF86	05-JUN-1997:9	V1076	WCCS SERVICING	40	PVD	Safety Management & Control	2.05.01
457	OPF86	05-JUN-1997:9	V1183	SEQ-53 ODS ELECTRICAL VERIFICATION	2	EPD	Power Management	2.05.01
458	OPF86	05-JUN-1997:9	V6012	ORB HYD SYSTEM INSPECTION (FWD)	16	HYD	Power Management	2.05.03
459	OPF86	05-JUN-1997:9	MPS-1282	TUBE SCRATCH REPAIR/DYE PEN	8	MPS	Propulsion	2.08.01
460	OPF86	05-JUN-1997:9	TFRC-009//TFRC-0809/0805	(MCR-18563)FRCS THRUSTER TILE PROCESS	8	TPS	Thermal Management	2.08.05
461	OPF86	05-JUN-1997:9	OHE-078	INSTL TRUSS PLTFRM	2	OHE	Structures , Mechanisms, Veh Handling	2.10.03
462	OPF86	05-JUN-1997:11	CM-665	(CHIT J5043)REMOVE EPS	4	PLO	Payload Accommodations	2.06.03
463	OPF86	05-JUN-1997:11	IPR-0006	EPS ELECTRICAL T/S	2	PLO	Structures , Mechanisms, Veh Handling	2.08.08
464	OPF86	05-JUN-1997:17	V1019VL1I1/VL2K/VL3K1	APU LEAK & FUNCTIONAL	8	APU	Power Management	2.05.01
465	OPF86	05-JUN-1997:17	V1291.004	HUMIDITY SEPARATOR FLOW PERFORMANCE TEST	6	ECL	Environmental Ctl & Life Spt	2.05.01
466	OPF86	05-JUN-1997:17	CM-677//COM-0233/34	CAMERA A/C REMOVAL	4	PLO	Payload Accommodations	2.06.03
467	OPF86	05-JUN-1997:17	CM-676	DECONFIGURE TETHERS	4	PLO	Payload Accommodations	2.06.03
468	OPF86	06-JUN-1997:5		SSME REMOVAL COMPLETE		SME	Propulsion	3.08.03
469	OPF86	06-JUN-1997:5	V5058	SSME REMOVAL COMPLETE		SME	Propulsion	3.08.03

No.	DATA SET	START TIME DD:MMM-YYYY:H	PROCEDURE NAME	DESCRIPTION	DURATION (Hours)	SYSTEM	DESIGN DISCIPLINE	GENERIC OPS FBS NUMBER (Appx A1)
470	OPF86	06-JUN-1997:6	PYR-115	ODS PYRO DEMATES & SAFING	3	PYR	Structures , Mechanisms, Veh Handling	2.03.02
471	OPF86	06-JUN-1997:8	V2258.001	HAZARDOUS WASTE REMOVAL (EG&G)	2	GSE	Ground Systems & Facilities	2.11.01
472	OPF86	06-JUN-1997:9	V5012.002	POST FLT DRAG CHUTE PYRO HARDWARE RMVL	8	PYR	Structures , Mechanisms, Veh Handling	2.03.02
473	OPF86	06-JUN-1997:9		POWER UP FROM: 86V000A - Note 1	8	OTC	Power Management	2.04.13
474	OPF86	06-JUN-1997:9	V1184	LOAD MMU 1	2	SOF	Cmd, Ctl & Health Mngmt	2.04.13
475	OPF86	06-JUN-1997:9	V1184	G9/P9/129/44/48		SOF	Cmd, Ctl & Health Mngmt	2.04.13
476	OPF86	06-JUN-1997:9	V30-14471/472/473/474	LH/RH ELEVON DRAIN HOLE INSP (BOROSCOPE)(NDE)	8	QC	Structures , Mechanisms, Veh Handling	2.05.02
477	OPF86	06-JUN-1997:9	V41-10018/10019/10020	ENGINE 1,2,3 MOUNTED HEAT SHIELD INSPECTION	16	NDE	Propulsion	2.05.02
478	OPF86	06-JUN-1997:9	V31-15115	VERT/RSB AREA INSP	8	QC	Structures , Mechanisms, Veh Handling	2.05.02
479	OPF86	06-JUN-1997:9	PLD-T101	TUNNEL INSPECTION(MDAC)(SPIKE)	4	PLO	Payload Accommodations	2.06.03
480	OPF86	06-JUN-1997:9	L6987C/PLD-T101	TUNNEL EXTERNAL CLEANING/INSP	8	PLO	Payload Accommodations	2.06.03
481	OPF86	06-JUN-1997:9	PLD-T101	REMOVE FWD EXT CATERPILLAR BLANKET (SPIKE REQ'D)(MDAC)	2	PLO	Payload Accommodations	2.06.03
482	OPF86	06-JUN-1997:9	V6063//V31-14199//V53-000	ODS INSPECTION	24	PLO	Payload Accommodations	2.06.09
483	OPF86	06-JUN-1997:9	V1270	ADP DEPLOY (FOR TPS INSPECTION)	2	MEQ	Structures , Mechanisms, Veh Handling	2.07.03
484	OPF86	06-JUN-1997:9	V1008.004	RADAR ALT TEST	4	COM	Communications	2.08.01
485	OPF86	06-JUN-1997:9	MPS-1282	TUBE SCRATCH REPAIR/DYE PEN FROM: 86V10S - Note 1	8	MPS	Propulsion	2.08.01
486	OPF86	06-JUN-1997:9	MPS-1284	E2 LOX GALLED THREADS	8	MPS	Propulsion	2.08.01
487	OPF86	06-JUN-1997:9	V5R01	POSU'S F/C 3 REMOVAL/INSTALLATION	12	FCP	Power Management	2.08.01
488	OPF86	06-JUN-1997:9	STR-XXXX	(CHIT J4243)RCC PINHOLE EVALUATION(NOT REQ'D)		STR	Structures , Mechanisms, Veh Handling	2.08.01
489	OPF86	06-JUN-1997:9	TFRC-009//TFRC-0809/0805	(MCR-18563)FRCS THRUSTER TILE PATTERN FAB FROM: 86V10M - Note 1	224	TPS	Thermal Management	2.08.05
490	OPF86	06-JUN-1997:9	V5058	INTERFACE TEST PLATE INSTL FROM: 86V5058 - Note 1	16	SME	Propulsion	3.08.02
491	OPF86	06-JUN-1997:9	V3581	CLOSE L/R XSLATOR PLATFORM AFTER SSME REMOVAL	2	GSE	Propulsion	3.08.03
492	OPF86	06-JUN-1997:11	V6028.003	POST FLT ORB REUSABLE SURFACE INSULATION INSP(ADP)	4	TPS	Thermal Management	2.05.02
493	OPF86	06-JUN-1997:13	V1019VL1I1/VL2K/VL3K1	APU LEAK & FUNCTIONAL FROM: 86V072 - Note 1	4	APU	Power Management	2.05.01
494	OPF86	06-JUN-1997:15	L5076B	FLOOR PAD REMOVAL	2	PLO	Payload Accommodations	2.06.02
495	OPF86	06-JUN-1997:17	CM-667	ODS FLEXSECTION INSPECTION(SPIKE REQ'D)	8	PLO	Payload Accommodations	2.06.09
496	OPF86	06-JUN-1997:21	FCP-A0018//V5R01	F/C 3 REMOVAL/INSTALLATION	6	FCP	Power Management	2.08.01
497	OPF86	07-JUN-1997:1	V1275.001	LOWER SURFACE/CARRIER PANEL/OMS LE WATERPROOFING	12	TPS	Thermal Management	2.04.05
498	OPF86	07-JUN-1997:9	SA086A	TCID LOAD	14	SOF	Cmd, Ctl & Health Mngmt	2.04.13

No.	DATA SET	START TIME DD:MMM-YYYY:H	PROCEDURE NAME	DESCRIPTION	DURATION (Hours)	SYSTEM	DESIGN DISCIPLINE	GENERIC OPS FBS NUMBER (Appx A1)
499	OPF86	07-JUN-1997:9		COM,ECL,EPD,INS,MEQ,OEL PROCESSING/HISTORICAL PWR DOWN PROCESSING(SAT. & SUN.) FROM: 86V350B - Note 1 FIXEDFLAG 2	4		Structures , Mechanisms, Veh Handling	2.08.04
500	OPF86	09-JUN-1997:9	V63-50056	TCS BLNKT,DRAG CHUTE R & R/INSPECTION	16	QC	Structures , Mechanisms, Veh Handling	2.02.05
501	OPF86	09-JUN-1997:9	V1165.008	NLG STRUT HYD LVL CHECKS(48 DECAY CK)	4	MEQ	Structures , Mechanisms, Veh Handling	2.04.01
502	OPF86	09-JUN-1997:9	V1184	LOAD MMU 2	2	SOF	Cmd, Ctl & Health Mngmt	2.04.13
503	OPF86	09-JUN-1997:9	V1184	G9/P9/129/44/48		SOF	Cmd, Ctl & Health Mngmt	2.04.13
504	OPF86	09-JUN-1997:9		POWER UP FROM: 86V000A - Note 1	12	OTC	Power Management	2.04.13
505	OPF86	09-JUN-1997:9	V1004	TASK 1 INITIAL MTU FREQ OSC (12 HR. WARM-UP)	12	INS	Cmd, Ctl & Health Mngmt	2.05.01
506	OPF86	09-JUN-1997:9	V30-14471/472/473/474	LH/RH ELEVON DRAIN HOLE INSP (BOROSCOPE)(NDE) FROM: 86V271G - Note 1	4	QC	Structures , Mechanisms, Veh Handling	2.05.02
507	OPF86	09-JUN-1997:9	V33-10001G//V33-00005	POST FLIGHT WINDOW INSPECT	16	NDE	Cockpit & Crew Cabin	2.05.02
508	OPF86	09-JUN-1997:9	V30-15381	DOME H/S COVER RING INSP(NOT REQ'D)		NDE	Structures , Mechanisms, Veh Handling	2.05.02
509	OPF86	09-JUN-1997:9	V6012	ORB HYD INSPECTION (MID)	16	HYD	Power Management	2.05.03
510	OPF86	09-JUN-1997:9	V1009.007	MPS LEAK & FUNCTIONAL TEST (GH2 SYSTEM)	7	MPS	Propulsion	2.05.03
511	OPF86	09-JUN-1997:9	V80-07700	RMV CADILLAC FTG FOR VERT BOLT INSP (30' PIC REQD)	4	STR	Structures , Mechanisms, Veh Handling	2.07.03
512	OPF86	09-JUN-1997:9	MPS-1281	PD3 CONTAMINATION INVEST FROM: 86V10K - Note 1	8	MPS	Propulsion	2.08.01
513	OPF86	09-JUN-1997:9		TILE PROCESSING	96	TPS	Thermal Management	2.08.05
514	OPF86	09-JUN-1997:11	FCP-A0018//V5R01	F/C 3 REMOVAL/INSTALLATION	4	FCP	Power Management	2.08.01
515	OPF86	09-JUN-1997:13	V1165.008	NLG STRUT HYD LVL CHECKS	2	MEQ	Structures , Mechanisms, Veh Handling	2.04.01
516	OPF86	09-JUN-1997:13	V1026.002	WMS POST FLIGHT SERVICING POI'S	8	FCP	Environmental Ctl & Life Spt	2.04.04
517	OPF86	09-JUN-1997:13	L5076B	FLOOR PAD INSTL	2	PLO	Payload Accommodations	2.06.02
518	OPF86	09-JUN-1997:13	V63-50001	RMV C-HATCH RING BLNKT	4	STR	Thermal Management	2.06.09
519	OPF86	09-JUN-1997:15	FCP-A0018//V5R01	F/C 3 REMOVAL/INSTALLATION	10	FCP	Power Management	2.08.01
520	OPF86	09-JUN-1997:17	V1180VL1	POSU 12 FLT CAP REMOVAL	8	OMS	Propulsion	2.02.05
521	OPF86	09-JUN-1997:17	V63-50001	OPEN BAY 1 AWT	4	STR	Thermal Management	2.02.05
522	OPF86	09-JUN-1997:17	FWD-3817/18/21/22/26/27	NLGD TPS TILE PROCESS/THERMAL BARRIERS FROM: 86V224B - Note 1	56	TPS	Thermal Management	2.04.06
523	OPF86	09-JUN-1997:17	V30-15401//V31-15134	LWR DECK DETAILED INSP	8	QC	Structures , Mechanisms, Veh Handling	2.05.02
524	OPF86	09-JUN-1997:21	V1004	TASK 1 MTU OSC FREQ/READ ADJ.(IF REQ'D)(CONT 24 HR PWR UP REQ'D IF ADJ MADE)	1	INS	Cmd, Ctl & Health Mngmt	2.05.01
525	OPF86	09-JUN-1997:23	V30-15207/209//V31-15113/	VERT/RSB AREA INSP	2	NDE	Structures , Mechanisms, Veh Handling	2.05.02
526	OPF86	10-JUN-1997:9	V5057	SEQ 23 INSTL ENG CVRS	4	SME	Propulsion	2.02.04
527	OPF86	10-JUN-1997:9		POWER UP FROM: 86V000A - Note 1	8	OTC	Power Management	2.04.13

No.	DATA SET	START TIME DD:MMM-YYYY:H	PROCEDURE NAME	DESCRIPTION	DURATION (Hours)	SYSTEM	DESIGN DISCIPLINE	GENERIC OPS FBS NUMBER (Appx A1)
528	OPF86	10-JUN-1997:9	V1291.004	HUMIDITY SEPARATOR FLOW PERFORMANCE TEST FROM: 86V104 - Note 1	8	ECL	Environmental Ctl & Life Spt	2.05.01
529	OPF86	10-JUN-1997:9	V30-15327	THRUST STRUCTURE INSP	20	QC	Structures , Mechanisms, Veh Handling	2.05.02
530	OPF86	10-JUN-1997:9	V30-15305	AFT INTERNAL DETAIL INSP FROM: 86V272E - Note 1	8	NDE	Structures , Mechanisms, Veh Handling	2.05.02
531	OPF86	10-JUN-1997:9	V30-15383	T-O UMB AREA INSP	8	QC	Structures , Mechanisms, Veh Handling	2.05.02
532	OPF86	10-JUN-1997:9	V30-15428	VERT TAIL FWD ATTACH BOLT INSPECTION (30' PIC BOARD REQD)	8	NDE	Structures , Mechanisms, Veh Handling	2.05.02
533	OPF86	10-JUN-1997:9	V30-15358//V31-15109	EXTERNAL AFT DETAILED INSP	8	NDE	Structures , Mechanisms, Veh Handling	2.05.02
534	OPF86	10-JUN-1997:9	V1009.006	(CHIT J4395)MPS LEAK & FUNCTIONAL TEST (G02 SYSTEM)	8	MPS	Propulsion	2.05.03
535	OPF86	10-JUN-1997:9	FCS-556	REMOVE SCAMP	2	FCS	Payload Accommodations	2.06.07
536	OPF86	10-JUN-1997:9	V9002.01/.05/.07/GNC-125	POSU'S DPF TEST	4	GNC	Propulsion	2.07.03
537	OPF86	10-JUN-1997:9	FCP-A0018//V5R01	F/C 3 REMOVAL/INSTALLATION	6	FCP	Power Management	2.08.01
538	OPF86	10-JUN-1997:9	DR HSS-457//PR STR-4600/0	HEATSHIELD REPAIR	48	STR	Thermal Management	2.08.05
539	OPF86	10-JUN-1997:9	V6051	ET DOOR INSP/T/B FROM: 86V222 - Note 1	32	TPS	Thermal Management	2.08.08
540	OPF86	10-JUN-1997:13	V1180VL1	POSU 12 FLT CAP REMOVAL FROM: 86V047B - Note 1	12	OMS	Propulsion	2.02.05
541	OPF86	10-JUN-1997:13	V1042.001	SMOKE DET FUNCT VERIF	4	ECL	Environmental Ctl & Life Spt	2.05.01
542	OPF86	10-JUN-1997:13	V33-10001G//V33-00005	POST FLIGHT WINDOW INSPECT FROM: 86V279 - Note 1	20	NDE	Cockpit & Crew Cabin	2.05.02
543	OPF86	10-JUN-1997:13	V9002.01/.05/.07	ET ACTUATOR CYCLING	4	GNC	Propulsion	2.07.03
544	OPF86	10-JUN-1997:15	V5R01	F/C 3 REMOVAL/INSTALLATION(ELECTRICAL/MECH MATES/BUBBLE LK CHK/FUNCTIONAL CHKS)	18	FCP	Power Management	2.08.01
545	OPF86	10-JUN-1997:17	V1180VL1	POSU 8/10 APS SPILL PROTECTION/SCUPPER INSTL	24	OMS	Propulsion	2.05.03
546	OPF86	10-JUN-1997:17	V1180VL1	POSU 7 APS ACCESS PANEL REMOVAL	8	OMS	Propulsion	2.05.03
547	OPF86	10-JUN-1997:17	V9021.001	OME TRICKLE PURGE MONITORING (OMS) FROM: 86V9021 - Note 1	448	GSE	Propulsion	2.07.01
548	OPF86	10-JUN-1997:17	V5069.006	JACK XFER TO AFT 999 HARD STRUTS (FOR MEQ-231)	4	OHE	Structures , Mechanisms, Veh Handling	2.07.03
549	OPF86	10-JUN-1997:17	OEL-1365	CONNECTOR R/R	16	OEL	Power Management	2.08.04
550	OPF86	10-JUN-1997:17	AFT-4894/95	AFT TE/BFLE TPS WORK	8	TPS	Thermal Management	2.08.05
551	OPF86	11-JUN-1997:1	V1275.001	B/F LE WATERPROOFING	8	TPS	Thermal Management	2.04.05
552	OPF86	11-JUN-1997:9		POWER UP FROM: 86V000A - Note 1	8	OTC	Power Management	2.04.13
553	OPF86	11-JUN-1997:9	V1076	R & R WCCS RACE TRACKS(REQ'S SPIKE)	8	PVD	Safety Management & Control	2.05.01
554	OPF86	11-JUN-1997:9	V1009.004	MPS LO2 SYSTEM LEAK & FUNCTIONAL	4	MPS	Propulsion	2.05.03
555	OPF86	11-JUN-1997:9	V80-97700	INSTL INSTALL CADILLAC FTG (30' PIC REQD)	8	STR	Structures , Mechanisms, Veh Handling	2.07.03
556	OPF86	11-JUN-1997:9	MPS-1288	PD3 INVESTIGATION	8	MPS	Propulsion	2.08.01

No.	DATA SET	START TIME DD:MMM-YYYY:H	PROCEDURE NAME	DESCRIPTION	DURATION (Hours)	SYSTEM	DESIGN DISCIPLINE	GENERIC OPS FBS NUMBER (Appx A1)
557	OPF86	11-JUN-1997:9	CM-671	REMOVE CAMERA FRAME	4	FCS	Cockpit & Crew Cabin	2.08.01
558	OPF86	11-JUN-1997:9	OEL-A0027	BKHD CONNECTOR INSPECTION	8	OEL	Power Management	2.08.04
559	OPF86	11-JUN-1997:9	V7253	ORB WINDOW POLISHING	68	STR	Structures , Mechanisms, Veh Handling	2.08.06
560	OPF86	11-JUN-1997:9	STR-A0251	(CHIT J5013)WING BOLT TORQUE CHKS	40	STR	Structures , Mechanisms, Veh Handling	2.08.06
561	OPF86	11-JUN-1997:9	STR-571	(CHIT J5048)VERT SEAL ENGR INSPECTION	8	STR	Structures , Mechanisms, Veh Handling	2.08.08
562	OPF86	11-JUN-1997:9	V5138.001	POSU 3 DRAG CHUTE PLTFM EXTENSION & TRACK INSTLN	8	OHE	Structures , Mechanisms, Veh Handling	3.03.03
563	OPF86	11-JUN-1997:9	V5138.001	POSU 2 DRAG CHUTE XFER FROM SHIPPING CRATE TO GSE(PARACHUTE FACILITY)	3	OHE	Structures , Mechanisms, Veh Handling	3.03.03
564	OPF86	11-JUN-1997:12	V5138.001	SEQ 03 SLING/CRANE CONNECT LIFT TO P/F 16	1	OHE	Structures , Mechanisms, Veh Handling	3.03.03
565	OPF86	11-JUN-1997:13	V5069.006	JACK XFER TO AFT 999 HARD STRUTS (FOR MEQ-231)	2	OHE	Structures , Mechanisms, Veh Handling	2.07.03
566	OPF86	11-JUN-1997:13	V5138.001	SEQ 03 LWR TO DRAG CHUTE PF WITH JIB CRANE	1	OHE	Structures , Mechanisms, Veh Handling	3.03.03
567	OPF86	11-JUN-1997:14	V5R01	LIFT FC3 TO CONTAINER	1	FCP	Power Management	2.08.01
568	OPF86	11-JUN-1997:17	MEQ-231	(CHIT J5047)ET DOOR LOAD TEST	16	MEQ	Structures , Mechanisms, Veh Handling	2.02.04
569	OPF86	11-JUN-1997:17	HYD-0759	LEAK CHECK	8	HYD	Power Management	2.08.02
570	OPF86	11-JUN-1997:17	IPR-0009	FLOODLIGHT 1 T/S	12	EPD	Power Management	2.08.04
571	OPF86	11-JUN-1997:17	AFT-4894/95	AFT TE/BFLE TPS WORK FROM: 86V10X - Note 1	8	TPS	Thermal Management	2.08.05
572	OPF86	11-JUN-1997:17	IPR-0008	OME PLUG PULLTEST/R&R	4	OMS	Structures , Mechanisms, Veh Handling	2.08.08
573	OPF86	12-JUN-1997:9	V3519	TASK 9 GSE AMMONIA CART SERVICING	4	ECL	Environmental Ctl & Life Spt	2.01.02
574	OPF86	12-JUN-1997:9	V5138.001	POSU 4 ORB PREPS	8	OHE	Ground Systems & Facilities	2.02.02
575	OPF86	12-JUN-1997:9	V5012.001	SEQ 02 INSTALL L/H AFT WIRE PROTECTION SHOP AIDE	1	PYR	Structures , Mechanisms, Veh Handling	2.04.02
576	OPF86	12-JUN-1997:9	V5012.003	POSU 1-3 OBTAIN SECTION 1.3 ITEMS SEQ 03-06	4	PYR	Structures , Mechanisms, Veh Handling	2.04.02
577	OPF86	12-JUN-1997:9	V80-97302/7401/7501/7502	RSB DOOR INSTALLATION	8	STR	Structures , Mechanisms, Veh Handling	2.04.05
578	OPF86	12-JUN-1997:9	FWD-3846	ADP TPS WORK	8	TPS	Thermal Management	2.04.06
579	OPF86	12-JUN-1997:9	V1053.005	PPO2 SENSOR REMOVAL	8	ECL	Environmental Ctl & Life Spt	2.04.12
580	OPF86	12-JUN-1997:9		POWER UP FROM: 86V000A - Note 1	12	OTC	Power Management	2.04.13
581	OPF86	12-JUN-1997:9	V33-10001G//V33-00005	POST FLIGHT WINDOW INSPECT FROM: 86V279 - Note 1	8	NDE	Cockpit & Crew Cabin	2.05.02
582	OPF86	12-JUN-1997:9	V9002.01	POSU'S B/F NULL	4	GNC	Propulsion	2.07.03
583	OPF86	12-JUN-1997:9	MPS-1288	PD3 INVESTIGATION FROM: 86V11D - Note 1	4	MPS	Propulsion	2.08.01
584	OPF86	12-JUN-1997:9	V5R01	F/C 3 REMOVAL/INSTALLATION(ELECTRICAL/MECH MATES/BUBBLE LK CHK/FUNCTIONAL CHKS) FROM: 86V5R01B - Note 1	12	FCP	Power Management	2.08.01
585	OPF86	12-JUN-1997:9		OMS LE TPS WORK	16	TPS	Thermal Management	2.08.05

No.	DATA SET	START TIME DD:MMM-YYYY:H	PROCEDURE NAME	DESCRIPTION	DURATION (Hours)	SYSTEM	DESIGN DISCIPLINE	GENERIC OPS FBS NUMBER (Appx A1)
586	OPF86	12-JUN-1997:9	IPR-0008	OME PLUG PULLTEST/R&R	2	OMS	Structures , Mechanisms, Veh Handling	2.08.08
587	OPF86	12-JUN-1997:9	IPR-0008	OME PLUG PULLTEST/R&R FROM: 86VIPR8 - Note 1	4	OMS	Structures , Mechanisms, Veh Handling	2.08.08
588	OPF86	12-JUN-1997:10	V5032.004	POSU 1-3-1.3 MERL- OFFLINE PYRO DISASSY & INVENTORY	7	PYR	Structures , Mechanisms, Veh Handling	2.03.02
589	OPF86	12-JUN-1997:13	V6054.003F	R/H OME NOZZLE INSPECT	4	NDE	Propulsion	2.05.03
590	OPF86	12-JUN-1997:13	V9002.01//GNC-125	POSITION B/F NULL/OPEN RSB/DPF TEST	4	HYD	Propulsion	2.07.03
591	OPF86	12-JUN-1997:13	HYD-0759	LEAK CHECK FROM: 86V11H - Note 1	4	HYD	Power Management	2.08.02
592	OPF86	12-JUN-1997:13	OEL-1371	MAIN C SENSOR R/R	12	OEL	Power Management	2.08.04
593	OPF86	12-JUN-1997:17	V5057	SEQ 12 REMOVE MID STOKE LOCKS	2	SME	Propulsion	2.02.04
594	OPF86	12-JUN-1997:17	V9021.001	OME TRICKLE PURGE ACTIVATION	4	OMS	Propulsion	2.04.08
595	OPF86	12-JUN-1997:17	V30-15368	VERT RSB INNER PNL INSP(RSB OPEN/WASH PNL)	24	NDE	Structures , Mechanisms, Veh Handling	2.05.02
596	OPF86	12-JUN-1997:17	V30-15358//V31-15109	EXTERNAL AFT DETAILED INSP FROM: 86V272T - Note 1	16	NDE	Structures , Mechanisms, Veh Handling	2.05.02
597	OPF86	12-JUN-1997:17	V30-15207/209//V31-15113/	VERT/RSB AREA INSP FROM: 86V272L - Note 1	24	NDE	Structures , Mechanisms, Veh Handling	2.05.02
598	OPF86	12-JUN-1997:17	V31-15115	VERT/RSB AREA INSP FROM: 86V272S - Note 1	8	QC	Structures , Mechanisms, Veh Handling	2.05.02
599	OPF86	12-JUN-1997:17	AFT-4894/95	AFT TE/BFLE TPS WORK FROM: 86V10X - Note 1	8	TPS	Thermal Management	2.08.05
600	OPF86	12-JUN-1997:21	V5R01	FC 3 REMOVAL/INSTALLATION COMPLETE		FCP	Power Management	2.08.01
601	OPF86	12-JUN-1997:21	V5R01	FC 3 REMOVAL/INSTALLATION COMPLETE		FCP	Power Management	2.08.01
602	OPF86	12-JUN-1997:21	STR-570	(CHIT J4990)TORQUE RE-CHECK OF B/F ACTUATOR ATTACH FITTING BOLTS	4	STR	Structures , Mechanisms, Veh Handling	2.08.06
603	OPF86	13-JUN-1997:9	V3519	TASK 9 POSU 1,2,3 GSE AMMONIA SAMPLING	8	ECL	Environmental Ctl & Life Spt	2.01.02
604	OPF86	13-JUN-1997:9	V5032.004	SEQ 02-04 OFFLINE PYRO DISASSY & INVENTORY (OSF)	8	PYR	Structures , Mechanisms, Veh Handling	2.03.02
605	OPF86	13-JUN-1997:9	V5012.003	SEQ 03,4 DRAG CHUTE MORTAR MECHANICAL INSTLN.	8	PYR	Structures , Mechanisms, Veh Handling	2.04.02
606	OPF86	13-JUN-1997:9	FWD-3846	ADP TPS WORK FROM: 86V11C - Note 1	8	TPS	Thermal Management	2.04.06
607	OPF86	13-JUN-1997:9	V1076	WCCS SERVICING FROM: 86V078 - Note 1	48	PVD	Safety Management & Control	2.05.01
608	OPF86	13-JUN-1997:9	COM-0235//COM-A0019	KU ANTENNA/INSPECTION/MR ID(NSLD)	8	COM	Communications	2.05.02
609	OPF86	13-JUN-1997:9	V30-13207//TES-0275	WINDOW 9 INSP/MOLD IMPRESSIONS	4	QC	Structures , Mechanisms, Veh Handling	2.05.02
610	OPF86	13-JUN-1997:9	V1180VL1	POSU 9 HARDWIRE SAFING CONFIG(VAB TECH TO WORK IN LCC)	2	OMS	Propulsion	2.05.03
611	OPF86	13-JUN-1997:9	V1009.003	MPS HE SYSTEM LK & FUNCT TEST	4	MPS	Propulsion	2.05.03
612	OPF86	13-JUN-1997:9	V1180VL1	POSU 2,3,4,5 GSE CB CONFIG	2	OMS	Propulsion	2.05.03
613	OPF86	13-JUN-1997:9	V1180VL1	POSU 8/10 APS SPILL PROTECTION/SCUPPER INSTL FROM: 86V204F - Note 1	8	OMS	Propulsion	2.05.03
614	OPF86	13-JUN-1997:9	V5K32.001//V80-01401/0140	G/F CP REMOVAL /POTTING REMOVAL	16	STR	Structures , Mechanisms, Veh Handling	2.08.01
615	OPF86	13-JUN-1997:9	STR-4586/87	ET CORROSION REPAIR	48	STR	Structures , Mechanisms, Veh Handling	2.08.01

No.	DATA SET	START TIME DD:MMM-YYYY:H	PROCEDURE NAME	DESCRIPTION	DURATION (Hours)	SYSTEM	DESIGN DISCIPLINE	GENERIC OPS FBS NUMBER (Appx A1)
616	OPF86	13-JUN-1997:9	V5R01	F/C 3 REMOVAL/INSTALLATION POI	8	FCP	Power Management	2.08.01
617	OPF86	13-JUN-1997:9	OEL-A0027	BKHD CONNECTOR INSPECTION FROM: 86V10U - Note 1	8	OEL	Power Management	2.08.04
618	OPF86	13-JUN-1997:9	OEL-1372	BROKEN WIRE T/S	16	OEL	Power Management	2.08.04
619	OPF86	13-JUN-1997:9	OEL-1371	MAIN C SENSOR R/R FROM: 86V11N - Note 1	8	OEL	Power Management	2.08.04
620	OPF86	13-JUN-1997:9		CRAWLER MOVE FROM CRAWLER PARKSITE TO MLP PARKSITE			Structures , Mechanisms, Veh Handling	4.01.02
621	OPF86	13-JUN-1997:13	V9021.001	SEQ 11/12 LRCS/RCRCS T/P DEACTIVATION	4	OMS	Propulsion	2.04.08
622	OPF86	13-JUN-1997:17		POWER UP FROM: 86V000A - Note 1	80	OTC	Power Management	2.04.13
623	OPF86	13-JUN-1997:17	V1180VL1P/VL2U/VL3I/VL4F	APS POD FUNCTIONAL C/O	80	OMS	Propulsion	2.05.03
624	OPF86	13-JUN-1997:17		OMS LE TPS WORK FROM: 86V11G - Note 1	32	TPS	Thermal Management	2.08.05
625	OPF86	13-JUN-1997:17		PADLEADER PAGE FOR OUTAGE			Structures , Mechanisms, Veh Handling	2.11.01
626	OPF86	14-JUN-1997:15	TFRC-009	R/R FRCS THRUSTER TILE	10	TPS	Thermal Management	2.08.05
627	OPF86	15-JUN-1997:1	V1275.001B	WATERPROOFING	12	TPS	Thermal Management	2.04.05
628	OPF86	15-JUN-1997:9		COM,ECL,EPD,INS,MEQ,OEL PROCESSING/HISTORICAL PWR DOWN PROCESSING(SAT. & SUN.) FROM: 86V350B - Note 1 FIXEDFLAG 2	4		Structures , Mechanisms, Veh Handling	2.08.04
629	OPF86	16-JUN-1997:9	PYR-A0008	UMB TORQUE CHECK	8	PYR	Structures , Mechanisms, Veh Handling	2.04.02
630	OPF86	16-JUN-1997:9	V6059.001A//.004	ELEVON COVE REWORK/C/P FROM: 86V198 - Note 1	24	TPS	Thermal Management	2.04.06
631	OPF86	16-JUN-1997:9	V5E29	GIMBAL BOLT R/R	32	SME	Propulsion	2.04.12
632	OPF86	16-JUN-1997:9	V1019VL1I1/VL2K/VL3K1	APU LK & FUNCTIONAL	8	APU	Power Management	2.05.01
633	OPF86	16-JUN-1997:9	V1093	POSU F/C SINGLE CELL VOLT TEST	24	FCP	Power Management	2.05.01
634	OPF86	16-JUN-1997:9	V6054.001E1/.002G/.003F	VERNIER INSP	8	NDE	Propulsion	2.05.03
635	OPF86	16-JUN-1997:9	V1009.003	MPS HE SYSTEM LK & FUNCT TEST FROM: 86V215C - Note 1	8	MPS	Propulsion	2.05.03
636	OPF86	16-JUN-1997:9	CM-A001//V63-50023//TCS-2	BOLT HOLE CONFIGURATION	32	PLO	Payload Accommodations	2.06.02
637	OPF86	16-JUN-1997:9	ECL-0954	A-HATCH LEAK CHECK	8	ECL	Thermal Management	2.08.02
638	OPF86	16-JUN-1997:9	V6051	ET DOOR T/B REWORK	8	TPS	Thermal Management	2.08.08
639	OPF86	16-JUN-1997:13	V1009.003	MPS HE SYSTEM LK & FUNCT TEST	2	MPS	Propulsion	2.05.03
640	OPF86	16-JUN-1997:13	STR-A0249	BAY 10 B/F DOUBLER BONDS	12	STR	Structures , Mechanisms, Veh Handling	2.08.06
641	OPF86	16-JUN-1997:17	V1019VL1I1/VL2K/VL3K1	APU LEAK & FUNCTIONAL	8	APU	Power Management	2.05.01
642	OPF86	16-JUN-1997:17	V30-15368	VERT RSB INNER PNL INSP(RSB OPEN/WASH PNL) FROM: 86V272 - Note 1	16	NDE	Structures , Mechanisms, Veh Handling	2.05.02

No.	DATA SET	START TIME DD:MMM-YYYY:H	PROCEDURE NAME	DESCRIPTION	DURATION (Hours)	SYSTEM	DESIGN DISCIPLINE	GENERIC OPS FBS NUMBER (Appx A1)
643	OPF86	16-JUN-1997:17	V30-15207/209//V31-15113/	VERT/RSB AREA INSP FROM: 86V272L - Note 1	24	NDE	Structures , Mechanisms, Veh Handling	2.05.02
644	OPF86	16-JUN-1997:17	V5K32.001//V80-01401/0140	G/F CP REMOVAL /POTTING REMOVAL FROM: 86V11Q - Note 1	8	STR	Structures , Mechanisms, Veh Handling	2.08.01
645	OPF86	16-JUN-1997:17	WWM-0034	CK VALVE RETORQUE	8	FCP	Power Management	2.08.07
646	OPF86	17-JUN-1997:9	V5012.003	SEQ 05 DRAG CHUTE ELECT. CONN	2	PYR	Structures , Mechanisms, Veh Handling	2.04.02
647	OPF86	17-JUN-1997:9	V1019VL1I1/VL2K/VL3K1	APU LEAK & FUNCTIONAL	12	APU	Power Management	2.05.01
648	OPF86	17-JUN-1997:9	V1058.006	'D' HATCH FUNCTIONAL VERIFICATION	8	MEQ	Structures , Mechanisms, Veh Handling	2.05.02
649	OPF86	17-JUN-1997:9	V30-15358//V31-15109	EXTERNAL AFT DETAILED INSP FROM: 86V272T - Note 1	8	NDE	Structures , Mechanisms, Veh Handling	2.05.02
650	OPF86	17-JUN-1997:9	V1009.003	MPS HE SYSTEM LK & FUNCT TEST FROM: 86V215C - Note 1	8	MPS	Propulsion	2.05.03
651	OPF86	17-JUN-1997:9	V5K32.001	WINDOW 1 R/R	42	STR	Structures , Mechanisms, Veh Handling	2.08.01
652	OPF86	17-JUN-1997:9	V6051	ET DOOR T/B REWORK FROM: 86V222B - Note 1	8	TPS	Thermal Management	2.08.08
653	OPF86	17-JUN-1997:11	V5012.003	SEQ 06 MORTAR & RETRACTOR CIR RESIST TEST	2	PYR	Structures , Mechanisms, Veh Handling	2.04.02
654	OPF86	17-JUN-1997:13	V3589	REMOVE TRUSS PICBOARDS	2	OHE	Structures , Mechanisms, Veh Handling	2.02.02
655	OPF86	17-JUN-1997:13	V5138.001	SEQ 04 DRAG CHUTE ALIGN & INSTALL INTO COMPARTMENT	4	OHE	Structures , Mechanisms, Veh Handling	3.03.03
656	OPF86	17-JUN-1997:15	V9048	ET DOORS TO 90	2	MEQ	Structures , Mechanisms, Veh Handling	2.07.03
657	OPF86	17-JUN-1997:17	V9046.004	PIN EVA SLIDEWIRES	1	PLO	Cockpit & Crew Cabin	2.04.04
658	OPF86	17-JUN-1997:17	V1093	F/C SINGLE CELL VOLT TEST	8	FCP	Power Management	2.05.01
659	OPF86	17-JUN-1997:17	V5K25.004//MEQ-0779/80/81	ET DOOR ENVIRONMENTAL SEAL R/R	8	MEQ	Thermal Management	2.08.08
660	OPF86	18-JUN-1997:9	FWD-3817/18/21/22/26/27	NLGD TPS TILE PROCESS/THERMAL BARRIERS FROM: 86V224B - Note 1	8	TPS	Thermal Management	2.04.06
661	OPF86	18-JUN-1997:9	V1093	F/C SINGLE CELL VOLT TEST	12	FCP	Power Management	2.05.01
662	OPF86	18-JUN-1997:9	V1086.001	POSU 2-7 FLT CABLE INSTL/V5012 POSU REM ACCESS PNL	24	EPD	Power Management	2.05.01
663	OPF86	18-JUN-1997:9	CE-193	AFD RECONFIG	24	PLO	Payload Accommodations	2.06.03
664	OPF86	18-JUN-1997:9	ECL-403	FES PLUG R/R	8	ECL	Environmental Ctl & Life Spt	2.08.01
665	OPF86	18-JUN-1997:9	FRC4-0474	F3D THRUSTER INSPECTION	4	OMS	Propulsion	2.08.01
666	OPF86	18-JUN-1997:9	IPR-0012	MD206 FLT CAP REMOVAL	8	OMS	Structures , Mechanisms, Veh Handling	2.08.08
667	OPF86	18-JUN-1997:9	V6051	ET DOOR T/B REWORK FROM: 86V222B - Note 1	8	TPS	Thermal Management	2.08.08
668	OPF86	18-JUN-1997:13	OEL-1371	MAIN C SENSOR R/R FROM: 86V11N - Note 1	12	OEL	Power Management	2.08.04
669	OPF86	18-JUN-1997:17	V6028.003	TPS POST FLT INSPECTION FROM: 86V191 - Note 1	48	TPS	Thermal Management	2.05.02
670	OPF86	18-JUN-1997:17	V5158	FLIGHT DOOR INSTALLATION	12	STR	Structures , Mechanisms, Veh Handling	2.10.02
671	OPF86	18-JUN-1997:21	V1093	PARTIAL OPEN/RESTORE CONTROLLED AREAS/GSE SECURING	4	FCP	Power Management	2.05.01

No.	DATA SET	START TIME DD:MMM-YYYY:H	PROCEDURE NAME	DESCRIPTION	DURATION (Hours)	SYSTEM	DESIGN DISCIPLINE	GENERIC OPS FBS NUMBER (Appx A1)
672	OPF86	18-JUN-1997:21	MEQ-0769	XRAY RH RAD 2	4	MEQ	Structures , Mechanisms, Veh Handling	2.05.02
673	OPF86	18-JUN-1997:21	V1015	POSU'S RADIATOR FUNCTIONAL	4	MEQ	Structures , Mechanisms, Veh Handling	2.05.02
674	OPF86	19-JUN-1997:9	FWD-3817/18/21/22/26/27	NLGD TPS TILE PROCESS/THERMAL BARRIERS FROM: 86V224B - Note 1	40	TPS	Thermal Management	2.04.06
675	OPF86	19-JUN-1997:9	V9021.001	OP 80,90 LRCS/RRCS T/P ACTIVATION	8	OMS	Propulsion	2.04.08
676	OPF86	19-JUN-1997:9	V1093	F/C SINGLE CELL VOLT TEST POI	4	FCP	Power Management	2.05.01
677	OPF86	19-JUN-1997:9	V1019VL1I1/VL2K/VL3K1	APU LEAK & FUNCTIONAL	4	APU	Power Management	2.05.01
678	OPF86	19-JUN-1997:9	V1022.001	POSU PRSD SYSTEM TEST	8	FCP	Power Management	2.05.01
679	OPF86	19-JUN-1997:9	V1062.001	POSU 1 AIR DATA SYSTEM FUNCT C/O	4	COM	Communications	2.05.01
680	OPF86	19-JUN-1997:9	V1015	POSU'S RADIATOR FUNCTIONAL(STRONGBACK/ZERO-G HOOKUP)	4	MEQ	Structures , Mechanisms, Veh Handling	2.05.02
681	OPF86	19-JUN-1997:9	V30-15368	VERT RSB INNER PNL INSP(RSB OPEN/WASH PNL) FROM: 86V272 - Note 1	8	NDE	Structures , Mechanisms, Veh Handling	2.05.02
682	OPF86	19-JUN-1997:9	V30-15207/209//V31-15113/	VERT/RSB AREA INSP FROM: 86V272L - Note 1	8	NDE	Structures , Mechanisms, Veh Handling	2.05.02
683	OPF86	19-JUN-1997:9	V1009.005	MPS LH2 SYSTEM LEAK & FUNCTIONAL TEST(CV13 X-RAY)	2	MPS	Propulsion	2.05.03
684	OPF86	19-JUN-1997:9	V1009.003	MPS HE SYSTEM LK & FUNCT TEST FROM: 86V215C - Note 1	4	MPS	Propulsion	2.05.03
685	OPF86	19-JUN-1997:9	APU-134	CONTROLLER C/O	4	APU	Power Management	2.08.01
686	OPF86	19-JUN-1997:9	MPS-1283	LH2 RECIRC MANIFOLD DENT DYE-PEN	8	NDE	Propulsion	2.08.01
687	OPF86	19-JUN-1997:9		TILE PROCESSING	256	TPS	Thermal Management	2.08.05
688	OPF86	19-JUN-1997:9	V7253	ORB WINDOW POLISHING FROM: 86V065 - Note 1	32	STR	Structures , Mechanisms, Veh Handling	2.08.06
689	OPF86	19-JUN-1997:9	V5K25.004//MEQ-0779/80/81	ET DOOR ENVIRONMENTAL SEAL R/R FROM: 86V15A - Note 1	16	MEQ	Thermal Management	2.08.08
690	OPF86	19-JUN-1997:13	V5138.001E//V5012.003D	INSTL CARRIER PNL/DRAG CHUTE P/F STOWAGE	12	TPS	Structures , Mechanisms, Veh Handling	2.04.05
691	OPF86	19-JUN-1997:13	V1015	RADIATOR FUNCTIONAL	8	MEQ	Structures , Mechanisms, Veh Handling	2.05.02
692	OPF86	19-JUN-1997:13	PLD-P001	PAINT CHIP REPAIR	4	PLO	Payload Accommodations	2.06.01
693	OPF86	19-JUN-1997:13	STR-4576/77	ELEVON HOIST PT FIT CHECK	4	STR	Structures , Mechanisms, Veh Handling	2.08.01
694	OPF86	19-JUN-1997:13	STR-4586/87	ET CORROSION REPAIR FROM: 86V272V - Note 1	28	STR	Structures , Mechanisms, Veh Handling	2.08.01
695	OPF86	19-JUN-1997:13	HYD-A0014	WSB 3 REG CYCLE	4	HYD	Power Management	2.08.02
696	OPF86	19-JUN-1997:21	V31-14195	P/L RADIATOR LOWER SURFACE INSP(WORK ONLY IF RADIATOR DEPLOYED)(NOT REQ'D)		QC	Structures , Mechanisms, Veh Handling	2.05.02
697	OPF86	19-JUN-1997:21	V1015	STOW RADIATORS/REMOVE RAD STRONGBACK	4	MEQ	Structures , Mechanisms, Veh Handling	2.05.02
698	OPF86	20-JUN-1997:9	V5126.001	WING TEMPILABELS INSTL	8	TPS	Thermal Management	2.04.05
699	OPF86	20-JUN-1997:9	TES-0275	WINDOW 9 MOLD IMPRESSIONS	2	NDE	Thermal Management	2.04.06

No.	DATA SET	START TIME DD:MMM-YYYY:H	PROCEDURE NAME	DESCRIPTION	DURATION (Hours)	SYSTEM	DESIGN DISCIPLINE	GENERIC OPS FBS NUMBER (Appx A1)
700	OPF86	20-JUN-1997:9	V6059.001A//.004	ELEVON COVE REWORK/C/P FROM: 86V198 - Note 1	32	TPS	Thermal Management	2.04.06
701	OPF86	20-JUN-1997:9	V1086.001	POSU 2-7 FLT CABLE INSTL/V5012 POSU REM ACCESS PNL FROM: 86V082A1 - Note 1	2	EPD	Power Management	2.05.01
702	OPF86	20-JUN-1997:9	V1062.001	POSU 1 AIR DATA SYSTEM FUNCT C/O FROM: 86V043A - Note 1	1	COM	Communications	2.05.01
703	OPF86	20-JUN-1997:9	V1019VL1I1/VL2K/VL3K 1	APU LEAK & FUNCTIONAL POI	4	APU	Power Management	2.05.01
704	OPF86	20-JUN-1997:9	V31-13121//V1058.005	SEAL SURF INSP/'C' HATCH FUNCTIONAL VERIFICATION	8	MEQ	Structures , Mechanisms, Veh Handling	2.05.02
705	OPF86	20-JUN-1997:9	V1015	INSTL RAD COVERS	2	MEQ	Structures , Mechanisms, Veh Handling	2.05.02
706	OPF86	20-JUN-1997:9	V1009.003	MPS HE SYSTEM LK & FUNCT TEST FROM: 86V215C - Note 1	8	MPS	Propulsion	2.05.03
707	OPF86	20-JUN-1997:9	HYD-A0014	WSB 3 REG CYCLE	16	HYD	Power Management	2.08.02
708	OPF86	20-JUN-1997:9	STR-A0249	BAY 10 B/F DOUBLER BONDS FROM: 86V37 - Note 1	64	STR	Structures , Mechanisms, Veh Handling	2.08.06
709	OPF86	20-JUN-1997:9	V6051	ET DOOR T/B REWORK FROM: 86V222B - Note 1	16	TPS	Thermal Management	2.08.08
710	OPF86	20-JUN-1997:10	V1062.001D1/IPR-0017	AIR DATA SYSTEM CK OUT	7	COM	Communications	2.05.01
711	OPF86	20-JUN-1997:11	V1086.001G1/.002/.004A 1/.	MEC/PIC TEST	6	EPD	Power Management	2.05.01
712	OPF86	20-JUN-1997:13	V9046.003	DEPIN EVA SLIDEWIRES	1	PLO	Cockpit & Crew Cabin	2.04.04
713	OPF86	20-JUN-1997:13	PLD-T001	PAINT CHIP REPAIR	4	PLO	Payload Accommodations	2.06.03
714	OPF86	20-JUN-1997:17	OEL-1372	BROKEN WIRE T/S FROM: 86V11O - Note 1	24	OEL	Power Management	2.08.04
715	OPF86	20-JUN-1997:17	OEL-1371	MAIN C SENSOR R/R FROM: 86V11N - Note 1	24	OEL	Power Management	2.08.04
716	OPF86	20-JUN-1997:17		OMS LE TPS WORK FROM: 86V11G - Note 1	40	TPS	Thermal Management	2.08.05
717	OPF86	20-JUN-1997:21	CM-668	ICAPC INSTALLATION(BAY 6 STBD)	4	PLO	Payload Accommodations	2.06.03
718	OPF86	22-JUN-1997:1	V1275.001	LOWER SURFACE/AFT SIDEWALL/VERT WATERPROOFING	10	TPS	Thermal Management	2.04.05
719	OPF86	23-JUN-1997:9	V6005.001	POSU 1 STARTRACKER CLEAN ROOM SETUP	8	COM	Communications	2.02.04
720	OPF86	23-JUN-1997:9		POWER UP FROM: 86V000A - Note 1	40	OTC	Power Management	2.04.13
721	OPF86	23-JUN-1997:9	V1086.001G1/.002/.004A 1/.	MEC/PIC TEST FROM: 86V082 - Note 1	8	EPD	Power Management	2.05.01
722	OPF86	23-JUN-1997:9	V1022.001	PRSD SYSTEM TEST/POI	24	FCP	Power Management	2.05.01
723	OPF86	23-JUN-1997:9	STR-568	HATCH WINDOW INSPECTION	8	QC	Structures , Mechanisms, Veh Handling	2.05.02
724	OPF86	23-JUN-1997:9	V1180VL1P/VL2U/VL3I/V L4F	APS POD FUNCTIONAL C/O FROM: 86V047 - Note 1	42	OMS	Propulsion	2.05.03
725	OPF86	23-JUN-1997:9	V1009.003	MPS HE SYSTEM LK & FUNCT TEST FROM: 86V215C - Note 1	8	MPS	Propulsion	2.05.03
726	OPF86	23-JUN-1997:9	CM-673	ICAPC INSTALLATION(BAY 6 PORT)	8	PLO	Payload Accommodations	2.06.03
727	OPF86	23-JUN-1997:9		INSTL ODS/TRUSS PLTFRM PIC BOARD	2	OHE	Structures , Mechanisms, Veh Handling	2.06.06
728	OPF86	23-JUN-1997:9	OEL-1371	MAIN C SENSOR RETEST	4	OEL	Power Management	2.08.04

No.	DATA SET	START TIME DD:MMM-YYYY:H	PROCEDURE NAME	DESCRIPTION	DURATION (Hours)	SYSTEM	DESIGN DISCIPLINE	GENERIC OPS FBS NUMBER (Appx A1)
729	OPF86	23-JUN-1997:9	V7253	ORB WINDOW POLISH	96	STR	Structures , Mechanisms, Veh Handling	2.08.06
730	OPF86	23-JUN-1997:9	V5K25.004//MEQ-0779/80/81	ET DOOR ENVIRONMENTAL SEAL R/R FROM: 86V15A - Note 1	8	MEQ	Thermal Management	2.08.08
731	OPF86	23-JUN-1997:9	S70-1201-02-003-0008	FILTER R/R	8	ECL	Propulsion	2.08.09
732	OPF86	23-JUN-1997:9	STR-XXXX	(CHIT J5005)RCC LAUNCH PAD CONTAMINATION(WORKED AT PAD)		STR	Structures , Mechanisms, Veh Handling	2.09.02
733	OPF86	23-JUN-1997:13	STR-4586/87	ET CORROSION REPAIR FROM: 86V272V - Note 1	28	STR	Structures , Mechanisms, Veh Handling	2.08.01
734	OPF86	24-JUN-1997:9	TES-0275	WINDOW 9 MOLD IMPRESSIONS FROM: 86V270K - Note 1	2	NDE	Thermal Management	2.04.06
735	OPF86	24-JUN-1997:9	V1037	POSU'S TASK 1 OR 3 NH3 SERVICING (PENDING FLT US-AGE)(START 7 DAYS PRIOR TO V1037 CTS)	8	ECL	Thermal Management	2.04.08
736	OPF86	24-JUN-1997:9	V1086.001G1/.002/.004A 1/.	MEC/PIC TEST FROM: 86V082 - Note 1	8	EPD	Power Management	2.05.01
737	OPF86	24-JUN-1997:9	V1076	WCCS SERVICING FROM: 86V078 - Note 1	32	PVD	Safety Management & Control	2.05.01
738	OPF86	24-JUN-1997:9	V6005.001	STARTRACKER LIGHTSHADE POST FLT INSPECTION	6	COM	Communications	2.05.01
739	OPF86	24-JUN-1997:9	V6028.003	TPS POST FLT INSPECTION FROM: 86V191 - Note 1	8	TPS	Thermal Management	2.05.02
740	OPF86	24-JUN-1997:9	V5067	CLOSE AFT HATCH	8	MEQ	Structures , Mechanisms, Veh Handling	2.07.03
741	OPF86	24-JUN-1997:9	FCP-XXX	FLIGHT CAP R/R	4	FCP	Power Management	2.08.01
742	OPF86	24-JUN-1997:9	V6051	ET DOOR T/B REWORK FROM: 86V222B - Note 1	8	TPS	Thermal Management	2.08.08
743	OPF86	24-JUN-1997:13	V1314	OP 10 POSITION ALL VENT DOORS	4	PVD	Safety Management & Control	2.05.01
744	OPF86	24-JUN-1997:13	V1314	POSITION LH VENT DR TO PURGE	4	PVD	Safety Management & Control	2.05.01
745	OPF86	24-JUN-1997:17	V5K32.001	WINDOW 1 R/R FROM: 86V11R - Note 1	8	STR	Structures , Mechanisms, Veh Handling	2.08.01
746	OPF86	24-JUN-1997:17	GNC-0152/53//V5B03.001J/.	ADTA 1/4 REMOVAL	8	GNC	Guid, Nav & Ctl	2.08.01
747	OPF86	25-JUN-1997:9	FWD-3817/18/21/22/26/27	NLGD TPS TILE PROCESS/THERMAL BARRIERS FROM: 86V224B - Note 1	8	TPS	Thermal Management	2.04.06
748	OPF86	25-JUN-1997:9	V1053.002	CABIN PRESSURE XDUCER C/O AMBIENT	6	ECL	Environmental Ctl & Life Spt	2.05.01
749	OPF86	25-JUN-1997:9	V30-15358//V31-15109	EXTERNAL AFT DETAILED INSP FROM: 86V272T - Note 1	8	NDE	Structures , Mechanisms, Veh Handling	2.05.02
750	OPF86	25-JUN-1997:9	V30/31	RADIATOR/BULB SEAL INSPECTIONS/ANOMALY REWORK(NOT REQUIRED 6/18/97)		MEQ	Structures , Mechanisms, Veh Handling	2.05.02
751	OPF86	25-JUN-1997:9	COM-0235//COM-A0019	KU ANTENNA/INSPECTION/MR ID(NSLD) FROM: 86V39 - Note 1	8	COM	Communications	2.05.02
752	OPF86	25-JUN-1997:9	V6028.003	TPS POST FLT INSPECTION FROM: 86V191 - Note 1	8	TPS	Thermal Management	2.05.02
753	OPF86	25-JUN-1997:9	V6051	ET DOOR T/B REWORK FROM: 86V222B - Note 1	16	TPS	Thermal Management	2.08.08
754	OPF86	25-JUN-1997:11	MEQ-231	(CHIT J5047)ET DOOR LOAD TEST FROM: 86V10V - Note 1	6	MEQ	Structures , Mechanisms, Veh Handling	2.02.04
755	OPF86	25-JUN-1997:11	STR-581//STR-4629//STR-46	(MCR 18866)PAYLOAD BAY SIDEWALL CARRIER MAIN FRAME ATTACH POINT CLIP CHANGEOUT	6	STR	Structures , Mechanisms, Veh Handling	2.06.08
756	OPF86	25-JUN-1997:11	V63-50016/V34-00008	REMOVE WING BLKS/INST STEP-UP	1	TCS	Thermal Management	2.06.09

No.	DATA SET	START TIME DD:MMM-YYYY:H	PROCEDURE NAME	DESCRIPTION	DURATION (Hours)	SYSTEM	DESIGN DISCIPLINE	GENERIC OPS FBS NUMBER (Appx A1)
757	OPF86	25-JUN-1997:12	V63-50016/V34-00008	REMOVE WING BLKS/INST STEP-UP	5	TCS	Thermal Management	2.06.09
758	OPF86	25-JUN-1997:13	IPR-0020	STARTRACKER T/S	4	COM	Structures , Mechanisms, Veh Handling	2.08.08
759	OPF86	25-JUN-1997:15	V1053.002	CABIN PRESSURE XDUCER C/O AMBIENT	2	ECL	Environmental Ctl & Life Spt	2.05.01
760	OPF86	25-JUN-1997:17	VITT-063	EPS SHARP EDGE WSP	2	FCS	Cockpit & Crew Cabin	2.04.04
761	OPF86	25-JUN-1997:17	DDC-0130	R/R FLOODLIGHT 7	8	DPS	Power Management	2.08.01
762	OPF86	25-JUN-1997:23	MEQ-0767	LH RAD 4 XRAY	2	NDE	Structures , Mechanisms, Veh Handling	2.05.02
763	OPF86	26-JUN-1997:1	V1275.001	C/P'S WATERPROOFING	8	TPS	Thermal Management	2.04.06
764	OPF86	26-JUN-1997:9	V1009.001	POSU TENT BUILDUP FOR MPS 17 INCH DISC INSPECT	2	MPS	Propulsion	2.02.04
765	OPF86	26-JUN-1997:9	FWD-3805/6/7//FWD-A0106	CREW HATCH T/B R/R	32	TPS	Thermal Management	2.04.06
766	OPF86	26-JUN-1997:9	FWD-3817/18/21/22/26/27	NLGD TPS TILE PROCESS/THERMAL BARRIERS FROM: 86V224B - Note 1	120	TPS	Thermal Management	2.04.06
767	OPF86	26-JUN-1997:9		POWER UP FROM: 86V000A - Note 1	8	OTC	Power Management	2.04.13
768	OPF86	26-JUN-1997:9	V1022.001	POI'S PRSD SYSTEM TEST	8	FCP	Power Management	2.05.01
769	OPF86	26-JUN-1997:9	V1053.002	CABIN PRESSURE XDUCER C/O AMBIENT FROM: 86V110 - Note 1	8	ECL	Environmental Ctl & Life Spt	2.05.01
770	OPF86	26-JUN-1997:9	V6028.003	TPS POST FLT INSPECTION FROM: 86V191 - Note 1	8	TPS	Thermal Management	2.05.02
771	OPF86	26-JUN-1997:9	CE-193	AFD RECONFIG FROM: 86VC193 - Note 1	8	PLO	Payload Accommodations	2.06.03
772	OPF86	26-JUN-1997:9	CM-665	INSTALL EPS(CUSTOMER REQ'D)	6	PLO	Payload Accommodations	2.06.03
773	OPF86	26-JUN-1997:9	V5067	OPEN UPPER HATCH	2	MEQ	Structures , Mechanisms, Veh Handling	2.07.03
774	OPF86	26-JUN-1997:9	V5K32.001	WINDOW 1 R/R FROM: 86V11R - Note 1	8	STR	Structures , Mechanisms, Veh Handling	2.08.01
775	OPF86	26-JUN-1997:9	TFRC-009//TFRC-0809/0805	(MCR-18563)FRCS THRUSTER TILE PATTERN FAB FROM: 86V10M - Note 1	126	TPS	Thermal Management	2.08.05
776	OPF86	26-JUN-1997:9	LWNG-3031/3058/59/LWNG-33	LESS -088 TILE BOND	56	TPS	Thermal Management	2.08.05
777	OPF86	26-JUN-1997:9	STR-A0249	BAY 10 B/F DOUBLER BONDS FROM: 86V37 - Note 1	16	STR	Structures , Mechanisms, Veh Handling	2.08.06
778	OPF86	26-JUN-1997:9	IPR-0020	STARTRACKER T/S FROM: 86V13G - Note 1	8	COM	Structures , Mechanisms, Veh Handling	2.08.08
779	OPF86	26-JUN-1997:11	V1009.001	MPS 17 INCH DISCONNECT INSPECTIONS	6	MPS	Propulsion	2.05.02
780	OPF86	26-JUN-1997:11	STR-A0263	CLEAN UPPER HATCH WINDOW	2	MEQ	Structures , Mechanisms, Veh Handling	2.08.06
781	OPF86	26-JUN-1997:13		PLB ELEC/MECH &AFD CONFIGURE(SUMMARY BAR)	160	PLO	Payload Accommodations	2.06.05
782	OPF86	26-JUN-1997:13	STR-581//STR-4629//STR-46	(MCR 18866)PAYLOAD BAY SIDEWALL CARRIER MAIN FRAME ATTACH POINT CLIP CHANGEOUT FROM: 86V18866 - Note 1	12	STR	Structures , Mechanisms, Veh Handling	2.06.08
783	OPF86	26-JUN-1997:15	CE-194	EPS MATES	2	PLO	Payload Accommodations	2.06.05
784	OPF86	26-JUN-1997:17	V5138.001E//V5012.003D	INSTL CARRIER PNL/DRAG CHUTE P/F STOWAGE FROM: 86V004E - Note 1	8	TPS	Structures , Mechanisms, Veh Handling	2.04.05
785	OPF86	26-JUN-1997:17	PT-041	INSTL TARGET/LOCKOUT CRANES	4	PLO	Payload Accommodations	2.06.06

No.	DATA SET	START TIME DD:MMM-YYYY:H	PROCEDURE NAME	DESCRIPTION	DURATION (Hours)	SYSTEM	DESIGN DISCIPLINE	GENERIC OPS FBS NUMBER (Appx A1)
786	OPF86	26-JUN-1997:17	DDC-0130	R/R FLOODLIGHT 7 FROM: 86V13F - Note 1	4	DPS	Power Management	2.08.01
787	OPF86	26-JUN-1997:17	RWNG-3611	LOWER LESS #10 TILE R&R	24	TPS	Thermal Management	2.08.05
788	OPF86	26-JUN-1997:17	V1017	POSU'S WSB CHECKOUT & SERVICING	8	HYD	Power Management	2.09.02
789	OPF86	27-JUN-1997:1	OPT-039	EPS OPTICAL MEASUREMENTS	8	PLO	Payload Accommodations	2.06.07
790	OPF86	27-JUN-1997:9	V5126.001	AFT TEMPILABELS INSTL	8	TPS	Thermal Management	2.04.05
791	OPF86	27-JUN-1997:9		POWER UP FROM: 86V000A - Note 1	8	OTC	Power Management	2.04.13
792	OPF86	27-JUN-1997:9	V1022.001	POI'S PRSD SYSTEM TEST FROM: 86V064Z - Note 1	4	FCP	Power Management	2.05.01
793	OPF86	27-JUN-1997:9	V1009.001	MPS 17 INCH DISCONNECT INSPECTIONS FROM: 86V215B - Note 1	4	MPS	Propulsion	2.05.02
794	OPF86	27-JUN-1997:9	PLD-T100//CM-667	INSTALL CATERPILLAR BLANKET	4	PLO	Payload Accommodations	2.06.03
795	OPF86	27-JUN-1997:9	OPT-039	EPS OPTICAL MEASUREMENTS	2	PLO	Payload Accommodations	2.06.07
796	OPF86	27-JUN-1997:9	V9002.05/V9002.01	POSITION B/F DWN/CLOSE RSB/ET ACTUATOR CYCLING	18	HYD	Propulsion	2.07.03
797	OPF86	27-JUN-1997:9	V5K32.001	WINDOW 1 R/R FROM: 86V11R - Note 1	8	STR	Structures , Mechanisms, Veh Handling	2.08.01
798	OPF86	27-JUN-1997:9	MPS-1281	PD3 FLAPPER SEAL INSP	8	MPS	Propulsion	2.08.01
799	OPF86	27-JUN-1997:9	OEL-1373	WIRE HARNESS C/O(PWR DWN)	16	OEL	Power Management	2.08.04
800	OPF86	27-JUN-1997:9	STR-A0249	BAY 10 B/F DOUBLER BONDS FROM: 86V37 - Note 1	16	STR	Structures , Mechanisms, Veh Handling	2.08.06
801	OPF86	27-JUN-1997:9	V5005	POSU'S SSME INSTALLATION	16	SME	Propulsion	3.03.01
802	OPF86	27-JUN-1997:9	PT-041	EPS CHECKOUT	4	PLO	Payload Accommodations	4.04.05
803	OPF86	27-JUN-1997:13		G2/S2/185/22/24(PT-041)		PLO	Structures , Mechanisms, Veh Handling	2.06.05
804	OPF86	27-JUN-1997:13	CE-0427/8	SPLIT KAPTON/WIRE REPAIR	4	PLO	Payload Accommodations	2.06.08
805	OPF86	27-JUN-1997:13	STR-581//STR-4629//STR-46	(MCR 18866)PAYLOAD BAY SIDEWALL CARRIER MAIN FRAME ATTACH POINT CLIP CHANGEOUT FROM: 86V18866 - Note 1	28	STR	Structures , Mechanisms, Veh Handling	2.06.08
806	OPF86	27-JUN-1997:13	PT-041	EPS CHECKOUT	4	PLO	Payload Accommodations	4.04.05
807	OPF86	27-JUN-1997:17	V6018.001	CABIN AIR RECIRC INSP/MAINT (PWR DWN REQD)	8	PLO	Payload Accommodations	2.04.04
808	OPF86	27-JUN-1997:17	V1291.001	CABIN POS PRESS RV OPS/NEG PRESS RV INSPT/AIR RECIRC SCREEN CLEANING (REQ PWR DWN)	4	ECL	Environmental Ctl & Life Spt	2.05.01
809	OPF86	27-JUN-1997:17	DDC-0130//STR-4630	R/R FLOODLIGHT	8	DPS	Power Management	2.08.01
810	OPF86	27-JUN-1997:17	IPR-0021	FC2 T/S	8	FCP	Structures , Mechanisms, Veh Handling	2.08.08
811	OPF86	28-JUN-1997:7	OEL-1373	WIRE HARNESS C/O(PWR DWN) FROM: 86V13W - Note 1	18	OEL	Power Management	2.08.04
812	OPF86	28-JUN-1997:7	STR-A0249	BAY 10 B/F DOUBLER BONDS FROM: 86V37 - Note 1	18	STR	Structures , Mechanisms, Veh Handling	2.08.06
813	OPF86	28-JUN-1997:7	V5005	POSU'S SSME INSTALLATION FROM: 86V119A - Note 1	38	SME	Propulsion	3.03.01
814	OPF86	28-JUN-1997:9	CM-672	EXTENDED APC INSTL INSTALLATION(BAY 2 PORT)	8	PLO	Payload Accommodations	2.06.03
815	OPF86	28-JUN-1997:9	STR-4626	RSB CLIP REPAIR	8	STR	Structures , Mechanisms, Veh Handling	2.08.01

No.	DATA SET	START TIME DD:MMM-YYYY:H	PROCEDURE NAME	DESCRIPTION	DURATION (Hours)	SYSTEM	DESIGN DISCIPLINE	GENERIC OPS FBS NUMBER (Appx A1)
816	OPF86	28-JUN-1997:9	V1017	POSU'S WSB CHECKOUT & SERVICING FROM: 86V051A - Note 1	8	HYD	Power Management	2.09.02
817	OPF86	29-JUN-1997:9	V6051	ET DOOR T/B REWORK FROM: 86V222B - Note 1	8	TPS	Thermal Management	2.08.08
818	OPF86	30-JUN-1997:7	V1017	POSU'S WSB CHECKOUT & SERVICING FROM: 86V051A - Note 1	2	HYD	Power Management	2.09.02
819	OPF86	30-JUN-1997:9	V5126.001	AFT TEMPILABELS INSTL FROM: 86V12U - Note 1	8	TPS	Thermal Management	2.04.05
820	OPF86	30-JUN-1997:9	V1037	POSU'S TASK 2 NH3 SERVICING (PENDING FLT USAGE)	38	ECL	Thermal Management	2.04.08
821	OPF86	30-JUN-1997:9	V1179.012	MADS I/F VERIFICATION	5	INS	Cmd, Ctl & Health Mngmt	2.05.01
822	OPF86	30-JUN-1997:9	V1291.001	CABIN POS PRESS RV OPS/NEG PRESS RV INSPT/AIR RECIRC SCREEN CLEANING (REQ PWR DWN) FROM: 86V255 - Note 1	4	ECL	Environmental Ctl & Life Spt	2.05.01
823	OPF86	30-JUN-1997:9	V31-15113	RSB ZONAL/CLOSE OUT INSP	2	NDE	Structures , Mechanisms, Veh Handling	2.05.02
824	OPF86	30-JUN-1997:9	V31-15105/114	AFT AV BAY INSP(CONT AT THE PAD)		QC	Structures , Mechanisms, Veh Handling	2.05.02
825	OPF86	30-JUN-1997:9	V9019	MPS VJ LINE CHECKS(SEQ 1 EVERY 30 DAYS)(DELETED NOT REQ'D PER J. KEATLEY)		MPS	Propulsion	2.05.03
826	OPF86	30-JUN-1997:9	V1028.001	DEDICATED DISPLAY & DISPLAY DRIVER UNIT C/O	4	DPS	Cmd, Ctl & Health Mngmt	2.05.04
827	OPF86	30-JUN-1997:9	V9048	ET DOOR TO 90	2	MEQ	Structures , Mechanisms, Veh Handling	2.07.03
828	OPF86	30-JUN-1997:9	V5K32.001	WINDOW 1 R/R FROM: 86V11R - Note 1	8	STR	Structures , Mechanisms, Veh Handling	2.08.01
829	OPF86	30-JUN-1997:9	TCS-1224//CM-666//CM-0343	EXTENDED APC INSTALLATION(BAY 2 STBD)	16	PLO	Thermal Management	2.08.05
830	OPF86	30-JUN-1997:9	V1017	WSB CHECKOUT & SERVICING	8	HYD	Power Management	2.09.02
831	OPF86	30-JUN-1997:11	V9002.05/V9002.01	POSITION B/F DWN/CLOSE RSB/ELEVON FUUL-UP/ET ACTUATOR CYCLING	2	HYD	Propulsion	2.07.03
832	OPF86	30-JUN-1997:11	V6037	ORB TPS-ROLLOUT INSP	2	TPS	Thermal Management	2.10.02
833	OPF86	30-JUN-1997:13	MEQ-231	(CHIT J5047)ET DOOR LOAD TEST FROM: 86V10V - Note 1	4	MEQ	Structures , Mechanisms, Veh Handling	2.02.04
834	OPF86	30-JUN-1997:17	MEQ-231	MOD 5(MCR 19023) ET DOOR BOLT MOD	8	MEQ	Structures , Mechanisms, Veh Handling	2.02.04
835	OPF86	30-JUN-1997:17	V9048	ET DOOR TO 90	2	MEQ	Structures , Mechanisms, Veh Handling	2.07.03
836	OPF86	30-JUN-1997:17	DDC-0130//STR-4630	R/R FLOODLIGHT FROM: 86V13K - Note 1	4	DPS	Power Management	2.08.01
837	OPF86	30-JUN-1997:17	STR-A0249	BAY 10 B/F DOUBLER BONDS FROM: 86V37 - Note 1	8	STR	Structures , Mechanisms, Veh Handling	2.08.06
838	OPF86	30-JUN-1997:17	V3599	ORBITER ALARM TEST	8	GSE	Ground Systems & Facilities	2.11.01
839	OPF86	01-JUL-1997:6	V6018.001	CABIN AIR RECIRC INSP/MAINT (PWR DWN REQD) FROM: 86V163 - Note 1	18	PLO	Payload Accommodations	2.04.04
840	OPF86	01-JUL-1997:6	V6018.002	CABIN AIR RECIRC. INSP/MAINT.(PWR UP OR DWN)(NIB)	10	PLO	Payload Accommodations	2.04.04
841	OPF86	01-JUL-1997:6	FWD-3818/21/26/27	NLGD TPS PATTERN FABS/THERMAL BARRIERS	3	TPS	Thermal Management	2.04.06
842	OPF86	01-JUL-1997:6	V1076	WCCS SERVICING FROM: 86V078 - Note 1	11	PVD	Safety Management & Control	2.05.01
843	OPF86	01-JUL-1997:6	LWNG-330/331	LESS TILE R&R	3	TPS	Thermal Management	2.08.05
844	OPF86	01-JUL-1997:6	STR-A0251	(CHIT J5013)WING BOLT TORQUE CHKS FROM: 86V251 - Note 1	10	STR	Structures , Mechanisms, Veh Handling	2.08.06

No.	DATA SET	START TIME DD:MMM-YYYY:H	PROCEDURE NAME	DESCRIPTION	DURATION (Hours)	SYSTEM	DESIGN DISCIPLINE	GENERIC OPS FBS NUMBER (Appx A1)
845	OPF86	01-JUL-1997:6	STR-A0249	BAY 10 B/F DOUBLER BONDS FROM: 86V37 - Note 1	19	STR	Structures , Mechanisms, Veh Handling	2.08.06
846	OPF86	01-JUL-1997:9	V30-14437	AFT HOIST SUPPORT FITTING	8	NDE	Ground Systems & Facilities	2.02.02
847	OPF86	01-JUL-1997:9	V5012.004	POSU 1-3-1.3 MERL-FWD SEP ASSY & NLG STRUT THRUSTER INSTALL/CONN	6	PYR	Structures , Mechanisms, Veh Handling	2.04.02
848	OPF86	01-JUL-1997:9	V1291.002	ECL TURNAROUND AFT/MID OPERATIONS	16	ECL	Thermal Management	2.05.01
849	OPF86	01-JUL-1997:9	EPD-294	POSU'S ODS PYRO SYSTEM VERIFICATION	16	EPD	Power Management	2.05.01
850	OPF86	01-JUL-1997:9	V30-15358//V31-15109	EXTERNAL AFT DETAILED INSP FROM: 86V272T - Note 1	2	NDE	Structures , Mechanisms, Veh Handling	2.05.02
851	OPF86	01-JUL-1997:9	CM-679	DESTOW PSA	2	PLO	Payload Accommodations	2.06.03
852	OPF86	01-JUL-1997:9	STR-4621/22	ET DR CORROSION REPAIR(SANDING)	16	STR	Structures , Mechanisms, Veh Handling	2.08.01
853	OPF86	01-JUL-1997:9	ECL-AXXXX	RE-ID FILTER ASSY(NOT REQ'D)		ECL	Thermal Management	2.08.01
854	OPF86	01-JUL-1997:9	V1017	WSB CHECKOUT & SERVICING FROM: 86V051 - Note 1	8	HYD	Power Management	2.09.02
855	OPF86	01-JUL-1997:9	V5005	POSU'S SSME INSTALLATION FROM: 86V119A - Note 1	16	SME	Propulsion	3.03.01
856	OPF86	01-JUL-1997:10	PLD-T100//CM-667	INSTALL CATERPILLAR BLANKET FROM: 86V13E - Note 1	2	PLO	Payload Accommodations	2.06.03
857	OPF86	01-JUL-1997:13	STR-581//STR-4629//STR-46	(MCR 18866)PAYLOAD BAY SIDEWALL CARRIER MAIN FRAME ATTACH POINT CLIP CHANGEOUT FROM: 86V18866 - Note 1	4	STR	Structures , Mechanisms, Veh Handling	2.06.08
858	OPF86	01-JUL-1997:13	V5138.001E//V5012.003D	REMOVE GSE FROM PLATFORM/LWR TO FLOOR	1	OHE	Structures , Mechanisms, Veh Handling	3.03.03
859	OPF86	01-JUL-1997:16	V9002	RECONFIGURE TO DEDICATED	2	HYD	Propulsion	2.07.03
860	OPF86	01-JUL-1997:17	V5K32.001	WINDOW 1 C/P	8	TPS	Cockpit & Crew Cabin	2.04.04
861	OPF86	01-JUL-1997:17	V9021.001	OME TRICKLE PURGE DEACTIVATION	4	OMS	Propulsion	2.04.08
862	OPF86	01-JUL-1997:17	MPS-1283	EPOXY LINE REPAIR	8	MPS	Propulsion	2.08.01
863	OPF86	01-JUL-1997:17	RWNG-3611	LOWER LESS #10 TILE R&R FROM: 86V223E - Note 1	23	TPS	Thermal Management	2.08.05
864	OPF86	01-JUL-1997:17	V1017	WSB 1,2,3 CORE FILL	8	HYD	Power Management	2.09.02
865	OPF86	01-JUL-1997:21	V1026.003	WCS INSTALL	12	FCP	Environmental Ctl & Life Spt	2.04.04
866	OPF86	02-JUL-1997:1	ECL-401	FES PURGE FROM: 86V284B - Note 1	492	ECL	Environmental Ctl & Life Spt	2.07.01
867	OPF86	02-JUL-1997:1		SSME HYSTER AVAIL		SME	Structures , Mechanisms, Veh Handling	3.08.02
868	OPF86	02-JUL-1997:9	V1158VL2X//IPR-0023	BALL VALVE LEAK CHECK	16	OMS	Propulsion	2.03.01
869	OPF86	02-JUL-1997:9	V5012.004	SEQ 03 FWD SEP BOLT ASSEMBLY INSTALLATION	4	PYR	Structures , Mechanisms, Veh Handling	2.04.02
870	OPF86	02-JUL-1997:9		POWER UP FROM: 86V000A - Note 1	88	OTC	Power Management	2.04.13
871	OPF86	02-JUL-1997:9	V1076	WCCS FUNCTIONAL POI'S	8	PVD	Safety Management & Control	2.05.01
872	OPF86	02-JUL-1997:9	EPD-294	ODS PYRO SYSTEM VERIFICATION	26	EPD	Power Management	2.05.01
873	OPF86	02-JUL-1997:9	CM-669//STR-4631/33	BRACKET INSTALLATION	4	PLO	Payload Accommodations	2.06.02
874	OPF86	02-JUL-1997:9	V02-70004	REMOVE PSA	8	PLO	Payload Accommodations	2.06.03

No.	DATA SET	START TIME DD:MMM-YYYY:H	PROCEDURE NAME	DESCRIPTION	DURATION (Hours)	SYSTEM	DESIGN DISCIPLINE	GENERIC OPS FBS NUMBER (Appx A1)
875	OPF86	02-JUL-1997:9	V9048	CYCLE ET DOORS	8	MEQ	Structures , Mechanisms, Veh Handling	2.07.03
876	OPF86	02-JUL-1997:9	TCS-1224//CM-666//CM-0343	EXTENDED APC INSTALLATION(BAY 2 STBD) FROM: 86V18S - Note 1	12	PLO	Thermal Management	2.08.05
877	OPF86	02-JUL-1997:9	AFT-XXX	TILE REPAIRS BASE HEATSHIELD	8	TPS	Thermal Management	2.08.05
878	OPF86	02-JUL-1997:9	V7253	ORB WINDOW POLISH FROM: 86V065B - Note 1	48	STR	Structures , Mechanisms, Veh Handling	2.08.06
879	OPF86	02-JUL-1997:9	STR-A0249	BAY 10 B/F DOUBLER BONDS FROM: 86V37 - Note 1	16	STR	Structures , Mechanisms, Veh Handling	2.08.06
880	OPF86	02-JUL-1997:9	V6051	ET DOOR T/B REWORK	8	TPS	Thermal Management	2.08.08
881	OPF86	02-JUL-1997:9	V1017	WSB POST SERV GN2 REG CREEP TEST(AFTER 48 HR DECAY CKS)	8	HYD	Power Management	2.09.02
882	OPF86	02-JUL-1997:13	V63-50023	BAY 13 LINER	2	PLO	Thermal Management	2.06.09
883	OPF86	02-JUL-1997:13	GNC-0152/53//V5B03.001J/.	ADTA 1/4 INSTALLATION	20	GNC	Guid, Nav & Ctl	2.08.01
884	OPF86	02-JUL-1997:15	V1037	TASK 2 NH3 SERVICING (PENDING FLT USAGE)	11	ECL	Thermal Management	2.04.08
885	OPF86	02-JUL-1997:17	V5K32.001	WINDOW 1 C/P FROM: 86V11U - Note 1	22	TPS	Cockpit & Crew Cabin	2.04.04
886	OPF86	02-JUL-1997:21	V5057	REMOVE HEATSHIELD ENG CVRS	4	SME	Propulsion	2.02.04
887	OPF86	03-JUL-1997:6	TES-0272	WINDOW 1 C/P	3	TPS	Thermal Management	2.04.06
888	OPF86	03-JUL-1997:6	FWD-3818/21/26/27	NLGD TPS PATTERN FABS/THERMAL BARRIERS FROM: 86V224C - Note 1	3	TPS	Thermal Management	2.04.06
889	OPF86	03-JUL-1997:6	V02-70004	REMOVE PSA FROM: 86V21C - Note 1	11	PLO	Payload Accommodations	2.06.03
890	OPF86	03-JUL-1997:6	TFRC-009	(MCR-18563)FRCS THRUSTER TILE PATTERN FAB	3	TPS	Thermal Management	2.08.05
891	OPF86	03-JUL-1997:6	LWNG-330/331	LESS TILE R&R FROM: 86V223F - Note 1	3	TPS	Thermal Management	2.08.05
892	OPF86	03-JUL-1997:6	V7253	ORB WINDOW POLISH	3	STR	Structures , Mechanisms, Veh Handling	2.08.06
893	OPF86	03-JUL-1997:6	V3581	OPEN L/R XSLATOR PLATFORM FOR SSME INSTALLATION	2	GSE	Propulsion	3.03.02
894	OPF86	03-JUL-1997:8	V5005	INSTALL SSME 1,2,3	16	SME	Propulsion	3.03.03
895	OPF86	03-JUL-1997:9	V1026.003	WCS INSTALL FROM: 86V033 - Note 1	16	FCP	Environmental Ctl & Life Spt	2.04.04
896	OPF86	03-JUL-1997:9	V1078	POSU 5,6,7 APU 1,2,3 SCUPPER INSTL	8	APU	Power Management	2.04.07
897	OPF86	03-JUL-1997:9	V1037	POI 2 VENT PORT FLEXHOSE REMOVAL(NOT REQ'D)		ECL	Thermal Management	2.04.08
898	OPF86	03-JUL-1997:9	V1238	OMS/RCS FLT CONTROLS C/O(NOT REQ'D PER ENGR)		GNC	Guid, Nav & Ctl	2.05.01
899	OPF86	03-JUL-1997:9	V1238	POSU'S FLT CONTROLS OMS/RCS C/O & POD I/F VERIF		GNC	Guid, Nav & Ctl	2.05.01
900	OPF86	03-JUL-1997:9	V1097	ORB/ET UMBILICAL DOOR FUNCTIONAL POSU	4	MEQ	Structures , Mechanisms, Veh Handling	2.05.02
901	OPF86	03-JUL-1997:9	V5R01	POSU FC 2 REMOVAL/INSTALLATION	2	FCP	Power Management	2.08.01
902	OPF86	03-JUL-1997:9	TCS-1224//CM-666//CM-0343	EXTENDED APC INSTALLATION(BAY 2 STBD) FROM: 86V18S - Note 1	8	PLO	Thermal Management	2.08.05
903	OPF86	03-JUL-1997:9	V6051	ET DOOR T/B REWORK FROM: 86V222G - Note 1	4	TPS	Thermal Management	2.08.08

No.	DATA SET	START TIME DD:MMM-YYYY:H	PROCEDURE NAME	DESCRIPTION	DURATION (Hours)	SYSTEM	DESIGN DISCIPLINE	GENERIC OPS FBS NUMBER (Appx A1)
904	OPF86	03-JUL-1997:9	V1032.003	ORBITER CLOSEOUT	152	MPS	Propulsion	2.10.03
905	OPF86	03-JUL-1997:11	V5R01	POSU FC 2 REMOVAL/INSTALLATION	6	FCP	Power Management	2.08.01
906	OPF86	03-JUL-1997:13	V1026.004	POSU WCS FUNCTIONAL	16	FCP	Environmental Ctl & Life Spt	2.05.01
907	OPF86	03-JUL-1997:13	V1097	ORB/ET UMBILICAL CLOSEOUT DOOR FUNCT TEST	8	MEQ	Structures , Mechanisms, Veh Handling	2.05.02
908	OPF86	03-JUL-1997:17	V1008.003	POSU'S MSBLS SYSTEM TEST	4	COM	Communications	2.05.01
909	OPF86	03-JUL-1997:17	DDC-0131	INSTL FLOODLIGHT	8	EPD	Power Management	2.08.01
910	OPF86	03-JUL-1997:21	CM-669/CM-0345	BAY 13 STBD GAS BEAM INSTALLATION	4	PLO	Payload Accommodations	2.06.02
911	OPF86	03-JUL-1997:23	STR-A0264	PAINT FWD FLIGHT DECK	2	STR	Structures , Mechanisms, Veh Handling	2.08.06
912	OPF86	03-JUL-1997:24	V5005	SSME INSTLN		SME	Propulsion	3.03.03
913	OPF86	03-JUL-1997:24		SSME INSTALLATION COMPLETE		SME	Propulsion	3.03.03
914	OPF86	03-JUL-1997:24	V3581	CLOSE L/R XSLATOR PLATFORM AFTER SSME INSTALLATION	1	GSE	Propulsion	3.03.03
915	OPF86	07-JUL-1997:9	FWD-3817/18/21/22/26/27	NLGD TPS TILE PROCESS/THERMAL BARRIERS FROM: 86V224B - Note 1	16	TPS	Thermal Management	2.04.06
916	OPF86	07-JUL-1997:9	V1008.003	MSBLS SYSTEM TEST	8	COM	Communications	2.05.01
917	OPF86	07-JUL-1997:9	EPD-294	ODS PYRO SYSTEM VERIFICATION FROM: 86V414E - Note 1	8	EPD	Power Management	2.05.01
918	OPF86	07-JUL-1997:9	CM-669	GROUND STRAP INSTALLATION	4	PLO	Payload Accommodations	2.06.05
919	OPF86	07-JUL-1997:9	MPS-1283	EPOXY LINE REPAIR FROM: 86V10AB - Note 1	8	MPS	Propulsion	2.08.01
920	OPF86	07-JUL-1997:9	V1062.001	POSU 1 AIR DATA SYSTEM FUNCT C/O	5	COM	Guid, Nav & Ctl	2.08.01
921	OPF86	07-JUL-1997:9	V5R01	POSU FC 2 REMOVAL/INSTALLATION FROM: 86V50A - Note 1	8	FCP	Power Management	2.08.01
922	OPF86	07-JUL-1997:9	COM-0235//COM-A0019//COM-	KU BAND REWORK INSPECTION(NSLD)	2	COM	Communications	2.08.04
923	OPF86	07-JUL-1997:9	TFRC-009//TFRC-0809/0805	(MCR-18563)FRCS THRUSTER TILE PATTERN FAB FROM: 86V10M - Note 1	80	TPS	Thermal Management	2.08.05
924	OPF86	07-JUL-1997:9	STR-A0249	BAY 10 B/F DOUBLER BONDS FROM: 86V37 - Note 1	40	STR	Structures , Mechanisms, Veh Handling	2.08.06
925	OPF86	07-JUL-1997:9	IPR-0025	RH ET DOOR T/S	8	MEQ	Structures , Mechanisms, Veh Handling	2.08.08
926	OPF86	07-JUL-1997:9	V1017	WSB POST SERV GN2 REG CREEP TEST(AFTER 48 HR DECAY CKS) FROM: 86V051C - Note 1	16	HYD	Power Management	2.09.02
927	OPF86	07-JUL-1997:9	V6055.001	ORBITER MODAL INSPECTION PREPS	96	STR	Structures , Mechanisms, Veh Handling	2.09.02
928	OPF86	07-JUL-1997:9	V5005	ENGINE SECURING	38	SME	Propulsion	3.03.05
929	OPF86	07-JUL-1997:11	PT-043	GAS PREMATE	2	PLO	Payload Accommodations	2.06.02
930	OPF86	07-JUL-1997:13	V1041.001	POSU 2,3,4 GSE CONFIG/GN2 SAMPLING(START 3 DAYS PRIOR TO CTS)	4	ECL	Environmental Ctl & Life Spt	2.04.08
931	OPF86	07-JUL-1997:13	V1026.004	WCS FUNCTIONAL	20	FCP	Environmental Ctl & Life Spt	2.05.01
932	OPF86	07-JUL-1997:13	CE-194	GAS MATES	4	PLO	Payload Accommodations	2.06.05

No.	DATA SET	START TIME DD:MMM-YYYY:H	PROCEDURE NAME	DESCRIPTION	DURATION (Hours)	SYSTEM	DESIGN DISCIPLINE	GENERIC OPS FBS NUMBER (Appx A1)
933	OPF86	07-JUL-1997:13	CE-194	SEED II WIRE HARNESS CONFIGURATION	4	PLO	Payload Accommodations	2.06.05
934	OPF86	07-JUL-1997:13	CM-670	BAY 13 STBD GAS CAN(SEEDS II) INSTALLATION(CUSTOMER REQ'D)	4	PLO	Payload Accommodations	2.06.07
935	OPF86	07-JUL-1997:13	DDC-0131	INSTL FLOODLIGHT FROM: 86V14H - Note 1	4	EPD	Power Management	2.08.01
936	OPF86	07-JUL-1997:14	V1062.001	AIR DATA SYSTEM CK OUT	3	COM	Guid, Nav & Ctl	2.08.01
937	OPF86	07-JUL-1997:17	V1055.002	POSU'S POTABLE H2O SERVICING	44	FCP	Power Management	2.04.08
938	OPF86	07-JUL-1997:17	V1008.003	MSBLS SYSTEM TEST POI	4	COM	Communications	2.05.01
939	OPF86	07-JUL-1997:17	V5067	CLOSE C-HATCH FOR FLT	4	MEQ	Structures , Mechanisms, Veh Handling	2.07.03
940	OPF86	07-JUL-1997:17	V5R01	FC 2 REMOVAL/INSTALLATION	10	FCP	Power Management	2.08.01
941	OPF86	07-JUL-1997:17	RWNG-3611	LOWER LESS #10 TILE R&R FROM: 86V223E - Note 1	8	TPS	Thermal Management	2.08.05
942	OPF86	07-JUL-1997:21	V1100	ADP DEPLOYMENT TEST	3	MEQ	Structures , Mechanisms, Veh Handling	2.05.02
943	OPF86	08-JUL-1997:9	FCS-XXX	ODS STOWAGE BAG INSTL(NOT REQ'D WORKED AT PAD)		FCS	Cockpit & Crew Cabin	2.04.04
944	OPF86	08-JUL-1997:9	V9046.015	VOLUME 'B' LINER INSTL	8	FCS	Cockpit & Crew Cabin	2.04.04
945	OPF86	08-JUL-1997:9	V1078	G.B. GN2 FLIGHT PRESSURE	8	APU	Power Management	2.04.07
946	OPF86	08-JUL-1997:9	V1037	POI 1 INSTL ECL NH3 SERV DOOR	2	ECL	Thermal Management	2.04.08
947	OPF86	08-JUL-1997:9	CE-194	SMCH RECONFIG	4	PLO	Payload Accommodations	2.06.05
948	OPF86	08-JUL-1997:9	CM-670	BAY 13 STBD GAS CAN(SEEDS II) INSTALLATION(CUSTOMER REQ'D) FROM: 86V20 - Note 1	4	PLO	Payload Accommodations	2.06.07
949	OPF86	08-JUL-1997:9	FCS-551	SNAP/CLIPS/VELCRO VERIF	8	FCS	Payload Accommodations	2.06.07
950	OPF86	08-JUL-1997:9	V63-50001	INSTL C HATCH COVER	4	STR	Thermal Management	2.06.09
951	OPF86	08-JUL-1997:9	FRC4-109	FRCS REG T/S	2	OMS	Propulsion	2.08.01
952	OPF86	08-JUL-1997:9	RWNG-3611	LOWER LESS #10 TILE R&R	8	TPS	Thermal Management	2.08.05
953	OPF86	08-JUL-1997:9	RSI-185	MODAL TEST TILE MARKING/LAYOUT	40	TPS	Thermal Management	2.08.05
954	OPF86	08-JUL-1997:9	STR-A0263	R/R UPPER HATCH WINDOW	8	STR	Structures , Mechanisms, Veh Handling	2.08.08
955	OPF86	08-JUL-1997:9	PT-043	GAS IVT	4	PLO	Payload Accommodations	4.04.05
956	OPF86	08-JUL-1997:11	V5R01	FC 2 REMOVAL/INSTALLATION	4	FCP	Power Management	2.08.01
957	OPF86	08-JUL-1997:11	FRC4-109	FRCS REG T/S	6	OMS	Propulsion	2.08.01
958	OPF86	08-JUL-1997:13	V34-00008	REMOVE WING STEPS UPS	4	TCS	Thermal Management	2.02.04
959	OPF86	08-JUL-1997:13	CM-0346	RAISED METAL ON PSA	4	PLO	Payload Accommodations	2.06.03
960	OPF86	08-JUL-1997:13	V1017	WSB POST SERV GN2 REG CREEP TEST(AFTER 48 HR DECAY CKS) FROM: 86V051C - Note 1	2	HYD	Power Management	2.09.02
961	OPF86	08-JUL-1997:13	V1011.03O1/.04O/.05M1//IP	SSME/MPS INTEGRATED TESTING (POST ENGINE INSTL)	12	SME	Propulsion	3.04.02
962	OPF86	08-JUL-1997:15	MPS-273	GSE LEAK CHECKS AT PD10/16	2	MPS	Propulsion	2.04.09

No.	DATA SET	START TIME DD:MMM-YYYY:H	PROCEDURE NAME	DESCRIPTION	DURATION (Hours)	SYSTEM	DESIGN DISCIPLINE	GENERIC OPS FBS NUMBER (Appx A1)
963	OPF86	08-JUL-1997:15	V5R01	FC 2 REMOVAL/INSTALLATION	2	FCP	Power Management	2.08.01
964	OPF86	08-JUL-1997:15	V1017	WSB CHECKOUT & SERVICING POI'S	2	HYD	Power Management	2.09.02
965	OPF86	08-JUL-1997:17	V5139	FWD RCC/S&G INSTALLATION	8	TPS	Thermal Management	2.04.06
966	OPF86	08-JUL-1997:17	V1078	APU LUBE OIL SERVICING POI	8	APU	Power Management	2.04.08
967	OPF86	08-JUL-1997:17	V5R01	FC 2 REMOVAL/INSTALLATION/INSTL INTO S/C	2	FCP	Power Management	2.08.01
968	OPF86	08-JUL-1997:17	TCS-1224//CM-666//CM-0343	EXTENDED APC INSTALLATION(BAY 2 STBD) FROM: 86V18S - Note 1	8	PLO	Thermal Management	2.08.05
969	OPF86	08-JUL-1997:17	LWNG-3031/3058/59/LWNG-33	LESS -088 TILE BOND FROM: 86V223D - Note 1	8	TPS	Thermal Management	2.08.05
970	OPF86	08-JUL-1997:19	V5R01	FC 2 REMOVAL/INSTALLATION/MANIFOLD PRESS/MASS SPEC	34	FCP	Power Management	2.08.01
971	OPF86	09-JUL-1997:1	RSI-186	REMOVE 6 TILES	16	TPS	Thermal Management	2.08.05
972	OPF86	09-JUL-1997:6	CM-674	MEEPS CONTAINER INSTL	11	PLO	Payload Accommodations	2.06.03
973	OPF86	09-JUL-1997:9	V5126.001	TEMPILABELS INSTL	8	TPS	Thermal Management	2.04.05
974	OPF86	09-JUL-1997:9	FWD-3817/18/21/22/26/27	NLGD TPS TILE PROCESS/THERMAL BARRIERS FROM: 86V224B - Note 1	8	TPS	Thermal Management	2.04.06
975	OPF86	09-JUL-1997:9	V1041.001	POSU 7,5 ENGR WALKDOWN/FR PREPS	4	ECL	Environmental Ctl & Life Spt	2.04.08
976	OPF86	09-JUL-1997:9	MPS-277	(CHIT J5054)LATCH SHAFT SEAL LEAK CHECK	16	MPS	Propulsion	2.05.03
977	OPF86	09-JUL-1997:9	STS-86-PLD-005	PAINT REPAIR	4	PLO	Payload Accommodations	2.06.01
978	OPF86	09-JUL-1997:9	CM-0346	RAISED METAL ON PSA FROM: 86V21H - Note 1	8	PLO	Payload Accommodations	2.06.03
979	OPF86	09-JUL-1997:9	V9002.08	SEQ 5,6,7,8 INSTL ELEVON LOCKS	8	HYD	Propulsion	2.07.03
980	OPF86	09-JUL-1997:9	V5069.005	JACK XFER TO AFT 570'S	2	OHE	Structures , Mechanisms, Veh Handling	2.07.03
981	OPF86	09-JUL-1997:9	MPS-1283	EPOXY LINE REPAIR FROM: 86V10AB - Note 1	8	MPS	Propulsion	2.08.01
982	OPF86	09-JUL-1997:9	OEL-1380	(V5005)CONNECTOR REPAIR	8	OEL	Power Management	2.08.04
983	OPF86	09-JUL-1997:9	RWNG-3611	LOWER LESS #10 TILE R&R FROM: 86V223E - Note 1	40	TPS	Thermal Management	2.08.05
984	OPF86	09-JUL-1997:9	AFT-5032/4913//STR-4636	DOME HEATSHIELD BLANKET R/R	12	TPS	Thermal Management	2.08.05
985	OPF86	09-JUL-1997:9	V7253	ORB WINDOW POLISH FROM: 86V065B - Note 1	88	STR	Structures , Mechanisms, Veh Handling	2.08.06
986	OPF86	09-JUL-1997:9	STR-A0249//V80-95859	BODY FLAP C/P INSTL	16	STR	Structures , Mechanisms, Veh Handling	2.08.06
987	OPF86	09-JUL-1997:11	V5069.005	JACK XFER TO AFT 570'S	2	OHE	Structures , Mechanisms, Veh Handling	2.07.03
988	OPF86	09-JUL-1997:13	V9028.002	POSU'S MLG STRUT HYD LEVEL CK AND GN2 LK CKS	16	MEQ	Structures , Mechanisms, Veh Handling	2.04.01
989	OPF86	09-JUL-1997:13	V1336.001	DOCKING MECHANISM FUNCTIONAL	20	MEQ	Structures , Mechanisms, Veh Handling	2.05.01
990	OPF86	09-JUL-1997:13	V1011.03O1/.04O/.05M1//IP	SSME/MPS INTEGRATED TESTING (POST ENGINE INSTL) FROM: 86V265 - Note 1	12	SME	Propulsion	3.04.02
991	OPF86	09-JUL-1997:17	V5139	FWD RCC/S&G INSTALLATION FROM: 86V118B - Note 1	56	TPS	Thermal Management	2.04.06

No.	DATA SET	START TIME DD:MMM-YYYY:H	PROCEDURE NAME	DESCRIPTION	DURATION (Hours)	SYSTEM	DESIGN DISCIPLINE	GENERIC OPS FBS NUMBER (Appx A1)
992	OPF86	09-JUL-1997:17	STR-4621/22	ET DR CORROSION REPAIR(SANDING) FROM: 86V272W - Note 1	8	STR	Structures , Mechanisms, Veh Handling	2.08.01
993	OPF86	09-JUL-1997:17	LWNG-3031/3058/59/LWNG-33	LESS -088 TILE BOND FROM: 86V223D - Note 1	40	TPS	Thermal Management	2.08.05
994	OPF86	10-JUL-1997:9	V5057	SEQ 4,5,8,10,11,31 PIN TVC'S	4	MPS	Propulsion	2.02.04
995	OPF86	10-JUL-1997:9	V1158VL2X//IPR-0023	BALL VALVE LEAK CHECK FROM: 86V204AH - Note 1	8	OMS	Propulsion	2.03.01
996	OPF86	10-JUL-1997:9	V5126.001	TEMPILABELS INSTL FROM: 86V12W - Note 1	8	TPS	Thermal Management	2.04.05
997	OPF86	10-JUL-1997:9	FWD-3817/18/21/22/26/27	NLGD TPS TILE PROCESS/THERMAL BARRIERS FROM: 86V224B - Note 1	8	TPS	Thermal Management	2.04.06
998	OPF86	10-JUL-1997:9	V1041.001	TASK 1 ECLSS GN2 SERVICE (SEQ 9 REQ'S 3 DAY DECAY TEST)	8	ECL	Environmental Ctl & Life Spt	2.04.08
999	OPF86	10-JUL-1997:9	V1291.003	ECL ROTARY EQUIPMENT C/O	8	ECL	Environmental Ctl & Life Spt	2.05.01
1000	OPF86	10-JUL-1997:9	CM-677//CM-680	CAMERA A/C INSTL/REMV B CAMERA	2	PLO	Payload Accommodations	2.06.03
1001	OPF86	10-JUL-1997:9	V63-50017	C/O AFT BLKHD	4	TCS	Thermal Management	2.06.09
1002	OPF86	10-JUL-1997:9	V9002.08	SEQ 5,6,7,8 INSTL ELEVON LOCKS FROM: 86V13N - Note 1	8	HYD	Propulsion	2.07.03
1003	OPF86	10-JUL-1997:9	STR-4621/22	ET DR CORROSION REPAIR	8	STR	Structures , Mechanisms, Veh Handling	2.08.01
1004	OPF86	10-JUL-1997:9	AFT-5032/4913//STR-4636	DOME HEATSHIELD BLANKET R/R FROM: 86V23C - Note 1	16	TPS	Thermal Management	2.08.05
1005	OPF86	10-JUL-1997:9	STR-A0249	BAY 10 B/F DOUBLER BONDS FROM: 86V37 - Note 1	24	STR	Structures , Mechanisms, Veh Handling	2.08.06
1006	OPF86	10-JUL-1997:9	V1017	WSB CHECKOUT & SERVICING(DECAY CHECK)		HYD	Power Management	2.09.02
1007	OPF86	10-JUL-1997:13	V1165.007	MLG STRUT HYD LVL CK & GN2 LK CKS	12	MEQ	Structures , Mechanisms, Veh Handling	2.04.10
1008	OPF86	10-JUL-1997:13	V63-50015/16	C/O WINGBOX	4	TCS	Thermal Management	2.06.09
1009	OPF86	10-JUL-1997:13	PT-044	SSV PREMATE TEST	4	PLO	Payload Accommodations	2.06.09
1010	OPF86	10-JUL-1997:13	TCS-1224//CM-666//CM-0343	EXTENDED APC INSTALLATION(BAY 2 STBD) FROM: 86V18S - Note 1	10	PLO	Thermal Management	2.08.05
1011	OPF86	10-JUL-1997:13	STR-A0251	(CHIT J5013)WING BOLT TORQUE CHKS FROM: 86V251 - Note 1	4	STR	Structures , Mechanisms, Veh Handling	2.08.06
1012	OPF86	10-JUL-1997:17	V1078	48 HR GN2 BOTTLE DECAY TEST(AFTER SEQ 29,30,31)		APU	Power Management	2.05.03
1013	OPF86	10-JUL-1997:17	V5165	INSTALL PRESSURE DOME/LEAK CHECK	8	MEQ	Structures , Mechanisms, Veh Handling	2.05.03
1014	OPF86	10-JUL-1997:17	V63-50050/51	PLBD BLNKT INSP	4	STR	Thermal Management	2.06.09
1015	OPF86	10-JUL-1997:17	V9002.07	POSU LOWER NLG STRUT	2	HYD	Propulsion	2.07.03
1016	OPF86	10-JUL-1997:17	V41-50021/22/23/24/25/26	INSTALL E1,E2,E3 MOUNTED H/S/DOME H/S	8	OHE	Propulsion	3.03.05
1017	OPF86	10-JUL-1997:17	V1011.05M1//MPS-1281	POI 2,3 SSME FOAM INSULATION INSTLN	40	SME	Propulsion	3.03.05
1018	OPF86	10-JUL-1997:19	V9002.07	LOWER NLG STRUT	2	HYD	Propulsion	2.07.03
1019	OPF86	10-JUL-1997:21	V1165.004	NLG WHEEL & TIRE ASSY INSTL	4	MEQ	Structures , Mechanisms, Veh Handling	2.04.01
1020	OPF86	10-JUL-1997:21	V5R01	FC 2 REMOVAL/INSTALLATION POI	4	FCP	Power Management	2.08.01

No.	DATA SET	START TIME DD:MMM-YYYY:H	PROCEDURE NAME	DESCRIPTION	DURATION (Hours)	SYSTEM	DESIGN DISCIPLINE	GENERIC OPS FBS NUMBER (Appx A1)
1021	OPF86	10-JUL-1997:23	V5R01	FC 2 REMOVAL/INSTALLATION COMPLETE		FCP	Power Management	2.08.01
1022	OPF86	10-JUL-1997:23	V5R01	FC 2 REMOVAL/INSTALLATION COMPLETE		FCP	Power Management	2.08.01
1023	OPF86	11-JUL-1997:9	G41-20017	POSITION DAVIT CRANE ON 19R		SME	Propulsion	2.02.02
1024	OPF86	11-JUL-1997:9	PYR-116/117	POSU'S ODS POWER OFF STRAY VOLTAGE CHKS/PYRO CON-NECTS(REQ'S PWR DWN/AFT HATCH CLOSED/TRUSS ACCESS)	16	PYR	Structures , Mechanisms, Veh Handling	2.03.02
1025	OPF86	11-JUL-1997:9	V1165.004	NLG WHEEL & TIRE ASSY INSTL	4	MEQ	Structures , Mechanisms, Veh Handling	2.04.01
1026	OPF86	11-JUL-1997:9	V5012.004	SEQ 04 NLG STRUT THRUSTER INSTALLATION	2	PYR	Structures , Mechanisms, Veh Handling	2.04.02
1027	OPF86	11-JUL-1997:9	V5012.007	POSU 1&2-1.3 MERL-FWD NSI CIR RESISTANCE TEST/FIREX ELECT CONN.	4	PYR	Structures , Mechanisms, Veh Handling	2.04.02
1028	OPF86	11-JUL-1997:9	V5012.004	POSU 3-1.3 MERL- FWD ET SEP BOLT & NLG STRUT THRUSTER ELECT CONN.	2	PYR	Structures , Mechanisms, Veh Handling	2.04.02
1029	OPF86	11-JUL-1997:9	FWD-3817/18/21/22/26/27	NLGD TPS TILE PROCESS/THERMAL BARRIERS FROM: 86V224B - Note 1	8	TPS	Thermal Management	2.04.06
1030	OPF86	11-JUL-1997:9	V1041.001	TASK 1 ECLSS GN2 SERVICE (SEQ 9 REQ'S 3 DAY DECAY TEST) FROM: 86V036 - Note 1	8	ECL	Environmental Ctl & Life Spt	2.04.08
1031	OPF86	11-JUL-1997:9		POWER UP FROM: 86V000A - Note 1	5	OTC	Power Management	2.04.13
1032	OPF86	11-JUL-1997:9	V1078	24 HR GB DECAY TEST(AFTER SEQ 26,27,28)		APU	Power Management	2.05.01
1033	OPF86	11-JUL-1997:9	V63-50001	PSA INSTL C/O INSPECTION	2	PLO	Thermal Management	2.06.09
1034	OPF86	11-JUL-1997:9	V9002.06/.07	RECONFIGURE HOSES TO FULL-UP	4	HYD	Propulsion	2.07.03
1035	OPF86	11-JUL-1997:9	DDC-0132	M051 PANEL REMOVAL	8	DPS	Power Management	2.08.01
1036	OPF86	11-JUL-1997:9	UA-A0046	GPS T/S	8	COM	Communications	2.08.04
1037	OPF86	11-JUL-1997:9	COM-336	SCAN RETEST	4	COM	Communications	2.08.04
1038	OPF86	11-JUL-1997:9	TCS-1224//CM-666//CM-0343	EXTENDED APC INSTALLATION(BAY 2 STBD) FROM: 86V18S - Note 1	4	PLO	Thermal Management	2.08.05
1039	OPF86	11-JUL-1997:9	STR-A0251	(CHIT J5013)WING BOLT TORQUE CHKS FROM: 86V251 - Note 1	4	STR	Structures , Mechanisms, Veh Handling	2.08.06
1040	OPF86	11-JUL-1997:9	V41-50021/22/23/24/25/26	INSTALL E1,E2,E3 MOUNTED H/S/DOME H/S	48	OHE	Propulsion	3.03.05
1041	OPF86	11-JUL-1997:11	V5012.004	SEQ 05 FWD ET SEP BOLT & NLG STRUT THRUSTER ELECT CONN.	2	PYR	Structures , Mechanisms, Veh Handling	2.04.02
1042	OPF86	11-JUL-1997:11	V1086.003A1	MEC/PIC TEST	4	EPD	Power Management	2.05.01
1043	OPF86	11-JUL-1997:11	STS-86-PLD-005	PAINT REPAIR	2	PLO	Payload Accommodations	2.06.01
1044	OPF86	11-JUL-1997:11	V02-70004	INSTALL PSA	12	PLO	Payload Accommodations	2.06.03
1045	OPF86	11-JUL-1997:17	CM-677//CM-680	CAMERA A/C INSTL/REMV B CAMERA FROM: 86V12T - Note 1	4	PLO	Payload Accommodations	2.06.03
1046	OPF86	11-JUL-1997:17	STR-4597	REBOND NLG BOOT STRAP	8	STR	Structures , Mechanisms, Veh Handling	2.08.01
1047	OPF86	11-JUL-1997:17	OEL-1381	RETERM WIRE	8	OEL	Power Management	2.08.04
1048	OPF86	11-JUL-1997:21	V1256.002	POSU'S CCTV COMM SYSTEM CKS	4	COM	Communications	2.05.01

No.	DATA SET	START TIME DD:MMM-YYYY:H	PROCEDURE NAME	DESCRIPTION	DURATION (Hours)	SYSTEM	DESIGN DISCIPLINE	GENERIC OPS FBS NUMBER (Appx A1)
1049	OPF86	12-JUL-1997:1		TCID LOAD(NO LPS)	16	SOF	Cmd, Ctl & Health Mngmt	2.04.13
1050	OPF86	12-JUL-1997:9	PYR-116/117	ODS POWER OFF STRAY VOLTAGE CHKS/PYRO CON-NECTS(REQ'S PWR DWN/AFT HATCH CLOSED/TRUSS ACCESS)	11	PYR	Structures , Mechanisms, Veh Handling	2.03.02
1051	OPF86	12-JUL-1997:9	FWD-3817/18/21/22/26/27	NLGD TPS TILE PROCESS/THERMAL BARRIERS FROM: 86V224B - Note 1	16	TPS	Thermal Management	2.04.06
1052	OPF86	12-JUL-1997:9	V6044.001	MSBLS DESSICANT CHECKS (RUN 2)	2	COM	Communications	2.04.11
1053	OPF86	12-JUL-1997:9		COM,ECL,EPD,INS,MEQ,OEL PROCESSING/HISTORICAL PWR DOWN PROCESSING(SAT. & SUN.) FROM: 86V350B - Note 1 FIXEDFLAG 2	32		Structures , Mechanisms, Veh Handling	2.08.04
1054	OPF86	13-JUL-1997:9	MEQ-0769	RADIATOR DOUBLER BOND	24	MEQ	Structures , Mechanisms, Veh Handling	2.05.02
1055	OPF86	13-JUL-1997:9	STR-A0249//V80-95859	BODY FLAP C/P INSTL FROM: 86V37A - Note 1	24	STR	Structures , Mechanisms, Veh Handling	2.08.06
1056	OPF86	14-JUL-1997:9	MEQ-238	ODS UPPER HATCH FUNCTIONAL	12	MEQ	Structures , Mechanisms, Veh Handling	2.02.04
1057	OPF86	14-JUL-1997:9	V1165.002	MLG WHEEL & TIRE ASSY INSTL	16	MEQ	Structures , Mechanisms, Veh Handling	2.04.01
1058	OPF86	14-JUL-1997:9	V5012.005	POSU 1-1.3 MERL-FWD SEP ET YOKE INSTALL	4	PYR	Structures , Mechanisms, Veh Handling	2.04.02
1059	OPF86	14-JUL-1997:9	FWD-3805/6/7//FWD-A0106	CREW HATCH T/B R/R FROM: 86V12V - Note 1	24	TPS	Thermal Management	2.04.06
1060	OPF86	14-JUL-1997:9	FWD-3817/18/21/22/26/27	NLGD TPS TILE PROCESS/THERMAL BARRIERS FROM: 86V224B - Note 1	16	TPS	Thermal Management	2.04.06
1061	OPF86	14-JUL-1997:9	V9021.001	ACTIVATE OMS TRICKLE PURGE	4	OMS	Propulsion	2.04.08
1062	OPF86	14-JUL-1997:9	V1077.001	ORB FUEL CELL COOLANT ULLAGE VERIFICATION	4	FCP	Power Management	2.05.03
1063	OPF86	14-JUL-1997:9	V1078	48 HR GB DECAY TEST(AFTER INITIAL 24 HR DECAY TEST)		APU	Power Management	2.05.03
1064	OPF86	14-JUL-1997:9	CM-675	BAY 5 PORT/STBD UFR INSTL(WORK AT PAD)		PLO	Payload Accommodations	2.06.03
1065	OPF86	14-JUL-1997:9		G2/S2/179/22/24(PT-042)	4	PLO	Structures , Mechanisms, Veh Handling	2.06.05
1066	OPF86	14-JUL-1997:9	V9002.06/.07	RECONFIGURE HOSES TO FULL-UP FROM: 86V13M - Note 1	16	HYD	Propulsion	2.07.03
1067	OPF86	14-JUL-1997:9	TFRC-009//TFRC-0809/0805	(MCR-18563)FRCS THRUSTER TILE PATTERN FAB FROM: 86V10M - Note 1	80	TPS	Thermal Management	2.08.05
1068	OPF86	14-JUL-1997:9		TILE PROCESSING	208	TPS	Thermal Management	2.08.05
1069	OPF86	14-JUL-1997:9	V6055.001	ORBITER MODAL INSPECTION	16	STR	Structures , Mechanisms, Veh Handling	2.09.02
1070	OPF86	14-JUL-1997:9	V6029.001	SEQ 10,11,12 THRUSTER ROLLOUT PREPS	16	OMS	Propulsion	2.10.02
1071	OPF86	14-JUL-1997:9	V6037	ORB TPS-ROLLOUT INSP FROM: 86V228 - Note 1	110	TPS	Thermal Management	2.10.02
1072	OPF86	14-JUL-1997:9	V41-50021/22/23/24/25/26/	INSTALL E1,E2,E3 MOUNTED H/S/DOME H/S/FLEX SEAL REPAIR	16	OHE	Propulsion	3.03.05
1073	OPF86	14-JUL-1997:9	PT-042	SSV IVT	4	PLO	Payload Accommodations	4.04.05
1074	OPF86	14-JUL-1997:10		POWER UP FROM: 86V000A - Note 1	31	OTC	Power Management	2.04.13
1075	OPF86	14-JUL-1997:13	V1055.002D1/WWM-0036/0037	POTABLE H20 SERVICING/R/R CV	16	FCP	Power Management	2.04.08

No.	DATA SET	START TIME DD:MMM-YYYY:H	PROCEDURE NAME	DESCRIPTION	DURATION (Hours)	SYSTEM	DESIGN DISCIPLINE	GENERIC OPS FBS NUMBER (Appx A1)
1076	OPF86	14-JUL-1997:13	V1041.001	3 DAY(MIN) LONG TERM DECAY TEST		ECL	Environmental Ctl & Life Spt	2.04.08
1077	OPF86	14-JUL-1997:13	V1256.002	//IPR-0031 CCTV COMM SYSTEM CKS	8	COM	Communications	2.05.01
1078	OPF86	14-JUL-1997:13	V1336.001/.002A1	ODS LUBE AND STOW FOR FLT	8	MEQ	Structures , Mechanisms, Veh Handling	2.05.01
1079	OPF86	14-JUL-1997:13	V1077.001	ORB FUEL CELL 1/3 COOLANT ULLAGE VERIFICATION	4	FCP	Power Management	2.05.03
1080	OPF86	14-JUL-1997:13	V02-70004//STR-4638/39//C	(CHIT K5056)INSTALL PSA/EVAL STRU	4	PLO	Payload Accommodations	2.06.03
1081	OPF86	14-JUL-1997:13		G9/129/42		PLO	Structures , Mechanisms, Veh Handling	2.06.05
1082	OPF86	14-JUL-1997:13	DDC-0131	INSTL FLOODLIGHT FROM: 86V14H - Note 1	6	EPD	Power Management	2.08.01
1083	OPF86	14-JUL-1997:13	UA-A0046	GPS T/S FROM: 86V46 - Note 1	4	COM	Communications	2.08.04
1084	OPF86	14-JUL-1997:17	V1065X1//V1048T1//V1063.0	POSU'S GNC INTEGRATED HYDRAULIC OPERATIONS	16	GNC	Guid, Nav & Ctl	2.05.01
1085	OPF86	14-JUL-1997:17	CM-678	REMOVE OSVS TARGET COVERS		PLO	Payload Accommodations	2.06.03
1086	OPF86	14-JUL-1997:17	V5050U2/V5050.002	MIDBODY CEIT START		PLO	Payload Accommodations	2.06.09
1087	OPF86	14-JUL-1997:17		MIDBODY CEIT START		PLO	Payload Accommodations	2.06.09
1088	OPF86	14-JUL-1997:17	V1264	SEQ 25 GSE CABLE DISCONNECT FROM DA	1	COM	Communications	2.10.02
1089	OPF86	14-JUL-1997:17	V5050.002	CEIT	4	PLO	Payload Accommodations	4.04.05
1090	OPF86	14-JUL-1997:18	V1276.001	OSSU 3 DA GSE CONNECTOR CLOSEOUT	1	COM	Communications	2.10.02
1091	OPF86	14-JUL-1997:24		HEAT SHIELD INSTALLATION COMPLETE		SME	Propulsion	3.03.03
1092	OPF86	15-JUL-1997:1	V5057	SEQ 12 REMOVE MID STOKE LOCKS	2	SME	Propulsion	2.02.04
1093	OPF86	15-JUL-1997:9	V1086.014	RMS NSI RESISTANCE TEST	2	EPD	Power Management	2.05.01
1094	OPF86	15-JUL-1997:9	V1065X-1//V1048T1//V1063.	OP 50 GNC INTEGRATED HYD OPS	16	GNC	Guid, Nav & Ctl	2.05.01
1095	OPF86	15-JUL-1997:9	MEQ-0769	RADIATOR DOUBLER BOND FROM: 86V11W - Note 1	4	MEQ	Structures , Mechanisms, Veh Handling	2.05.02
1096	OPF86	15-JUL-1997:9	V1011.05	PNEUMATIC LEAK CHECKS	8	SME	Propulsion	2.05.03
1097	OPF86	15-JUL-1997:9	V1009.005	SEQ 4 LH2 MANIFOLD DECAY TEST	4	MPS	Propulsion	2.05.03
1098	OPF86	15-JUL-1997:9	L6987	MDS&DS TNL EXTERNAL C/O INSPECTION	4	PLO	Payload Accommodations	2.06.03
1099	OPF86	15-JUL-1997:9	CM-674	MEEPS CONTAINER INSTL FROM: 86V27 - Note 1	8	PLO	Payload Accommodations	2.06.03
1100	OPF86	15-JUL-1997:9	V1176.002	POSU'S FINAL PLB CLEAN	16	PLO	Payload Accommodations	2.06.07
1101	OPF86	15-JUL-1997:9	V9002.06	HYD QD LEAK CHECKS	8	HYD	Propulsion	2.07.03
1102	OPF86	15-JUL-1997:9	STR-A0251	(CHIT J5013)WING BOLT TORQUE CHKS FROM: 86V251 - Note 1	12	STR	Structures , Mechanisms, Veh Handling	2.08.06
1103	OPF86	15-JUL-1997:9	V6055.001	ORBITER MODAL INSPECTION	80	STR	Structures , Mechanisms, Veh Handling	2.09.02
1104	OPF86	15-JUL-1997:9	V1032.002	SEQ 5 CLOSE UPPER HATCH FOR FLT	4	PLO	Structures , Mechanisms, Veh Handling	2.10.02
1105	OPF86	15-JUL-1997:13	V1041.001	POI SCUPPER REMOVAL	2	ECL	Environmental Ctl & Life Spt	2.04.08

No.	DATA SET	START TIME DD:MMM-YYYY:H	PROCEDURE NAME	DESCRIPTION	DURATION (Hours)	SYSTEM	DESIGN DISCIPLINE	GENERIC OPS FBS NUMBER (Appx A1)
1106	OPF86	15-JUL-1997:13	V02-70004//STR-4638/39//C	(CHIT K5056)INSTALL PSA/EVAL STRU FROM: 86V21J - Note 1	12	PLO	Payload Accommodations	2.06.03
1107	OPF86	15-JUL-1997:13	DDC-0130	INT/EXT LIGHTING FUNCTIONAL	2	EPD	Power Management	2.08.01
1108	OPF86	15-JUL-1997:13	DDC-0133	PILOT OVERHEAD FLOODLIGHT C/O	4	OEL	Power Management	2.08.01
1109	OPF86	15-JUL-1997:15	V1276.001	SEQ 12 FINAL SELF TEST	1	COM	Communications	2.05.01
1110	OPF86	15-JUL-1997:16	V1276.001	SEQ 13 DA CABLE UNWRAP	2	COM	Communications	2.10.02
1111	OPF86	15-JUL-1997:17	MEQ-0766/0772	BULB SEAL REPAIRS	8	MEQ	Structures , Mechanisms, Veh Handling	2.05.02
1112	OPF86	15-JUL-1997:17	CM-671	INSTALL CAMERA BRACKET	2	PLO	Payload Accommodations	2.06.02
1113	OPF86	15-JUL-1997:17	V5050U-2	CEIT POI	4	FCS	Payload Accommodations	4.04.05
1114	OPF86	15-JUL-1997:18	V1264.001	SEQ 30 DA COUNTERWEIGHT REMOVAL	2	COM	Communications	2.10.02
1115	OPF86	15-JUL-1997:21	V5067	OPEN AFT HATCH	4	MEQ	Structures , Mechanisms, Veh Handling	2.07.03
1116	OPF86	15-JUL-1997:21	MEQ-A0074	DRY FILM LUBE APPLIC	4	MEQ	Structures , Mechanisms, Veh Handling	2.08.08
1117	OPF86	16-JUL-1997:1	V5057	SEQ 11 INSTL MID STOKE LOCKS	2	SME	Propulsion	2.02.04
1118	OPF86	16-JUL-1997:1		INTEGRATED HYD/FLIGHT CONTROL OPS COMPLETE		GNC	Guid, Nav & Ctl	2.05.01
1119	OPF86	16-JUL-1997:5	V5012.007	SEQ 03-FWD NSI RESIST TEST-NLG,FWD SEP,MLG,KU ANT. & FIREX	2	PYR	Structures , Mechanisms, Veh Handling	2.04.02
1120	OPF86	16-JUL-1997:5	V63-50014/23	C/O BAY 11	4	PLO	Thermal Management	2.06.09
1121	OPF86	16-JUL-1997:7	V5012.007	SEQ 04-FIREX ELECTRICAL CONN	2	PYR	Structures , Mechanisms, Veh Handling	2.04.02
1122	OPF86	16-JUL-1997:9	V1165.011	MLG SEP HARNESS INSTALLATION	4	INS	Cmd, Ctl & Health Mngmt	2.04.01
1123	OPF86	16-JUL-1997:9	V80-95907/933/935//AFT-49	INSTALL E1,E2,E3 SSME CARRIER PANELS	48	STR	Thermal Management	2.04.05
1124	OPF86	16-JUL-1997:9	FWD-3817/18/21/22/26/27	NLGD TPS TILE PROCESS/THERMAL BARRIERS FROM: 86V224B - Note 1	16	TPS	Thermal Management	2.04.06
1125	OPF86	16-JUL-1997:9	V1065X1//V1048T1//V1063A-	GNC INTEGRATED HYDRAULIC OPERATIONS POI'S	4	GNC	Guid, Nav & Ctl	2.05.01
1126	OPF86	16-JUL-1997:9	MEQ-0777	1307 BLKHD FITTING INSPECTION	2	MEQ	Structures , Mechanisms, Veh Handling	2.05.02
1127	OPF86	16-JUL-1997:9	MEQ-0766/0772	BULB SEAL REPAIRS FROM: 86V32A - Note 1	20	MEQ	Structures , Mechanisms, Veh Handling	2.05.02
1128	OPF86	16-JUL-1997:9	L5067B	TUNNEL PAD INSTALLATION(MDS&DS)	2	PLO	Payload Accommodations	2.06.03
1129	OPF86	16-JUL-1997:9	V1176.002	POSU'S FINAL PLB CLEAN FROM: 86V418A - Note 1	64	PLO	Payload Accommodations	2.06.07
1130	OPF86	16-JUL-1997:9	CM-671	TARGET CROSSHAIR VERIF.	4	PLO	Payload Accommodations	2.06.09
1131	OPF86	16-JUL-1997:9	ECL-A0061//V63-50003	FWD BLKHD INSP	10	STR	Thermal Management	2.08.01
1132	OPF86	16-JUL-1997:9	HYD-0771	O-RING REPAIR	16	HYD	Power Management	2.08.02
1133	OPF86	16-JUL-1997:9	RWNG-3603,4,5,23	WING SPAR TILE INSTL	24	TPS	Thermal Management	2.08.05
1134	OPF86	16-JUL-1997:9	STR-AXXXX	(MCR 19040)AFT ATTCH PLATE MODIFICATION(DEFER TO OMDP)		STR	Structures , Mechanisms, Veh Handling	2.09.01

No.	DATA SET	START TIME DD:MMM-YYYY:H	PROCEDURE NAME	DESCRIPTION	DURATION (Hours)	SYSTEM	DESIGN DISCIPLINE	GENERIC OPS FBS NUMBER (Appx A1)
1135	OPF86	16-JUL-1997:9	V1032.002B	MIDBODY CLOSEOUTS	40	STR	Structures , Mechanisms, Veh Handling	2.10.02
1136	OPF86	16-JUL-1997:9	V1032.003	ORBITER CLOSEOUT FROM: 86V112 - Note 1	48	MPS	Propulsion	2.10.03
1137	OPF86	16-JUL-1997:9	V1011.05M1//MPS-1281	POI 2,3 SSME FOAM INSULATION INSTLN FROM: 86V265Z - Note 1	48	SME	Propulsion	3.03.05
1138	OPF86	16-JUL-1997:11		POWER UP FROM: 86V000A - Note 1	10	OTC	Power Management	2.04.13
1139	OPF86	16-JUL-1997:11	V1080	MULTIFUNCTION CRT DISPLAY SYSTEM CHECKOUT	4	DPS	Cmd, Ctl & Health Mngmt	2.05.01
1140	OPF86	16-JUL-1997:11	CM-335/381	ODS CAMERA TURNAROUND CHECKOUT	4	PLO	Payload Accommodations	2.06.09
1141	OPF86	16-JUL-1997:13	V1004	TASK 5 MTU VERIF (REQ'S OPS 0)	1	INS	Cmd, Ctl & Health Mngmt	2.05.01
1142	OPF86	16-JUL-1997:13	V02-70004//STR-4638/39//C	(CHIT K5056)INSTALL PSA/EVAL STRU FROM: 86V21J - Note 1	20	PLO	Payload Accommodations	2.06.03
1143	OPF86	16-JUL-1997:17	MEQ-237	AFT HATCH FUNCTIONAL	8	MEQ	Structures , Mechanisms, Veh Handling	2.02.04
1144	OPF86	16-JUL-1997:17	V1055.002D1/WWM-0036/0037	POTABLE H20 SERVICING/R/R CV FROM: 86V039 - Note 1	12	FCP	Power Management	2.04.08
1145	OPF86	16-JUL-1997:17	V00-10072	FINAL PLB INSPECTION	40	QC	Payload Accommodations	2.05.02
1146	OPF86	16-JUL-1997:17	MEQ-0771/0768	KOROPON REPAIRS	8	FCP	Structures , Mechanisms, Veh Handling	2.05.02
1147	OPF86	16-JUL-1997:17	CM-676	RE-INSTALL TETHERS	4	PLO	Payload Accommodations	2.06.03
1148	OPF86	16-JUL-1997:17	CM-0349/0350/0351/0352/VI	SHARP EDGE REPAIRS	8	PLO	Payload Accommodations	2.06.03
1149	OPF86	16-JUL-1997:17	V6042E//CM-0358/0359	OIL/VERIFY LATCHES	24	PLO	Payload Accommodations	2.06.03
1150	OPF86	16-JUL-1997:17	V5050U-2	CEIT POI FROM: 86V414L - Note 1	8	FCS	Payload Accommodations	4.04.05
1151	OPF86	16-JUL-1997:21	MEQ-A0074	DRY FILM LUBE APPLIC FROM: 86V422C - Note 1	2	MEQ	Structures , Mechanisms, Veh Handling	2.08.08
1152	OPF86	16-JUL-1997:21	V1032.002	SEQ 6 ODS EXTERNAL C/O	8	VPL	Payload Accommodations	2.10.02
1153	OPF86	17-JUL-1997:1	V1275.001	C/P'S WATERPROOFING	8	TPS	Thermal Management	2.04.05
1154	OPF86	17-JUL-1997:9	FWD-3805/6/7//FWD-A0106	CREW HATCH T/B R/R FROM: 86V12V - Note 1	16	TPS	Thermal Management	2.04.06
1155	OPF86	17-JUL-1997:9		POWER UP FROM: 86V000A - Note 1	8	OTC	Power Management	2.04.13
1156	OPF86	17-JUL-1997:9	V1004	TASK 3 OI PCMMU C/O (REQ'S OPS 0)	1	INS	Cmd, Ctl & Health Mngmt	2.05.01
1157	OPF86	17-JUL-1997:9	V1058.002	AIRLOCK HATCH 'A' FUNCTIONAL VERIFICATION	6	MEQ	Structures , Mechanisms, Veh Handling	2.05.02
1158	OPF86	17-JUL-1997:9	MEQ-0766/0772	BULB SEAL REPAIRS FROM: 86V32A - Note 1	38	MEQ	Structures , Mechanisms, Veh Handling	2.05.02
1159	OPF86	17-JUL-1997:9	V1097	ET DOOR TO CENTERLINE	4	MEQ	Structures , Mechanisms, Veh Handling	2.05.02
1160	OPF86	17-JUL-1997:9	V9019	FH17 PUMP DOWN	32	MPS	Propulsion	2.05.03
1161	OPF86	17-JUL-1997:9	V9002	CONFIGURE FOR MODAL	8	HYD	Propulsion	2.07.03
1162	OPF86	17-JUL-1997:9	HYD-0771	PD27 LEAK CHECK	4	HYD	Power Management	2.08.02
1163	OPF86	17-JUL-1997:9	OHE-078	RMV TRUSS PLTFRM(POST PYR STRAY VOLTAGE)	4	OHE	Structures , Mechanisms, Veh Handling	2.10.03
1164	OPF86	17-JUL-1997:10	V1004	TASK 4 PCMMU FORMAT C/O (REQ'S OPS 0)	1	INS	Cmd, Ctl & Health Mngmt	2.05.01

No.	DATA SET	START TIME DD:MMM-YYYY:H	PROCEDURE NAME	DESCRIPTION	DURATION (Hours)	SYSTEM	DESIGN DISCIPLINE	GENERIC OPS FBS NUMBER (Appx A1)
1165	OPF86	17-JUL-1997:13	V9046.004	PIN EVA SLIDEWIRES	4	PLO	Cockpit & Crew Cabin	2.04.04
1166	OPF86	17-JUL-1997:13	ECL-414	CABIN LEAK CHECK	2	ECL	Environmental Ctl & Life Spt	2.05.04
1167	OPF86	17-JUL-1997:13	ECL-A0061//V63-50003	FWD BLKHD INSP FROM: 86V080I - Note 1	4	STR	Thermal Management	2.08.01
1168	OPF86	17-JUL-1997:15	ECL-414	CABIN LEAK CHECK	6	ECL	Environmental Ctl & Life Spt	2.05.04
1169	OPF86	17-JUL-1997:17		POWER UP	2	OTC	Power Management	2.04.13
1170	OPF86	17-JUL-1997:17	MEQ-0771/0768	KOROPON REPAIRS FROM: 86V16F - Note 1	4	FCP	Structures , Mechanisms, Veh Handling	2.05.02
1171	OPF86	17-JUL-1997:17	CM-676	ROUTE TETHERS	4	PLO	Payload Accommodations	2.06.03
1172	OPF86	17-JUL-1997:17	CM-679	STOW PSA	2	PLO	Payload Accommodations	2.06.03
1173	OPF86	17-JUL-1997:17	CM-674	INSTL MEEP BOND JUMPER	4	PLO	Payload Accommodations	2.06.05
1174	OPF86	17-JUL-1997:17	V9002	CONFIGURE FOR MODAL	2	HYD	Propulsion	2.07.03
1175	OPF86	17-JUL-1997:19	MEQ-0777	1307 BLKHD FITTING INSPECTION	2	MEQ	Structures , Mechanisms, Veh Handling	2.05.02
1176	OPF86	17-JUL-1997:21	CM-665	EPS C/O PHOTOS	2	PLO	Payload Accommodations	2.06.07
1177	OPF86	18-JUL-1997:1	V63-50005	CLOSEOUT BAY 2	8	STR	Thermal Management	2.06.09
1178	OPF86	18-JUL-1997:1	ECL-A0061//V63-50003	FWD BLKHD INSP FROM: 86V080I - Note 1	4	STR	Thermal Management	2.08.01
1179	OPF86	18-JUL-1997:7	FWD-3805/6/7//FWD-A0106	CREW HATCH T/B R/R FROM: 86V12V - Note 1	18	TPS	Thermal Management	2.04.06
1180	OPF86	18-JUL-1997:9	V1165.013	NLG SEP HARNESS INSTALLATION	8	INS	Cmd, Ctl & Health Mngmt	2.04.01
1181	OPF86	18-JUL-1997:9	FWD-3817/18/21/22/26/27	NLGD TPS TILE PROCESS/THERMAL BARRIERS FROM: 86V224B - Note 1	8	TPS	Thermal Management	2.04.06
1182	OPF86	18-JUL-1997:9	ECL-414	CABIN LEAK CHECK FROM: 86V14S - Note 1	10	ECL	Environmental Ctl & Life Spt	2.05.04
1183	OPF86	18-JUL-1997:9	V9002	RECONFIGURE TO DEDICATED	4	HYD	Propulsion	2.07.03
1184	OPF86	18-JUL-1997:9	V9002.07	PIN RH/LH MLG STRUT	4	HYD	Propulsion	2.07.03
1185	OPF86	18-JUL-1997:9	STR-4625	RH BAGGIE RETAINER FIT CHECK	8	STR	Structures , Mechanisms, Veh Handling	2.08.01
1186	OPF86	18-JUL-1997:10	V6042	REMOVE LUMALLOY		PLO	Payload Accommodations	2.06.09
1187	OPF86	18-JUL-1997:13	MEQ-0777	MR ID PLB FTG	2	MEQ	Structures , Mechanisms, Veh Handling	2.05.02
1188	OPF86	18-JUL-1997:13	V5067	CLOSE AFT HATCH	2	MEQ	Structures , Mechanisms, Veh Handling	2.07.03
1189	OPF86	18-JUL-1997:17	V11-00001/00002	RCC CVR REMOVAL/INSTL	10	STR	Structures , Mechanisms, Veh Handling	2.02.04
1190	OPF86	18-JUL-1997:17	V34-60001/60002	TEMPORARY CLOSE FLIPPER DOORS	8	STR	Structures , Mechanisms, Veh Handling	2.02.06
1191	OPF86	18-JUL-1997:17	MEQ-0769	RAD #2 RH TAPE PAINT	10	MEQ	Structures , Mechanisms, Veh Handling	2.05.02
1192	OPF86	18-JUL-1997:17	V63-50005	CLOSEOUT BAY 2 FROM: 86V080H - Note 1	4	STR	Thermal Management	2.06.09
1193	OPF86	18-JUL-1997:17	STR-4643	BORON STRUT PAINT	4	STR	Structures , Mechanisms, Veh Handling	2.08.01
1194	OPF86	18-JUL-1997:21	V5139	FWD RCC/S&G INSTALLATION FROM: 86V118B - Note 1	4	TPS	Thermal Management	2.04.06

No.	DATA SET	START TIME DD:MMM-YYYY:H	PROCEDURE NAME	DESCRIPTION	DURATION (Hours)	SYSTEM	DESIGN DISCIPLINE	GENERIC OPS FBS NUMBER (Appx A1)
1195	OPF86	18-JUL-1997:21	V63-50006//PVD-204	CLOSEOUT BAY 3	6	STR	Thermal Management	2.06.09
1196	OPF86	18-JUL-1997:23	V00-10072	FINAL PLB INSPECTION FROM: 86V080A - Note 1	2	QC	Payload Accommodations	2.05.02
1197	OPF86	19-JUL-1997:5	V5018	POSU 1 PLBD FINAL CLOSING PREPS	4	MEQ	Structures , Mechanisms, Veh Handling	2.10.02
1198	OPF86	19-JUL-1997:9	FWD-3817/18/21/22/26/27	NLGD TPS TILE PROCESS/THERMAL BARRIERS FROM: 86V224B - Note 1	16	TPS	Thermal Management	2.04.06
1199	OPF86	19-JUL-1997:9		POWER UP(SAT & SUN) FROM: 86V000B - Note 1	4	OTC	Power Management	2.04.13
1200	OPF86	19-JUL-1997:9	V1037	POI 3 NH3 SYSTEM 10 DAY DECAY CKS (MIN)		ECL	Thermal Management	2.05.03
1201	OPF86	19-JUL-1997:9	V63-50023	PLB C/O INSP/PHOTOS	2	STR	Thermal Management	2.06.09
1202	OPF86	19-JUL-1997:9		COM,ECL,EPD,INS,MEQ,OEL PROCESSING/HISTORICAL PWR DOWN PROCESSING(SAT. & SUN.) FROM: 86V350B - Note 1 FIXEDFLAG 2	32		Structures , Mechanisms, Veh Handling	2.08.04
1203	OPF86	19-JUL-1997:11	V1176.002	FINAL PLB CLEAN	6	PLO	Payload Accommodations	2.06.07
1204	OPF86	19-JUL-1997:17	V5018	FINAL PAYLOAD BAY DOOR CLOSING/STRONGBACK REMOVAL	4	MEQ	Structures , Mechanisms, Veh Handling	2.10.02
1205	OPF86	19-JUL-1997:18	V1276.001	SEQ 14 KU FINAL STOW	1	COM	Communications	2.10.02
1206	OPF86	19-JUL-1997:21	V5018	FINAL PLBD CLOSURE		MEQ	Structures , Mechanisms, Veh Handling	2.07.03
1207	OPF86	19-JUL-1997:21		FINAL PLB DOORS CLOSED		MEQ	Structures , Mechanisms, Veh Handling	2.10.02
1208	OPF86	19-JUL-1997:21	V5018	POI 1 ZERO-G FIXTURE & STRONGBACK REMOVAL	4	MEQ	Structures , Mechanisms, Veh Handling	2.10.02
1209	OPF86	19-JUL-1997:23		TERMINATE PURGE CIRCUIT 1,2,3		PVD	Structures , Mechanisms, Veh Handling	2.07.01
1210	OPF86	20-JUL-1997:9	V6055.001	ORBITER MODAL INSPECTION	16	STR	Structures , Mechanisms, Veh Handling	2.09.02
1211	OPF86	21-JUL-1997:1		REACTIVATE PURGE CIRCUIT 1,2,3		PVD	Structures , Mechanisms, Veh Handling	2.07.01
1212	OPF86	21-JUL-1997:4	FWD-3817/18/21/22/26/27	NLGD TPS TILE PROCESS/THERMAL BARRIERS FROM: 86V224B - Note 1	21	TPS	Thermal Management	2.04.06
1213	OPF86	21-JUL-1997:9	FWD-3805/6/7//FWD-A0106	CREW HATCH T/B R/R FROM: 86V12V - Note 1	8	TPS	Thermal Management	2.04.06
1214	OPF86	21-JUL-1997:9	PVD-207	PAYLOAD PURGE VERIFICATION		PVD	Safety Management & Control	2.05.03
1215	OPF86	21-JUL-1997:9	ECL-414//ECL-0956	CABIN LEAK CHECK	12	ECL	Environmental Ctl & Life Spt	2.05.04
1216	OPF86	21-JUL-1997:9	V9002.07	DEPIN MLG STRUT	4	HYD	Propulsion	2.07.03
1217	OPF86	21-JUL-1997:9	TFRC-009//TFRC-0809/0805	(MCR-18563)FRCS THRUSTER TILE PATTERN FAB FROM: 86V10M - Note 1	32	TPS	Thermal Management	2.08.05
1218	OPF86	21-JUL-1997:9	RSI-186	ORBITER MODAL TILE REPLACEMENTS	80	TPS	Thermal Management	2.08.05
1219	OPF86	21-JUL-1997:9	LWNG-3031/3058/59/LWNG-33	LESS -088 TILE BOND FROM: 86V223D - Note 1	8	TPS	Thermal Management	2.08.05
1220	OPF86	21-JUL-1997:9	V6055.001	ORBITER MODAL INSPECTION	4	STR	Structures , Mechanisms, Veh Handling	2.09.02
1221	OPF86	21-JUL-1997:9	PLBD-0708//MID-2484/85/87	ORB PLBD TILE CLOSEOUT	48	TPS	Thermal Management	2.10.02

No.	DATA SET	START TIME DD:MMM-YYYY:H	PROCEDURE NAME	DESCRIPTION	DURATION (Hours)	SYSTEM	DESIGN DISCIPLINE	GENERIC OPS FBS NUMBER (Appx A1)
1222	OPF86	21-JUL-1997:13	V80-95907/933/935//AFT-49	INSTALL E1,E2,E3 SSME CARRIER PANELS FROM: 86V105B - Note 1	68	STR	Thermal Management	2.04.05
1223	OPF86	21-JUL-1997:13	V1058.003	CM HATCH & DOOR FUNCTIONAL	4	MEQ	Structures , Mechanisms, Veh Handling	2.05.02
1224	OPF86	21-JUL-1997:13	STR-4625	RH BAGGIE RETAINER FIT CHECK FROM: 86V14K - Note 1	4	STR	Structures , Mechanisms, Veh Handling	2.08.01
1225	OPF86	21-JUL-1997:13	V6055.001	ORBITER MODAL INSPECTION POI	4	STR	Structures , Mechanisms, Veh Handling	2.09.02
1226	OPF86	21-JUL-1997:13	V6055.001	ORBITER MODAL INSPECTION COMPLETE		STR	Structures , Mechanisms, Veh Handling	2.09.02
1227	OPF86	21-JUL-1997:13	V6055.001	ORBITER MODAL INSPECTION COMPLETE		STR	Structures , Mechanisms, Veh Handling	2.09.02
1228	OPF86	21-JUL-1997:13	V35-50002/STR-4581	INSTL ET ATTACH LINER	20	STR	Structures , Mechanisms, Veh Handling	2.10.02
1229	OPF86	21-JUL-1997:13	V1032.003	ORBITER CLOSEOUT FROM: 86V112 - Note 1	20	MPS	Propulsion	2.10.03
1230	OPF86	21-JUL-1997:13	V1011.05M1//MPS-1281	POI 2,3 SSME FOAM INSULATION INSTLN FROM: 86V265Z - Note 1	4	SME	Propulsion	3.03.05
1231	OPF86	22-JUL-1997:1		TPS TILE REPAIR	32	TPS	Thermal Management	2.08.05
1232	OPF86	22-JUL-1997:9	V5012.006	POSU 1-1.3 MERL-ET UMB SEP BLADE VLV VERIFY & AFT SEP NUT INSTLN.	8	PYR	Structures , Mechanisms, Veh Handling	2.04.02
1233	OPF86	22-JUL-1997:9	FWD-3823/94/95	CHIN PANEL C/P REWORK & INSTLN	16	TPS	Thermal Management	2.04.06
1234	OPF86	22-JUL-1997:9	FWD-3817/18/21/22/26/27	NLGD TPS TILE PROCESS/THERMAL BARRIERS FROM: 86V224B - Note 1	16	TPS	Thermal Management	2.04.06
1235	OPF86	22-JUL-1997:9	V1055.002D1/WWM-0036/0037	POTABLE H20 SERVICING/R/R CV FROM: 86V039 - Note 1	4	FCP	Power Management	2.04.08
1236	OPF86	22-JUL-1997:9	V1239G//V1240F//V1230I	F/M/A CLEANING FROM: 86V1239 - Note 1	8	GSE	Structures , Mechanisms, Veh Handling	2.04.09
1237	OPF86	22-JUL-1997:9	V6029.001	RCS THRUSTER DESICCANT INSPECTION (OMS) FROM: 86V6029 - Note 1	8	GSE	Propulsion	2.04.09
1238	OPF86	22-JUL-1997:9		POWER UP FROM: 86V000A - Note 1	4	OTC	Power Management	2.04.13
1239	OPF86	22-JUL-1997:9	V1161	ORBITER BUS REDUNDANCY TEST	2	OTC	Power Management	2.05.01
1240	OPF86	22-JUL-1997:9	INS-XXX	ET SEP CAMERA CHECKOUT(DELETED AS DUP V254)		INS	Cmd, Ctl & Health Mngmt	2.05.01
1241	OPF86	22-JUL-1997:9	V76-10001/2	L02/LH2 ET UMB MONOBALL CONNECTORS DETAIL INSPECTION	8	STR	Structures , Mechanisms, Veh Handling	2.05.02
1242	OPF86	22-JUL-1997:9	MEQ-0784	TUNNEL ADAPTER BUMPER REPAIR	8	MEQ	Structures , Mechanisms, Veh Handling	2.05.02
1243	OPF86	22-JUL-1997:9	V9019	FH17 PUMP DOWN FROM: 86V099C - Note 1	8	MPS	Propulsion	2.05.03
1244	OPF86	22-JUL-1997:9	ECL-414//ECL-0956	CABIN LEAK CHECK FROM: 86V14R - Note 1	2	ECL	Environmental Ctl & Life Spt	2.05.04
1245	OPF86	22-JUL-1997:9	V9021.001	OME TRICKLE PURGE MONITORING (OMS) FROM: 86V9021 - Note 1	8	GSE	Propulsion	2.07.01
1246	OPF86	22-JUL-1997:9	V9002.07	RETRACT NLG/MLG	2	HYD	Propulsion	2.07.03
1247	OPF86	22-JUL-1997:9	V5067	OPEN AFT HATCH	4	MEQ	Structures , Mechanisms, Veh Handling	2.07.03
1248	OPF86	22-JUL-1997:9	MPS-A0047	TORQUE CK	2	MPS	Propulsion	2.08.01
1249	OPF86	22-JUL-1997:9	STR-4625	RH BAGGIE RETAINER FIT CHECK FROM: 86V14K - Note 1	4	STR	Structures , Mechanisms, Veh Handling	2.08.01

No.	DATA SET	START TIME DD:MMM-YYYY:H	PROCEDURE NAME	DESCRIPTION	DURATION (Hours)	SYSTEM	DESIGN DISCIPLINE	GENERIC OPS FBS NUMBER (Appx A1)
1250	OPF86	22-JUL-1997:9	LWNG-3031/3058/59/LWNG-33	LESS -088 TILE BOND FROM: 86V223D - Note 1	16	TPS	Thermal Management	2.08.05
1251	OPF86	22-JUL-1997:9	V6055.001	ORBITER MODAL INSPECTION POI FROM: 86V90F - Note 1	8	STR	Structures , Mechanisms, Veh Handling	2.09.02
1252	OPF86	22-JUL-1997:9	V1032.002B	MIDBODY CLOSEOUTS FROM: 86V112A - Note 1	24	STR	Structures , Mechanisms, Veh Handling	2.10.02
1253	OPF86	22-JUL-1997:9	V1032.003	SEQ 03 MPS R/O PREPS	4	MPS	Propulsion	2.10.03
1254	OPF86	22-JUL-1997:9	V1011.05M1//MPS-1281	POI 2,3 SSME FOAM INSULATION INSTLN FROM: 86V265Z - Note 1	4	SME	Propulsion	3.03.05
1255	OPF86	22-JUL-1997:9	V5050	CEIT POSU'S	24	FCS	Cockpit & Crew Cabin	4.04.05
1256	OPF86	22-JUL-1997:11	V9002.05	RETRACT NLG	2	HYD	Propulsion	2.07.03
1257	OPF86	22-JUL-1997:13	V5012.005	SEQ 02-FWD SEP ET YOKE INSTALLATION	4	PYR	Structures , Mechanisms, Veh Handling	2.04.02
1258	OPF86	22-JUL-1997:13	MEQ-0783	D-HATCH SEAL REPAIR	4	MEQ	Structures , Mechanisms, Veh Handling	2.05.02
1259	OPF86	22-JUL-1997:13	DDC-0134	COMMANDER FLOODLIGHT R/R	4	OEL	Power Management	2.08.01
1260	OPF86	22-JUL-1997:13	DIG-0263//V5L04	(IPR-0010)MMU 1 R/R	4	DPS	Cmd, Ctl & Health Mngmt	2.08.01
1261	OPF86	22-JUL-1997:13	DDC-0132	INSTL M051F PANEL	4	OEL	Power Management	2.08.01
1262	OPF86	23-JUL-1997:9	V5012.006	SEQ 02-ET UMBILICAL SEP BLADE VALVE VERIFY AND AFT SEP NUT INSTALL	4	PYR	Structures , Mechanisms, Veh Handling	2.04.02
1263	OPF86	23-JUL-1997:9	V5012.007	SEQ 05-FWD SEP CLOSEOUT	2	PYR	Structures , Mechanisms, Veh Handling	2.04.02
1264	OPF86	23-JUL-1997:9	V5012.005	SEQ 02-FWD SEP ET YOKE INSTLN (12 HR RETORQUE/INSPECT & BOND CKS)	2	PYR	Structures , Mechanisms, Veh Handling	2.04.02
1265	OPF86	23-JUL-1997:9	FWD-3823/94/95	CHIN PANEL C/P REWORK & INSTLN FROM: 86V080B - Note 1	88	TPS	Thermal Management	2.04.06
1266	OPF86	23-JUL-1997:9	FWD-3817/18/21/22/26/27	NLGD TPS TILE PROCESS/THERMAL BARRIERS FROM: 86V224B - Note 1	16	TPS	Thermal Management	2.04.06
1267	OPF86	23-JUL-1997:9	V1055.002D1/WWM-0036/0037	POTABLE H20 SERVICING/R/R CV FROM: 86V039 - Note 1	4	FCP	Power Management	2.04.08
1268	OPF86	23-JUL-1997:9	V1239G//V1240F//V1230I	F/M/A CLEANING FROM: 86V1239 - Note 1	8	GSE	Structures , Mechanisms, Veh Handling	2.04.09
1269	OPF86	23-JUL-1997:9	V1063.001	OP 100 SSME MARKING TAPE REMOVAL	2	SME	Propulsion	2.04.09
1270	OPF86	23-JUL-1997:9	V6029.001	RCS THRUSTER DESICCANT INSPECTION (OMS) FROM: 86V6029 - Note 1	8	GSE	Propulsion	2.04.09
1271	OPF86	23-JUL-1997:9	V1053.006	PPO2 SENSOR INSTALLATION	8	ECL	Environmental Ctl & Life Spt	2.04.12
1272	OPF86	23-JUL-1997:9		POWER UP FROM: 86V000A - Note 1	24	OTC	Power Management	2.04.13
1273	OPF86	23-JUL-1997:9	V76-10001/2	L02/LH2 ET UMB MONOBALL CONNECTORS DETAIL INSPECTION FROM: 86V504JC - Note 1	4	STR	Structures , Mechanisms, Veh Handling	2.05.02
1274	OPF86	23-JUL-1997:9	V9019	FH17 PUMP DOWN FROM: 86V099C - Note 1	8	MPS	Propulsion	2.05.03
1275	OPF86	23-JUL-1997:9	V9021.001	OME TRICKLE PURGE MONITORING (OMS) FROM: 86V9021 - Note 1	8	GSE	Propulsion	2.07.01
1276	OPF86	23-JUL-1997:9	V9023	POSU 1 PLBD FINAL CLOSING PREPS	8	MEQ	Structures , Mechanisms, Veh Handling	2.07.03
1277	OPF86	23-JUL-1997:9	MPS-1295	LO2 FEEDLINE INSPECTION	4	NDE	Propulsion	2.08.01

No.	DATA SET	START TIME DD:MMM-YYYY:H	PROCEDURE NAME	DESCRIPTION	DURATION (Hours)	SYSTEM	DESIGN DISCIPLINE	GENERIC OPS FBS NUMBER (Appx A1)
1278	OPF86	23-JUL-1997:9	DDC-0132	INSTL M051F PANEL FROM: 86V26U - Note 1	4	OEL	Power Management	2.08.01
1279	OPF86	23-JUL-1997:9	DIG-0263//V5L04	(IPR-0010)MMU 1 R/R FROM: 86V15J - Note 1	4	DPS	Cmd, Ctl & Health Mngmt	2.08.01
1280	OPF86	23-JUL-1997:9	HYD-0772	HYD DRAIN LINE 3 REPAIR	4	HYD	Power Management	2.08.02
1281	OPF86	23-JUL-1997:9	TFRC-009//TFRC-0809/0805	(MCR-18563)FRCS THRUSTER TILE PATTERN FAB FROM: 86V10M - Note 1	16	TPS	Thermal Management	2.08.05
1282	OPF86	23-JUL-1997:9	LWNG-3031/3058/59/LWNG-33	LESS -088 TILE BOND FROM: 86V223D - Note 1	16	TPS	Thermal Management	2.08.05
1283	OPF86	23-JUL-1997:9	V6055.001	ORBITER MODAL INSPECTION POI FROM: 86V90F - Note 1	8	STR	Structures , Mechanisms, Veh Handling	2.09.02
1284	OPF86	23-JUL-1997:9	V6055.001	ORBITER MODAL INSPECTION POI	2	STR	Structures , Mechanisms, Veh Handling	2.09.02
1285	OPF86	23-JUL-1997:9	V41-20003	SSME R/O PREPS	8	SME	Propulsion	2.10.02
1286	OPF86	23-JUL-1997:13	V9002.07	EXTEND NLG STRUT	1	HYD	Propulsion	2.07.03
1287	OPF86	23-JUL-1997:14	V9002.07	PIN MLG	1	HYD	Propulsion	2.07.03
1288	OPF86	23-JUL-1997:15	V9028.003	SEQ 20 INSTL DR ROLLERS	1	HYD	Power Management	2.04.01
1289	OPF86	23-JUL-1997:16	V9002.05	RETRACT MLG/NLG	1	HYD	Propulsion	2.07.03
1290	OPF86	23-JUL-1997:17	V6049.004	MLGD EVALUATION & REWORK	8	TPS	Thermal Management	2.04.06
1291	OPF86	24-JUL-1997:9	V6049.004	MLGD EVALUATION & REWORK		TPS	Thermal Management	2.04.06
1292	OPF86	24-JUL-1997:9	V6049.004	MLGD EVALUATION & REWORK(NOT REQ'D)		TPS	Thermal Management	2.04.06
1293	OPF86	24-JUL-1997:9	V1239G//V1240F//V1230I	F/M/A CLEANING FROM: 86V1239 - Note 1	8	GSE	Structures , Mechanisms, Veh Handling	2.04.09
1294	OPF86	24-JUL-1997:9	V6029.001	RCS THRUSTER DESICCANT INSPECTION (OMS) FROM: 86V6029 - Note 1	8	GSE	Propulsion	2.04.09
1295	OPF86	24-JUL-1997:9	V6029.001	RCS THRUSTER DESICCANT INSPECTION (OMS)	2	GSE	Propulsion	2.04.09
1296	OPF86	24-JUL-1997:9	V9021.001	OME TRICKLE PURGE MONITORING (OMS) FROM: 86V9021 - Note 1	8	GSE	Propulsion	2.07.01
1297	OPF86	24-JUL-1997:9	V9023	CYCLE PLBDS/STRONGBACK REMOVAL	4	MEQ	Structures , Mechanisms, Veh Handling	2.07.03
1298	OPF86	24-JUL-1997:9	V9002.05	EXTEND MLG/NLG	2	HYD	Propulsion	2.07.03
1299	OPF86	24-JUL-1997:9	PFS-001	INSTL FAN PACKAGE	8	FCP	Power Management	2.08.08
1300	OPF86	24-JUL-1997:9	V1032.003	FINAL ROLL PREPS/CONFIG AFT CLOSEOUT	16	VPL	Structures , Mechanisms, Veh Handling	2.10.03
1301	OPF86	24-JUL-1997:13	FWD-3817/18/21/22/26/27	NLGD TPS TILE PROCESS/THERMAL BARRIERS FROM: 86V224B - Note 1	54	TPS	Thermal Management	2.04.06
1302	OPF86	24-JUL-1997:13	ECL-XXX	POSU'S FAN PACKAGE FLOW TEST	4	ECL	Environmental Ctl & Life Spt	2.05.01
1303	OPF86	24-JUL-1997:13	V9023	POI 1 ZERO-G FIXTURE & STRONGBACK REMOVAL	4	MEQ	Structures , Mechanisms, Veh Handling	2.07.03
1304	OPF86	24-JUL-1997:13	V9002.07	DEPIN/PIN MLG FOR FLT	2	HYD	Propulsion	2.07.03
1305	OPF86	24-JUL-1997:13	LWNG-3031/3058/59/LWNG-33	LESS -088 TILE BOND FROM: 86V223D - Note 1	28	TPS	Thermal Management	2.08.05

No.	DATA SET	START TIME DD:MMM-YYYY:H	PROCEDURE NAME	DESCRIPTION	DURATION (Hours)	SYSTEM	DESIGN DISCIPLINE	GENERIC OPS FBS NUMBER (Appx A1)
1306	OPF86	24-JUL-1997:13	PLBD-0708//MID-2484/85/87	ORB PLBD TILE CLOSEOUT FROM: 86V193 - Note 1	2	TPS	Thermal Management	2.10.02
1307	OPF86	24-JUL-1997:17	FCS-555	INSTL NET BAGS	8	FCS	Payload Accommodations	2.06.07
1308	OPF86	24-JUL-1997:17	FCS-552	INSTL ODS CABLES	8	FCS	Cockpit & Crew Cabin	2.07.03
1309	OPF86	24-JUL-1997:17	TFRC-009//TFRC-0809/0805	(MCR-18563)FRCS THRUSTER TILE PATTERN FAB FROM: 86V10M - Note 1	40	TPS	Thermal Management	2.08.05
1310	OPF86	24-JUL-1997:17	FCS-553	CEIT STOW	8	FCS	Cockpit & Crew Cabin	4.04.05
1311	OPF86	25-JUL-1997:5	FWD-3818/21/22/26/27	NLGD TPS TILE PROCESS/THERMAL BARRIERS	4	TPS	Thermal Management	2.04.06
1312	OPF86	25-JUL-1997:9	V1055.002D1/WWM-0036/0037	POTABLE H20 SERVICING/R/R CV FROM: 86V039 - Note 1	8	FCP	Power Management	2.04.08
1313	OPF86	25-JUL-1997:9	V1041.001	10 DAY(MIN) DECAY TEST(LPS)		ECL	Environmental Ctl & Life Spt	2.04.08
1314	OPF86	25-JUL-1997:9	V1239G//V1240F//V1230I	F/M/A CLEANING FROM: 86V1239 - Note 1	8	GSE	Structures , Mechanisms, Veh Handling	2.04.09
1315	OPF86	25-JUL-1997:9	V6029.001	RCS THRUSTER DESICCANT INSPECTION (OMS) FROM: 86V6029 - Note 1	8	GSE	Propulsion	2.04.09
1316	OPF86	25-JUL-1997:9	V1184	LOAD MMU 1	2	SOF	Cmd, Ctl & Health Mngmt	2.04.13
1317	OPF86	25-JUL-1997:9	V1098.001	POSU'S LANDING GEAR DOOR FUNCTIONAL	8	MEQ	Structures , Mechanisms, Veh Handling	2.05.01
1318	OPF86	25-JUL-1997:9	FCS-569	LOCKER REMOVAL/VELCRO INSTL(FOR COM-334)	8	FCS	Payload Accommodations	2.06.07
1319	OPF86	25-JUL-1997:9	V9021.001	OME TRICKLE PURGE MONITORING (OMS) FROM: 86V9021 - Note 1	8	GSE	Propulsion	2.07.01
1320	OPF86	25-JUL-1997:9	V9023	POI 1 ZERO-G FIXTURE & STRONGBACK REMOVAL FROM: 86V03B - Note 1	8	MEQ	Structures , Mechanisms, Veh Handling	2.07.03
1321	OPF86	25-JUL-1997:9	FCS-554	CEIT CABLE STOW	8	FCS	Cockpit & Crew Cabin	2.07.03
1322	OPF86	25-JUL-1997:9	HYD-0773	PD14 REPAIR	8	HYD	Power Management	2.08.02
1323	OPF86	25-JUL-1997:9	OEL-1384/85//DR OEL-1382/	CONNECTOR/WIRE REPAIR	8	OEL	Power Management	2.08.04
1324	OPF86	25-JUL-1997:9	PFS-001	INSTL FAN PACKAGE FROM: 86V01B - Note 1	4	FCP	Power Management	2.08.08
1325	OPF86	25-JUL-1997:13	ECL-413	FAN PACKAGE FLOW TEST	4	ECL	Environmental Ctl & Life Spt	2.05.01
1326	OPF86	25-JUL-1997:13	V1059	DPS COMP COMPLEX C/O (OPS MODE 0)	4	DPS	Cmd, Ctl & Health Mngmt	2.05.04
1327	OPF86	25-JUL-1997:13	DDC-0134	FLOODLIGHT RETEST	4	EPD	Power Management	2.08.01
1328	OPF86	25-JUL-1997:13	INS-134	(CHIT K5037)ET SEP CAMERA WIRING CHECKOUT SCHEDULE	4	INS	Cmd, Ctl & Health Mngmt	2.08.04
1329	OPF86	26-JUL-1997:9		COM,ECL,EPD,INS,MEQ,OEL PROCESSING/HISTORICAL PWR DOWN PROCESSING(SAT. & SUN.) FROM: 86V350B - Note 1 FIXEDFLAG 2	24		Structures , Mechanisms, Veh Handling	2.08.04
1330	OPF86	26-JUL-1997:9	AFT-XXXX	BHS REPAIR	8	TPS	Thermal Management	2.08.05
1331	OPF86	26-JUL-1997:9		CREW MODULE CEIT START		PLO	Structures , Mechanisms, Veh Handling	4.04.05
1332	OPF86	26-JUL-1997:9		CREW MODULE CEIT START		PLO	Structures , Mechanisms, Veh Handling	4.04.05

No.	DATA SET	START TIME DD:MMM-YYYY:H	PROCEDURE NAME	DESCRIPTION	DURATION (Hours)	SYSTEM	DESIGN DISCIPLINE	GENERIC OPS FBS NUMBER (Appx A1)
1333	OPF86	26-JUL-1997:9	V5050	CEIT	8	FCS	Payload Accommodations	4.04.05
1334	OPF86	26-JUL-1997:13	MEQ-236	CEIT A HATCH LATCH DEMO	2	MEQ	Structures , Mechanisms, Veh Handling	2.02.04
1335	OPF86	26-JUL-1997:17	FCS-552	ODS CABLE CHECK	4	FCS	Payload Accommodations	2.06.07
1336	OPF86	26-JUL-1997:17	FCS-553	CEIT DESTOW	8	FCS	Payload Accommodations	2.06.07
1337	OPF86	26-JUL-1997:17	FCS-554	CEIT CABLE DESTOW	8	FCS	Cockpit & Crew Cabin	2.07.03
1338	OPF86	27-JUL-1997:9	STR-603	(MCR 18933)GPS BRACKET BONDS/INSTL(FOR COM-334)	72	STR	Structures , Mechanisms, Veh Handling	2.09.01
1339	OPF86	27-JUL-1997:23	V3555	DEACTIVATE ORBITER PURGE AIR	2	PVD	Safety Management & Control	2.10.01
1340	OPF86	28-JUL-1997:9	V1239G//V1240F//V1230I	F/M/A CLEANING FROM: 86V1239 - Note 1	8	GSE	Structures , Mechanisms, Veh Handling	2.04.09
1341	OPF86	28-JUL-1997:9	V6029.001	RCS THRUSTER DESICCANT INSPECTION (OMS) FROM: 86V6029 - Note 1	24	GSE	Propulsion	2.04.09
1342	OPF86	28-JUL-1997:9	V6044.001	MSBLS DESSICANT CHECKS (RUN 3)	2	COM	Communications	2.05.01
1343	OPF86	28-JUL-1997:9	V3555	MONITOR ORBITER PURGE AIR (PVD) FROM: 86V3555 - Note 1	208	GSE	Safety Management & Control	2.07.01
1344	OPF86	28-JUL-1997:9	V9021.001	OME TRICKLE PURGE MONITORING (OMS) FROM: 86V9021 - Note 1	8	GSE	Propulsion	2.07.01
1345	OPF86	28-JUL-1997:9	V3555	ACTIVATE ORBITER PURGE AIR	2	PVD	Safety Management & Control	2.07.03
1346	OPF86	28-JUL-1997:9	V9002	ROUTE HOSES FOR CAMERA DR OPS	8	HYD	Propulsion	2.07.03
1347	OPF86	28-JUL-1997:9	OEL-1384/85//DR OEL-1382/	CONNECTOR/WIRE REPAIR FROM: 86V14M - Note 1	8	OEL	Power Management	2.08.04
1348	OPF86	28-JUL-1997:9	RSI-186	ORBITER MODAL TILE REPLACEMENTS FROM: 86V90D - Note 1	80	TPS	Thermal Management	2.08.05
1349	OPF86	28-JUL-1997:9	LWNG-3031/3058/59/LWNG-33	LESS -088 TILE BOND FROM: 86V223D - Note 1	32	TPS	Thermal Management	2.08.05
1350	OPF86	28-JUL-1997:9	PFS-001	FAN PACKAGE DUCT RECONFIG	8	ECL	Environmental Ctl & Life Spt	2.08.07
1351	OPF86	28-JUL-1997:9	V5018	POI PLBD FINAL CLOSING	8	MEQ	Structures , Mechanisms, Veh Handling	2.10.02
1352	OPF86	28-JUL-1997:9	V1032.003	FINAL ROLL PREPS/CONFIG AFT CLOSEOUT FROM: 86V112B - Note 1	8	VPL	Structures , Mechanisms, Veh Handling	2.10.03
1353	OPF86	28-JUL-1997:9	V9002	CONFIG HEATERS FOR ROLLOUT(CONTINGENCY)(NOT REQ'D THIS FLOW)		OMS	Propulsion	2.10.03
1354	OPF86	28-JUL-1997:9		WORK DAY BEAN COUNT TO SUPPORT OPF ROLLOUT REVIEW		OSO	Structures , Mechanisms, Veh Handling	4.01.01
1355	OPF86	28-JUL-1997:13	V1098.001	POSU'S LANDING GEAR DOOR FUNCTIONAL FROM: 86V054A - Note 1	2	MEQ	Structures , Mechanisms, Veh Handling	2.05.01
1356	OPF86	28-JUL-1997:13	HYD-0773	PD14 REPAIR FROM: 86V15W - Note 1	2	HYD	Power Management	2.08.02
1357	OPF86	29-JUL-1997:1	V3575	BRIDGE BUCKET CHECKOUT FROM: 86V3575 - Note 1	40	GSE	Structures , Mechanisms, Veh Handling	2.06.06
1358	OPF86	29-JUL-1997:1	V3576.003	BRIDGE BUCKET CLEANING FROM: 86V3576 - Note 1	40	GSE	Structures , Mechanisms, Veh Handling	2.11.01
1359	OPF86	29-JUL-1997:9	V1239G//V1240F//V1230I	F/M/A CLEANING FROM: 86V1239 - Note 1	8	GSE	Structures , Mechanisms, Veh Handling	2.04.09
1360	OPF86	29-JUL-1997:9	V1098.001	POSU'S LANDING GEAR DOOR FUNCTIONAL FROM: 86V054A - Note 1	4	MEQ	Structures , Mechanisms, Veh Handling	2.05.01

No.	DATA SET	START TIME DD:MMM-YYYY:H	PROCEDURE NAME	DESCRIPTION	DURATION (Hours)	SYSTEM	DESIGN DISCIPLINE	GENERIC OPS FBS NUMBER (Appx A1)
1361	OPF86	29-JUL-1997:9	V9021.001	OME TRICKLE PURGE MONITORING (OMS) FROM: 86V9021 - Note 1	8	GSE	Propulsion	2.07.01
1362	OPF86	29-JUL-1997:9	V9002.05	RETRACT NLG/MLG	1	HYD	Propulsion	2.07.03
1363	OPF86	29-JUL-1997:9	HYD-0771	PD27 LEAK CHECK	4	HYD	Power Management	2.08.02
1364	OPF86	29-JUL-1997:9	V1032.003	FINAL ROLL PREPS/CONFIG AFT CLOSEOUT FROM: 86V112B - Note 1	8	VPL	Structures , Mechanisms, Veh Handling	2.10.03
1365	OPF86	29-JUL-1997:9	V3599	ORBITER ALARM TEST	8	GSE	Ground Systems & Facilities	2.11.01
1366	OPF86	29-JUL-1997:10		POWER UP FROM: 86V000A - Note 1	7	OTC	Power Management	2.04.13
1367	OPF86	29-JUL-1997:10	UA-A0045	GPS P-CODE TEST(BFS OPS 0)	3	COM	Communications	2.05.01
1368	OPF86	29-JUL-1997:10	V9002.05	RETRACT NLG/MLG	1	HYD	Propulsion	2.07.03
1369	OPF86	29-JUL-1997:11	V9002	RECONFIGURE TO FULL UP HYD	2	HYD	Propulsion	2.07.03
1370	OPF86	29-JUL-1997:11	V6049.002	MYLAR PULL/ENGR WALKDOWN	2	TPS	Thermal Management	2.08.08
1371	OPF86	29-JUL-1997:13	V1098.001	LANDING GEAR FUNCTIONAL	2	MEQ	Structures , Mechanisms, Veh Handling	2.05.01
1372	OPF86	29-JUL-1997:13	INS-134	(CHIT K5037)ET SEP CAMERA WIRING CHECKOUT SCHEDULE FROM: 86V254 - Note 1	4	INS	Cmd, Ctl & Health Mngmt	2.08.04
1373	OPF86	29-JUL-1997:15	V1308F-2	OPS 40 FLT CONTROLS FINAL CYCLING	2	GNC	Guid, Nav & Ctl	2.04.08
1374	OPF86	29-JUL-1997:17	V1273.001B1//V5139C	NLGD STEP & GAP PERPHERY TILE MEASUREMENT	24	TPS	Thermal Management	2.04.06
1375	OPF86	30-JUL-1997:9	V5057	SEQ 13 THRUSTER CHAMBER DESICCANT CK	8	SME	Propulsion	2.02.04
1376	OPF86	30-JUL-1997:9	V80-40045	DOOR 45 INSTALL	8	STR	Structures , Mechanisms, Veh Handling	2.02.06
1377	OPF86	30-JUL-1997:9	V9002.09	ORB/HYD DISC/CONFIG TO DED HYD SYS	4	HYD	Power Management	2.04.01
1378	OPF86	30-JUL-1997:9	V5139	S & G ACROSS T/B'S	4	TPS	Thermal Management	2.04.06
1379	OPF86	30-JUL-1997:9	V1055.002D1/WWM-0036/0037	POTABLE H20 SERVICING/R/R CV FROM: 86V039 - Note 1	8	FCP	Power Management	2.04.08
1380	OPF86	30-JUL-1997:9	V1239G//V1240F//V1230I	F/M/A CLEANING FROM: 86V1239 - Note 1	8	GSE	Structures , Mechanisms, Veh Handling	2.04.09
1381	OPF86	30-JUL-1997:9	V6029.001	RCS THRUSTER DESICCANT INSPECTION (OMS) FROM: 86V6029 - Note 1	8	GSE	Propulsion	2.04.09
1382	OPF86	30-JUL-1997:9		POWER UP FROM: 86V000A - Note 1	8	OTC	Power Management	2.04.13
1383	OPF86	30-JUL-1997:9	ME2019-0820	GIMBAL MEASUREMENTS	8	SME	Propulsion	2.05.01
1384	OPF86	30-JUL-1997:9	CE-193	AFD RECONFIG	4	PLO	Payload Accommodations	2.06.03
1385	OPF86	30-JUL-1997:9	V9021.001	OME TRICKLE PURGE MONITORING (OMS) FROM: 86V9021 - Note 1	8	GSE	Propulsion	2.07.01
1386	OPF86	30-JUL-1997:9	V5067	SEQ 23 OPEN AFT HATCH	4	MEQ	Structures , Mechanisms, Veh Handling	2.07.03
1387	OPF86	30-JUL-1997:9	V9002	MATE HOSES FOR CAMERA DR OPS	8	HYD	Propulsion	2.07.03
1388	OPF86	30-JUL-1997:9	TFRC-009//TFRC-0809/0805	(MCR-18563)FRCS THRUSTER TILE PATTERN FAB FROM: 86V10M - Note 1	96	TPS	Thermal Management	2.08.05

No.	DATA SET	START TIME DD:MMM-YYYY:H	PROCEDURE NAME	DESCRIPTION	DURATION (Hours)	SYSTEM	DESIGN DISCIPLINE	GENERIC OPS FBS NUMBER (Appx A1)
1389	OPF86	30-JUL-1997:9	AFT-4880	LET DOOR TILE BOND	48	TPS	Thermal Management	2.08.05
1390	OPF86	30-JUL-1997:9	TRPO4-1597	ROMS TILE BOND	48	TPS	Thermal Management	2.08.05
1391	OPF86	30-JUL-1997:9	AFT-4925	BHS TILE BOND	80	TPS	Thermal Management	2.08.05
1392	OPF86	30-JUL-1997:9	AFT-4928	C/P TILE BOND	98	TPS	Thermal Management	2.08.05
1393	OPF86	30-JUL-1997:9	V1032.003	FINAL ROLL PREPS/CONFIG AFT CLOSEOUT FROM: 86V112B - Note 1	8	VPL	Structures , Mechanisms, Veh Handling	2.10.03
1394	OPF86	30-JUL-1997:13	L6987	MDS&DS FINAL TUNNEL INSPECTION/PAD REMOVAL	4	PLO	Payload Accommodations	2.06.03
1395	OPF86	30-JUL-1997:17	CAP-0043	INSPECT OX X-FEED LINE/HEATER RETEST	24	OMS	Propulsion	2.09.01
1396	OPF86	31-JUL-1997:1		TPS TILE REPAIR FROM: 86V223I - Note 1	48	TPS	Thermal Management	2.08.05
1397	OPF86	31-JUL-1997:9	TES-0276	WINDOW #3 C/P CLEANOUT	8	STR	Thermal Management	2.04.06
1398	OPF86	31-JUL-1997:9	V1055.002D1/WWM-0036/0037	POTABLE H20 SERVICING/R/R CV FROM: 86V039 - Note 1	26	FCP	Power Management	2.04.08
1399	OPF86	31-JUL-1997:9	V1239G//V1240F//V1230I	F/M/A CLEANING FROM: 86V1239 - Note 1	8	GSE	Structures , Mechanisms, Veh Handling	2.04.09
1400	OPF86	31-JUL-1997:9	V6029.001	RCS THRUSTER DESICCANT INSPECTION (OMS) FROM: 86V6029 - Note 1	8	GSE	Propulsion	2.04.09
1401	OPF86	31-JUL-1997:9		POWER UP FROM: 86V000A - Note 1	12	OTC	Power Management	2.04.13
1402	OPF86	31-JUL-1997:9	V1262.011	VENT DOOR ENVELOPE INTRUSION INSPECTION(NOT REQ'D PER S. RUSSELL)		PVD	Safety Management & Control	2.05.01
1403	OPF86	31-JUL-1997:9	V1077.001	(RUN 2)POSU'S ORB FUEL CELL 2 COOLANT ULLAGE VERIFICA-TION	8	FCP	Power Management	2.05.03
1404	OPF86	31-JUL-1997:9	CE-193	AFD RECONFIG FROM: 86VC193A - Note 1	4	PLO	Payload Accommodations	2.06.03
1405	OPF86	31-JUL-1997:9	FCS-552	POST CEIT ODS ROUTING VERIF	2	FCS	Payload Accommodations	2.06.07
1406	OPF86	31-JUL-1997:9	V9021.001	OME TRICKLE PURGE MONITORING (OMS) FROM: 86V9021 - Note 1	8	GSE	Propulsion	2.07.01
1407	OPF86	31-JUL-1997:9	V9002.01	EXTEND NLG/MLG	2	HYD	Propulsion	2.07.03
1408	OPF86	31-JUL-1997:9	FCS-0513	TEHM MOUNTING PLATE R/R	4	FCS	Cockpit & Crew Cabin	2.08.01
1409	OPF86	31-JUL-1997:9	STR-4655	LOCK TITE SCREW	8	STR	Structures , Mechanisms, Veh Handling	2.08.01
1410	OPF86	31-JUL-1997:9	FCS-0512	TEHM DATA CABLE CONNECTOR	4	FCS	Cockpit & Crew Cabin	2.08.01
1411	OPF86	31-JUL-1997:9	COM-334//OEL-164	(MCR 18933)GPS/SIGI MOD	48	COM	Communications	2.09.01
1412	OPF86	31-JUL-1997:9	V1032.002	SEQ 5 ODS/TA C/O	8	PLO	Payload Accommodations	2.10.02
1413	OPF86	31-JUL-1997:9	V1032.003	FINAL ROLL PREPS/CONFIG AFT CLOSEOUT FROM: 86V112B - Note 1	32	VPL	Structures , Mechanisms, Veh Handling	2.10.03
1414	OPF86	31-JUL-1997:9	FCS-561//VT-STS-86-PLD-T1	TEHM INSTL/IVT	4	FCS	Payload Accommodations	4.04.05
1415	OPF86	31-JUL-1997:11	V1053.004O1//ECL-0957	CO2 SENSOR INSTALL/CHECKOUT	6	ECL	Environmental Ctl & Life Spt	2.04.12
1416	OPF86	31-JUL-1997:11	V9002.01	EXTEND NLG/MLG	2	HYD	Propulsion	2.07.03

No.	DATA SET	START TIME DD:MMM-YYYY:H	PROCEDURE NAME	DESCRIPTION	DURATION (Hours)	SYSTEM	DESIGN DISCIPLINE	GENERIC OPS FBS NUMBER (Appx A1)
1417	OPF86	31-JUL-1997:13	V1165.007	MLG STRUT HYD LVL CK & GN2 LK CKS(48 DECAY CK)	12	MEQ	Cmd, Ctl & Health Mngmt	2.04.10
1418	OPF86	31-JUL-1997:17	S70-952-03-004-0053	R/R O-RING HYD SYSTEM2 SUPPLY HOSE	8	HYD	Ground Systems & Facilities	2.08.07
1419	OPF86	01-AUG-1997:9		OHE-079 TILE CONSTRAINTS	80	TPS	Structures , Mechanisms, Veh Handling	2.02.02
1420	OPF86	01-AUG-1997:9	V1158VL6	OMS CLOSEOUT	16	OMS	Propulsion	2.04.07
1421	OPF86	01-AUG-1997:9	V1239G//V1240F//V1230I	F/M/A CLEANING FROM: 86V1239 - Note 1	8	GSE	Structures , Mechanisms, Veh Handling	2.04.09
1422	OPF86	01-AUG-1997:9	V6029.001	RCS THRUSTER DESICCANT INSPECTION (OMS) FROM: 86V6029 - Note 1	8	GSE	Propulsion	2.04.09
1423	OPF86	01-AUG-1997:9		POWER UP FROM: 86V000A - Note 1	10	OTC	Power Management	2.04.13
1424	OPF86	01-AUG-1997:9	V1077.001	/IPR-0036 (RUN 2)ORB FUEL CELL 2 COOLANT ULLAGE VERIFICA-TION	8	FCP	Power Management	2.05.03
1425	OPF86	01-AUG-1997:9	V9021.001	OME TRICKLE PURGE MONITORING (OMS) FROM: 86V9021 - Note 1	8	GSE	Propulsion	2.07.01
1426	OPF86	01-AUG-1997:9	STR-603	(MCR 18933)GPS BRACKET BONDS/INSTL(FOR COM-334) FROM: 86V14J - Note 1	8	STR	Structures , Mechanisms, Veh Handling	2.09.01
1427	OPF86	01-AUG-1997:9	V1032.002	SEQ 5 ODS/TA C/O FROM: 86V112E - Note 1	8	PLO	Payload Accommodations	2.10.02
1428	OPF86	01-AUG-1997:13	MEQ-0787	NLG DOOR ENVIRONMENTAL SEAL REPAIR	4	MEQ	Structures , Mechanisms, Veh Handling	2.05.02
1429	OPF86	02-AUG-1997:9		COM,ECL,EPD,INS,MEQ,OEL PROCESSING/HISTORICAL PWR DOWN PROCESSING(SAT. & SUN.) FROM: 86V350B - Note 1 FIXEDFLAG 2	16		Structures , Mechanisms, Veh Handling	2.08.04
1430	OPF86	02-AUG-1997:9	V1032.001	FWD COMPT CL OUT	24	VPL	Structures , Mechanisms, Veh Handling	2.10.02
1431	OPF86	04-AUG-1997:9	V9021.001	OP 160,170 FINAL DEACTIVATE RCS TRICKLE PURGE/MASS SPEC	4	OMS	Propulsion	2.04.08
1432	OPF86	04-AUG-1997:9	V1239G//V1240F//V1230I	F/M/A CLEANING FROM: 86V1239 - Note 1	48	GSE	Structures , Mechanisms, Veh Handling	2.04.09
1433	OPF86	04-AUG-1997:9	V6029.001	RCS THRUSTER DESICCANT INSPECTION (OMS) FROM: 86V6029 - Note 1	40	GSE	Propulsion	2.04.09
1434	OPF86	04-AUG-1997:9	V1053.004O1//ECL-0957	CO2 SENSOR INSTALL/CHECKOUT FROM: 86V113 - Note 1	8	ECL	Environmental Ctl & Life Spt	2.04.12
1435	OPF86	04-AUG-1997:9	V34-00001/7/11/18//V34-90	MIDBODY C/O/TS FC3	16	FCP	Power Management	2.04.13
1436	OPF86	04-AUG-1997:9	V1007.006	POSU'S AFT STRUCT LK TEST/ORB COMPART POS PRESS TEST (LOW FLOW)	16	PVD	Safety Management & Control	2.05.02
1437	OPF86	04-AUG-1997:9	V9021.001	OME TRICKLE PURGE MONITORING (OMS) FROM: 86V9021 - Note 1	32	GSE	Propulsion	2.07.01
1438	OPF86	04-AUG-1997:9	STR-4655	LOCK TITE SCREW FROM: 86V18C - Note 1	8	STR	Structures , Mechanisms, Veh Handling	2.08.01
1439	OPF86	04-AUG-1997:9	COM-334//OEL-164	(MCR 18933)GPS/SIGI MOD FROM: 86V14D - Note 1	16	COM	Communications	2.09.01
1440	OPF86	04-AUG-1997:9	V1032.003	FINAL ROLL PREPS/CONFIG AFT CLOSEOUT FROM: 86V112B - Note 1	16	VPL	Structures , Mechanisms, Veh Handling	2.10.03
1441	OPF86	04-AUG-1997:13	V9021.001	OP 160,170,120 FINAL DEACTIVATE RCS TRICKLE PURGE/MASS SPEC	4	OMS	Propulsion	2.04.08
1442	OPF86	04-AUG-1997:13		POWER UP FROM: 86V000A - Note 1	4	OTC	Power Management	2.04.13

No.	DATA SET	START TIME DD:MMM-YYYY:H	PROCEDURE NAME	DESCRIPTION	DURATION (Hours)	SYSTEM	DESIGN DISCIPLINE	GENERIC OPS FBS NUMBER (Appx A1)
1443	OPF86	04-AUG-1997:13	V1077.001	(RUN 2)ORB FUEL CELL 2 COOLANT ULLAGE VERIFICATION POI'S	4	FCP	Power Management	2.05.03
1444	OPF86	04-AUG-1997:17	V9002.01	RETRACT NLG/MLG (DEDICATED)(SUPPORT V1007.006)	4	HYD	Propulsion	2.07.03
1445	OPF86	05-AUG-1997:1	COM-334//OEL-164	(MCR 18933)GPS/SIGI MOD FROM: 86V14D - Note 1	40	COM	Communications	2.09.01
1446	OPF86	05-AUG-1997:7	V80-94400	INSTALL DR 44(V1007.006)	2	STR	Structures , Mechanisms, Veh Handling	2.02.06
1447	OPF86	05-AUG-1997:9	V80-95907/933/935//AFT-49	INSTALL E1,E2,E3 SSME CARRIER PANELS FROM: 86V105B - Note 1	16	STR	Thermal Management	2.04.05
1448	OPF86	05-AUG-1997:9	V1055.002	POTABLE WATER SERVICING POI'S	8	FCP	Power Management	2.04.08
1449	OPF86	05-AUG-1997:9	V1055.002	POTABLE WATER 3 DAY MINIMUM DECAY CK		FCP	Power Management	2.04.08
1450	OPF86	05-AUG-1997:9		POWER UP FROM: 86V000A - Note 1	4	OTC	Power Management	2.04.13
1451	OPF86	05-AUG-1997:9	V9001	MADS SNAPSHOT		INS	Power Management	2.04.13
1452	OPF86	05-AUG-1997:9	V1007.006	AFT STRUCT LK TEST/ORB COMPT POS PRESS TEST	8	PVD	Safety Management & Control	2.05.02
1453	OPF86	05-AUG-1997:9	V1032.001	FWD COMPT CL OUT FROM: 86V268A - Note 1	8	VPL	Structures , Mechanisms, Veh Handling	2.10.02
1454	OPF86	05-AUG-1997:13	V80-04400	REMOVE DR 44(V1007.006)	2	STR	Structures , Mechanisms, Veh Handling	2.02.05
1455	OPF86	05-AUG-1997:13	V9002.01	EXTEND LG (DEDICATED)	4	HYD	Power Management	2.07.03
1456	OPF86	05-AUG-1997:13	V1032.002B//V80-94400//V6	C/O DR 44	4	STR	Structures , Mechanisms, Veh Handling	2.10.02
1457	OPF86	05-AUG-1997:17	V9028.002	FINAL TIRE PRESSURE FOR FLIGHT	16	MEQ	Structures , Mechanisms, Veh Handling	2.04.01
1458	OPF86	05-AUG-1997:17	V34-00001/7/11/18//V34-90	MIDBODY C/O/TS FC3 FROM: 86V1077D - Note 1	8	FCP	Power Management	2.04.13
1459	OPF86	05-AUG-1997:17	S5023.001	POSU'S ORB/OTS TRANSPORT OPS	8	OSO	Structures , Mechanisms, Veh Handling	2.10.03
1460	OPF86	06-AUG-1997:9	V1284	UPPER BODY FLAP BLACK RTV COATING		TPS	Thermal Management	2.04.05
1461	OPF86	06-AUG-1997:9	V9021.001	DEACTIVATE OMS TRICKLE PURGE	4	OMS	Propulsion	2.04.08
1462	OPF86	06-AUG-1997:9	V1007.006	AFT STRUCT LK TEST/ORB COMPT POS PRESS TEST POI	8	PVD	Safety Management & Control	2.05.02
1463	OPF86	06-AUG-1997:9	V9002.09	OP 140 DEMATE 1338 HOSE	2	HYD	Propulsion	2.07.03
1464	OPF86	06-AUG-1997:9	V1032.001	FWD COMPT CL OUT FROM: 86V268A - Note 1	8	VPL	Structures , Mechanisms, Veh Handling	2.10.02
1465	OPF86	06-AUG-1997:13	WWM-0039	POTABLE WATER SERVICING(BIONETICS)	4	FCP	Power Management	2.04.07
1466	OPF86	06-AUG-1997:13		POWER UP FROM: 86V000A - Note 1	10	OTC	Power Management	2.04.13
1467	OPF86	07-AUG-1997:6	V1032.001	FWD COMPT CL OUT/CLOSE HATCH	11	VPL	Structures , Mechanisms, Veh Handling	2.10.02
1468	OPF86	07-AUG-1997:9	V1098.002	POSU'S LDG BUNGEE LITE VERIF	2	MEQ	Structures , Mechanisms, Veh Handling	2.04.01
1469	OPF86	07-AUG-1997:9	V9046.010	VOL B REMOVAL	2	PLO	Cockpit & Crew Cabin	2.04.04
1470	OPF86	07-AUG-1997:9	V76-10001/2	L02/LH2 ET UMB MONOBALL FINAL R/O PREPS	8	STR	Structures , Mechanisms, Veh Handling	2.05.02
1471	OPF86	07-AUG-1997:9	V60-40002	DISCONNECT ORBITER GSE FREON SYSTEM	2	ECL	Thermal Management	2.10.01
1472	OPF86	07-AUG-1997:9	V76-40004J/5G	DISCONNECT T-O UMB CABLES	2	EPD	Power Management	2.10.01

No.	DATA SET	START TIME DD:MMM-YYYY:H	PROCEDURE NAME	DESCRIPTION	DURATION (Hours)	SYSTEM	DESIGN DISCIPLINE	GENERIC OPS FBS NUMBER (Appx A1)
1473	OPF86	07-AUG-1997:9	V6029.001	RUN 1/2 FINAL DESICCANT C/O	8	OMS	Propulsion	2.10.02
1474	OPF86	07-AUG-1997:9	S5023.001	POSU'S ORB/OTS TRANSPORT OPS FROM: 86V161A - Note 1	8	OSO	Structures , Mechanisms, Veh Handling	2.10.03
1475	OPF86	07-AUG-1997:9	OHE-079	POSU'S ORB JACKDOWN WT&CG	8	OHE	Structures , Mechanisms, Veh Handling	2.10.03
1476	OPF86	07-AUG-1997:11	V1098.002	LDG BUNGEE LITE VERIF (1660 BOX CABLE ROUTED THRU SIDE HATCH)	4	MEQ	Structures , Mechanisms, Veh Handling	2.04.01
1477	OPF86	07-AUG-1997:11	V3555	SEQ 6,9 DISCONNECT ORBITER PURGE AIR	2	PVD	Safety Management & Control	2.10.01
1478	OPF86	07-AUG-1997:13		REMOVE LH2/L02 T-O CARRIER FRAME/INSPECTION	4	STR	Structures , Mechanisms, Veh Handling	2.10.01
1479	OPF86	07-AUG-1997:13	V6050	CREW MODULE CLOSEOUT PHOTOS	4	OHE	Structures , Mechanisms, Veh Handling	2.10.02
1480	OPF86	07-AUG-1997:17	FWD-3806/7	CREW HATCH EVAL		TPS	Thermal Management	2.04.06
1481	OPF86	07-AUG-1997:17	OHE-079	ORB JACKDOWN WT/CG	8	OHE	Structures , Mechanisms, Veh Handling	2.10.03
1482	OPF86	08-AUG-1997:1	V5101	WGT/CG COMPLETE		OHE	Structures , Mechanisms, Veh Handling	2.10.03
1483	OPF86	08-AUG-1997:1		WGT/CG COMPLETE		OHE	Structures , Mechanisms, Veh Handling	2.10.03
1484	OPF86	08-AUG-1997:9		DAILY SUPV/EMPLOYEE BRIEFING			Structures , Mechanisms, Veh Handling	2.07.02
1485	OPF86	08-AUG-1997:9	S5023.001	POSU'S ORB/OTS TRANSPORT OPS FROM: 86V161A - Note 1	2	OSO	Structures , Mechanisms, Veh Handling	2.10.03
1486	OPF86	08-AUG-1997:11	S5023.001	ORBITER/OTS TRANSPORT OPERATIONS - OPF TO VAB	10	OSO	Structures , Mechanisms, Veh Handling	2.10.03
1487	OPF86	08-AUG-1997:17	FWD-3823/94/95	CHIN PANEL C/P REWORK & INSTLN FROM: 86V080B - Note 1	8	TPS	Thermal Management	2.04.06
1488	OPF86	08-AUG-1997:17	V3555	RECONNECT ORBITER PURGE AIR	4	PVD	Safety Management & Control	2.07.01
1489	OPF86	08-AUG-1997:17		DAILY SUPV/EMPLOYEE BRIEFING			Structures , Mechanisms, Veh Handling	2.07.02
1490	OPF86	08-AUG-1997:19		LPS C8(S5023)			Structures , Mechanisms, Veh Handling	2.01.01
1491	OPF86	09-AUG-1997:1		DAILY SUPV/EMPLOYEE BRIEFING			Structures , Mechanisms, Veh Handling	2.07.02
1492	OPF86	11-AUG-1997:1	FWD-3823/94/95	CHIN PANEL C/P REWORK & INSTLN FROM: 86V080B - Note 1	8	TPS	Thermal Management	2.04.06
1493	OPF86	11-AUG-1997:9	V3555	SEQ 6,9 DISCONNECT ORBITER PURGE AIR	2	PVD	Safety Management & Control	2.07.01
1494	OPF86	11-AUG-1997:9	V6029.001	RUN 1/2 FINAL DESICCANT C/O FROM: 86V090C - Note 1	1	OMS	Propulsion	2.10.02
1495	OPF86	11-AUG-1997:9	S5023.001	ORBITER/OTS TRANSPORT OPERATIONS - OPF TO VAB FROM: 86V161 - Note 1	2	OSO	Structures , Mechanisms, Veh Handling	2.10.03
1496	OPF86	11-AUG-1997:11	S5023.001	ROLLOUT FIRST MOTION		OSO	Structures , Mechanisms, Veh Handling	2.10.03
1497	PAD86	07-Aug-97	S0007 VL-2	T-0 LAUNCH STS-085	0	MIL	Cmd, Ctl & Health Mngmt	5.09.05
1498	PAD86	07-Aug-97	S2005	LAUNCH PAD SECURING	25	MIL	Ground Systems & Facilities	5.10.01
1499	PAD86	08-AUG-1997:8	T2445	ET/IT ARM VALID (EVERY FLOW)	7	ARM	Ground Systems & Facilities	5.01.03
1500	PAD86	08-AUG-1997:9	G3153	LOX SYSTEM WASHDOWN (EVERY FLOW)	6	LOX	Propulsion	5.01.02
1501	PAD86	08-AUG-1997:9	G6204	WASHDOWN PIPING (NIB)	7	LH2	Propulsion	5.01.02
1502	PAD86	08-AUG-1997:9		LAUNCH DMG ASSESSMENT PCR & PCR DOORS	4	PCR	Payload Accommodations	5.01.03

No.	DATA SET	START TIME DD:MMM-YYYY:H	PROCEDURE NAME	DESCRIPTION	DURATION (Hours)	SYSTEM	DESIGN DISCIPLINE	GENERIC OPS FBS NUMBER (Appx A1)
1503	PAD86	08-AUG-1997:9	A3204/V3592	DRAIN MLP FREON LOOP	15		Safety Management & Control	5.01.04
1504	PAD86	08-AUG-1997:9	G5107	LOX/MLP DEMATE (EVERY FLOW)	5	LOX	Propulsion	5.01.04
1505	PAD86	08-AUG-1997:9	G5207	LH2/MLP DEMATE (EVERY FLOW)	5	LH2	Propulsion	5.01.04
1506	PAD86	08-AUG-1997:9	G6223	WASHDOWN PIPING (NIB) EVERY FLOW	4	LH2	Propulsion	5.10.01
1507	PAD86	11-AUG-1997:8	G2108	LOX TANK REPLENISH (4 WAVES DAILY)	8	LOX	Propulsion	5.01.04
1508	PAD86	11-AUG-1997:8	G2272	VENT LINE PURGE PANEL C/O	5	LH2	Propulsion	5.03.01
1509	PAD86	11-AUG-1997:9	G6305	REWRAP GH2 VENT LINE CABLES (EVERY FLOW)	8	ELC	Power Management	5.01.01
1510	PAD86	11-AUG-1997:9	G2297	GN2/LH2 FIRE DETECT (EVERY FLOW)	72	HGS	Safety Management & Control	5.01.03
1511	PAD86	11-AUG-1997:9	V7192	WINDOW PURGE PANEL VAL	5	ECS	Safety Management & Control	5.01.03
1512	PAD86	11-AUG-1997:9	G2162	LOX VAPORIZER LEAK CHECK (OPTIONAL-EVERY FLOW)	24	LOX	Propulsion	5.01.03
1513	PAD86	11-AUG-1997:9	V2172	C70-1226-03-003 CREW CABIN INTEGRITY TEST PANEL VAL (EVERY FLOW)	6	ECL	Environmental Ctl & Life Spt	5.01.03
1514	PAD86	11-AUG-1997:9	V3541	HYD LK & FIRE DETECT (EVERY FLOW)	72	HGS	Safety Management & Control	5.01.03
1515	PAD86	11-AUG-1997:9	G6152	GOX VENT SEAL INSP (EVERY FLOW)	7	ARM	Ground Systems & Facilities	5.01.03
1516	PAD86	11-AUG-1997:9	G2160	TANK FILL HEADER LEAK CHECK (EVERY FLOW)	22	LOX	Propulsion	5.01.03
1517	PAD86	11-AUG-1997:9	G2278	S72-0697-13 VENT LINE GHE PNL C/O	16	LH2	Propulsion	5.01.03
1518	PAD86	11-AUG-1997:9	V2161.001	LOCAL ECS VALS (EVERY FLOW)	40	ECS	Safety Management & Control	5.01.03
1519	PAD86	11-AUG-1997:9	G2168	LOX PURGE SYS LEAK CHECK	23	LOX	Propulsion	5.03.02
1520	PAD86	11-AUG-1997:9	V3542	HAZ GAS VAPOR DETECT (EVERY FLOW)	72	HGS	Safety Management & Control	5.03.03
1521	PAD86	11-AUG-1997:9	V5084	PGHM RECONFIGURATION	8	PCR	Payload Accommodations	5.05.03
1522	PAD86	12-AUG-1997:8	T2404	LK CK ET VENT/PURGE SYS (LPS)(EVERY FLOW)	5	LOX	Propulsion	5.01.03
1523	PAD86	12-AUG-1997:8	G2159	GOX VENT VALID (EVERY FLOW)	8	ARM	Ground Systems & Facilities	5.01.03
1524	PAD86	12-AUG-1997:8	G2208	LH2 TANK REPLENISH	5	LH2	Propulsion	5.01.04
1525	PAD86	12-AUG-1997:8	V2263	NICKEL CAD. BATTERY MAINT (LOSS OF BATTERY SUPPORT)	5	ELC	Power Management	5.01.04
1526	PAD86	12-AUG-1997:9	V2073	OAA VALID (EVERY FLOW)	4	ARM	Ground Systems & Facilities	5.01.03
1527	PAD86	12-AUG-1997:9	V2074	HCU VALID (EVERY FLOW)	5	ARM	Ground Systems & Facilities	5.01.03
1528	PAD86	12-AUG-1997:9	T3445	LOAD CELL CHANGEOUT (EVERY FLOW)	6	ARM	Ground Systems & Facilities	5.01.04
1529	PAD86	12-AUG-1997:9	G2271	STOR AREA LANTERN RING PGE PNL	5	LH2	Propulsion	5.03.01
1530	PAD86	12-AUG-1997:9		MLP OFF PAD	3	MIL	Ground Systems & Facilities	5.10.02
1531	PAD86	13-AUG-1997:8	V2201	OX TANKER OFFLOAD/CIRC & SAMPLE	6	OMS	Propulsion	5.01.04
1532	PAD86	13-AUG-1997:9	V5084	PRELIMINARY LEAK CHECKS	8	PCR	Payload Accommodations	5.01.03
1533	PAD86	13-AUG-1997:9	G2273	S72-0697-2 VENT LINE C/O	4	LH2	Propulsion	5.01.03

No.	DATA SET	START TIME DD:MMM-YYYY:H	PROCEDURE NAME	DESCRIPTION	DURATION (Hours)	SYSTEM	DESIGN DISCIPLINE	GENERIC OPS FBS NUMBER (Appx A1)
1534	PAD86	13-AUG-1997:9	T2403	INTERTANK PURGE SYS VAL (EVERY FLOW)	8	LH2	Propulsion	5.01.03
1535	PAD86	13-AUG-1997:9	V2200	FUEL TANKER OFFLOAD/CICR & SAMPLE	1	OMS	Propulsion	5.01.04
1536	PAD86	13-AUG-1997:9	V2516	VACUUM VENT DUCT PURGE (EVERY FLOW)	4	ECS	Safety Management & Control	5.01.04
1537	PAD86	13-AUG-1997:9	T2405	ET NOSE CONE PURGE PNL (EVERY FLOW)	8	LOX	Propulsion	5.01.04
1538	PAD86	13-AUG-1997:9	S0044	SIM RUN (NO LPS)	5	TCO	Cmd, Ctl & Health Mngmt	5.03.04
1539	PAD86	14-AUG-1997:8	T1052	ET VAS VERIF (EVERY FLOW)	5	ELC	Power Management	5.01.03
1540	PAD86	14-AUG-1997:8	G2398	ET HEATER SYSTEM VAL (EVERY FLOW)	7	ELC	Power Management	5.01.03
1541	PAD86	14-AUG-1997:8	T2444	GH2 VENT LINE VALID (EVERY FLOW)	8	ARM	Ground Systems & Facilities	5.01.03
1542	PAD86	14-AUG-1997:8	G2108	LOX TANK REPLENISH	10	LOX	Propulsion	5.01.04
1543	PAD86	14-AUG-1997:9	V2161.002	ECS REMOTE VAL (LPS C-6)(EVERY FLOW)	8	ECS	Safety Management & Control	5.01.03
1544	PAD86	14-AUG-1997:9	V3521	RELOCATE/SERVICE & VALIDATE 0508 UNITS (EVERY FLOW)	6	ECL	Environmental Ctl & Life Spt	5.01.04
1545	PAD86	15-AUG-1997:8	G2208	LH2 TANK REPLENISH	5	LH2	Propulsion	5.01.04
1546	PAD86	15-AUG-1997:9	N5085	PGHM FINAL HYD LEAK CHK	6	PCR	Payload Accommodations	5.05.02
1547	PAD86	15-AUG-1997:9		PCR FINAL CUSTOMER WALKDOWN(0900)	2	PCR	Payload Accommodations	5.05.02
1548	PAD86	15-AUG-1997:16	V5161	LIFT S70-1317 CARTS TO RSS 135' LEVEL (EVERY FLOW)	5	OMS	Propulsion	5.06.02
1549	PAD86	15-AUG-1997:20	V7234	REINSTALL FLOWMETERS (EVERY FLOW)	3	OMS	Propulsion	5.01.04
1550	PAD86	15-AUG-1997:22	S70-0700-0-1-0277/0278	HINGE COLUMN FLEXHOSE REPLACEMENT	3	OMS	Propulsion	5.04.01
1551	PAD86	18-Aug-97	A5214	SSV AT PAD	0	MIL	Structures , Mechanisms, Veh Handling	5.02.01
1552	PAD86	18-Aug-97	S0009	LAUNCH PAD VALIDATION	25		Cmd, Ctl & Health Mngmt	5.03.02
1553	PAD86	18-AUG-1997:9	G2108	LOX TANK REPLENISH	10	LOX	Propulsion	5.01.04
1554	PAD86	18-AUG-1997:10	T1450	GUCP LEAK CHECK	4	LH2	Propulsion	5.03.02
1555	PAD86	19-AUG-1997:8	T2408	BLOW DOWN/SAMPLE (FSS) (EVERY FLOW)	6	LOX	Propulsion	5.01.02
1556	PAD86	19-AUG-1997:8	G2208	LH2 TANK REPLENISH	5	LH2	Propulsion	5.01.04
1557	PAD86	22-AUG-1997:8	G2208	LH2 TANK REPLENISH	5	LH2	Propulsion	5.01.04
1558	PAD86	26-AUG-1997:8	S0024VL-1	PROPELLANT LOAD	36	MIL	Propulsion	5.06.01
1559	PAD86	04-SEP-1997:9		PAYLOAD TO PAD	0		Structures , Mechanisms, Veh Handling	4.04.01
1560	PAD86	25-SEP-1997:9	SOOO7	STS-086 LAUNCH	0		Cmd, Ctl & Health Mngmt	5.09.05
1561	PAD86	14-OCT-1997:9	G2111	LOX TANK PRESSURIZATION TEST	8	LOX	Propulsion	5.01.02
1562	PAD86	14-OCT-1997:9	G2115	LOX TANK TEST	8	LOX	Propulsion	5.01.03
1563	PAD86	14-OCT-1997:9	G2211	TANK PRESURIZATION TEST	8	LH2	Propulsion	5.01.03
1564	PAD86	14-OCT-1997:9	V2161.002	ECS GN2 FLOWS	8	ECS	Safety Management & Control	5.01.03

No.	DATA SET	START TIME DD:MMM-YYYY:H	PROCEDURE NAME	DESCRIPTION	DURATION (Hours)	SYSTEM	DESIGN DISCIPLINE	GENERIC OPS FBS NUMBER (Appx A1)
1565	PAD86	14-OCT-1997:9	G2108	LOX TANK REPLENISH	12	LOX	Propulsion	5.01.04
1566	PAD86	14-OCT-1997:9	G2208	LH2 TANK REPLENISH	8	LH2	Propulsion	5.01.04
1567	PAD86	14-OCT-1997:9	S70-0700-0-1-0277/0278	REPLACE HINGE COLUMN OX. FILL/DRAIN HOSES	16	OMS	Propulsion	5.04.01
1568	PAD86	15-OCT-1997:9	G2108	LOX TANK REPLENISH	12	LOX	Propulsion	5.01.04
1569	PSF86	25-APR-1997:9		LEFT FORWARD OFFLOAD	59		Structures , Mechanisms, Veh Handling	6.01.03
1570	PSF86	29-APR-1997:9		LEFT FORWARD CENTER OFFLOAD	62		Structures , Mechanisms, Veh Handling	6.01.03
1571	PSF86	29-APR-1997:17	B5308	T15 REC INSP AFT SKIRT	5	SRM	Structures , Mechanisms, Veh Handling	3.01.03
1572	PSF86	30-APR-1997:17	B5308	T14 INSTALL AFT SKIRT INTO STAND	4	SRM	Structures , Mechanisms, Veh Handling	3.01.03
1573	PSF86	01-MAY-1997:18	B5309	T09 STIFF RING PREPS	4	SRM	Structures , Mechanisms, Veh Handling	3.02.01
1574	PSF86	01-MAY-1997:21	B5309	T09 STIFF RING PREPS	4	SRM	Structures , Mechanisms, Veh Handling	3.02.01
1575	PSF86	02-MAY-1997:13	B5308	T01 POSITION RAILCAR IN RPSF	2	SRM	Structures , Mechanisms, Veh Handling	6.01.02
1576	PSF86	02-MAY-1997:14	B5308	T01 REM INSTRUMENTATION	2	SRM	Structures , Mechanisms, Veh Handling	6.02.03
1577	PSF86	02-MAY-1997:14	B5308	T01 BEAM PREPS/REMOVE RAILCAR COVER	2	SRM	Structures , Mechanisms, Veh Handling	6.02.03
1578	PSF86	02-MAY-1997:15	B5308	T03 BEAM PREPS/REMOVE END COVER	2	SRM	Structures , Mechanisms, Veh Handling	6.02.03
1579	PSF86	02-MAY-1997:16	B5308	T05 GRAIN/INHIBITOR INSPECTION	4	SRM	Structures , Mechanisms, Veh Handling	6.03.01
1580	PSF86	02-MAY-1997:19	B5308	T05 NOZZLE INSPECTION	2	SRM	Structures , Mechanisms, Veh Handling	6.03.01
1581	PSF86	02-MAY-1997:20	B5308	T04 INSTALL END COVER	2	SRM	Structures , Mechanisms, Veh Handling	6.02.02
1582	PSF86	02-MAY-1997:21	B5308	T10 BEAM PREPS/ROTATE / SET SEGMENT ON 386 STAND	4	SRM	Structures , Mechanisms, Veh Handling	3.01.03
1583	PSF86	05-MAY-1997:9	B5308	LEFT AFT CENTER OFFLOAD	69	SRM	Structures , Mechanisms, Veh Handling	6.01.03
1584	PSF86	05-MAY-1997:20	B5309	T01 AFT SKIRT CLEVIS INSPECTION	5	SRM	Structures , Mechanisms, Veh Handling	3.04.01
1585	PSF86	06-MAY-1997:9	B5309	T01 BEAM PREPS/REMOVE AFT END RING/LIFT	3	SRM	Structures , Mechanisms, Veh Handling	3.02.01
1586	PSF86	06-MAY-1997:11	B5309	T01 TANG METAL PARTS INSPECTION	3	SRM	Structures , Mechanisms, Veh Handling	3.01.03
1587	PSF86	06-MAY-1997:13	B5309	T01 MATE AFT SEG TO AFT SKIRT/VIDEO	8	SRM	Structures , Mechanisms, Veh Handling	3.02.02
1588	PSF86	06-MAY-1997:20	B5309	T02 INSTALL INTERNAL ACCESS KIT	2	SRM	Structures , Mechanisms, Veh Handling	3.02.01
1589	PSF86	07-MAY-1997:12	B5309	T03 AFT SEG/NOZ SHIP LINK INSP	18	SRM	Structures , Mechanisms, Veh Handling	3.03.01
1590	PSF86	08-MAY-1997:9		NOZZLE OFFLOAD	42		Structures , Mechanisms, Veh Handling	6.01.03
1591	PSF86	09-MAY-1997:15	B5309	T05 SINE BAR INSP/ETA RING INSTL'N PREPS	25	SRM	Structures , Mechanisms, Veh Handling	3.03.04
1592	PSF86	12-MAY-1997:9		RIGHT FORWARD OFFLOAD	44		Structures , Mechanisms, Veh Handling	6.01.03
1593	PSF86	12-MAY-1997:17	B5309	T11 STIFF RING INSP/INSTALLATION	53	SRM	Structures , Mechanisms, Veh Handling	3.03.03
1594	PSF86	12-MAY-1997:21	B5309	T06 ETA RING INSTALLATION/K5NA CURE	74	SRM	Structures , Mechanisms, Veh Handling	3.03.03
1595	PSF86	14-MAY-1997:9		RIGHT FORWARD CENTER OFFLOAD	78		Structures , Mechanisms, Veh Handling	6.01.03

No.	DATA SET	START TIME DD:MMM-YYYY:H	PROCEDURE NAME	DESCRIPTION	DURATION (Hours)	SYSTEM	DESIGN DISCIPLINE	GENERIC OPS FBS NUMBER (Appx A1)
1596	PSF86	14-MAY-1997:20	B5309	T17 PIN RETAINER BAND INSULATION/K5NA CURE	96	SRM	Structures , Mechanisms, Veh Handling	3.03.05
1597	PSF86	15-MAY-1997:10	B5309	T42 REMOVE FWD HANDLING RING	6	SRM	Structures , Mechanisms, Veh Handling	3.06.01
1598	PSF86	15-MAY-1997:15	B5305	T01 CABLE WRAP	8	SRE	Cmd, Ctl & Health Mngmt	3.03.03
1599	PSF86	15-MAY-1997:17	B5309	T13 K5NA APPL TO STIFF RING/K5NA CURE	81	SRM	Structures , Mechanisms, Veh Handling	3.03.05
1600	PSF86	16-MAY-1997:17	B5309	T33 ETA RING FOAMING OPS	45	SRM	Structures , Mechanisms, Veh Handling	3.03.05
1601	PSF86	19-MAY-1997:9	B5309	T24 STRUT PREPS	3	SRM	Structures , Mechanisms, Veh Handling	3.03.01
1602	PSF86	19-MAY-1997:13	B5305	T01 CABLE WRAP	12	SRE	Cmd, Ctl & Health Mngmt	3.03.03
1603	PSF86	19-MAY-1997:16	B5309	T18 SYS TUN FLR PLATE ASSY INSTL/RTV CURE	43	SRM	Structures , Mechanisms, Veh Handling	3.03.03
1604	PSF86	19-MAY-1997:17	B5309	T38 DIAGONAL STRUT PRE-MOLD FABRICATION	29	SRM	Structures , Mechanisms, Veh Handling	3.02.01
1605	PSF86	20-MAY-1997:13	B5308	T15 REC INSP AFT SKIRT	10	SRM	Structures , Mechanisms, Veh Handling	6.01.03
1606	PSF86	21-MAY-1997:10	B5308	T14 INSTALL AFT SKIRT INTO STAND	5	SRM	Structures , Mechanisms, Veh Handling	3.01.03
1607	PSF86	21-MAY-1997:13	B5308	T01 POSITION RAILCAR IN RPSF	2	SRM	Structures , Mechanisms, Veh Handling	6.01.02
1608	PSF86	21-MAY-1997:13	B5308	T01 BEAM PREPS/REMOVE RAILCAR COVER	2	SRM	Structures , Mechanisms, Veh Handling	6.02.03
1609	PSF86	21-MAY-1997:14	B5308	T01 REM INSTRUMENTATION	2	SRM	Structures , Mechanisms, Veh Handling	6.02.03
1610	PSF86	21-MAY-1997:15	B5308	T03 BEAM PREPS/REMOVE END COVER	2	SRM	Structures , Mechanisms, Veh Handling	6.02.03
1611	PSF86	21-MAY-1997:16	B5305	T06 ETA RING CABLE INSTALLATION(STRUT)	14	SRE	Cmd, Ctl & Health Mngmt	3.03.03
1612	PSF86	21-MAY-1997:16	B5309	T07 AFT IEA INSTALLATION	9	SRM	Structures , Mechanisms, Veh Handling	3.03.03
1613	PSF86	21-MAY-1997:16	B5308	T05 GRAIN/INHIBITOR INSPECTION	5	SRM	Structures , Mechanisms, Veh Handling	6.03.01
1614	PSF86	21-MAY-1997:19	B5309	T08 UPPER AND DIAGONAL STRUT INSTALLATION	5	SRM	Structures , Mechanisms, Veh Handling	3.03.03
1615	PSF86	21-MAY-1997:20	B5309	T25 LOWER STRUT INSTALLATION	5	SRM	Structures , Mechanisms, Veh Handling	3.03.03
1616	PSF86	21-MAY-1997:20	B5308	T05 NOZZLE INSPECTION	2	SRM	Structures , Mechanisms, Veh Handling	6.03.01
1617	PSF86	21-MAY-1997:21	B5305	T02 SYS TUNNEL CABLE INSTALLATION	42	SRE	Cmd, Ctl & Health Mngmt	3.03.03
1618	PSF86	21-MAY-1997:21	B5308	T04 INSTALL END COVER	2	SRM	Structures , Mechanisms, Veh Handling	6.02.02
1619	PSF86	22-MAY-1997:14	B5309	T04 INSTALL NOZZLE	10	SRM	Structures , Mechanisms, Veh Handling	3.03.03
1620	PSF86	22-MAY-1997:14	B5305	T04 O/I STRUT CBL TEST/VAL	9	SRE	Cmd, Ctl & Health Mngmt	3.04.01
1621	PSF86	22-MAY-1997:16	B5308	T10 BEAM PREPS/ROTATE / SET SEGMENT ON STAND	9	SRM	Structures , Mechanisms, Veh Handling	3.01.03
1622	PSF86	22-MAY-1997:19	B5309	T14 NOZZLE MASS FLOW LEAK CHECK	18	SRM	Structures , Mechanisms, Veh Handling	3.04.01
1623	PSF86	23-MAY-1997:9	B5309	T01 AFT SKIRT CLEVIS INSPECTION	3	SRM	Structures , Mechanisms, Veh Handling	3.02.01
1624	PSF86	23-MAY-1997:11	B5309	T01 BEAM PREPS/REMOVE AFT END RING/LIFT	3	SRM	Structures , Mechanisms, Veh Handling	3.02.02
1625	PSF86	23-MAY-1997:13	B5309	T01 TANG METAL INSPECTION	2	SRM	Structures , Mechanisms, Veh Handling	3.02.01
1626	PSF86	23-MAY-1997:14	B5309	T01 MATE AFT SEG TO AFT SKIRT/ VIDEO	8	SRM	Structures , Mechanisms, Veh Handling	3.02.02

No.	DATA SET	START TIME DD:MMM-YYYY:H	PROCEDURE NAME	DESCRIPTION	DURATION (Hours)	SYSTEM	DESIGN DISCIPLINE	GENERIC OPS FBS NUMBER (Appx A1)
1627	PSF86	23-MAY-1997:21	B5309	T03 AFT SEG/NOZ SHIP LINK INSP	17	SRM	Structures , Mechanisms, Veh Handling	3.02.01
1628	PSF86	23-MAY-1997:21	B5309	T02 INSTALL INTERNAL ACCESS KIT	2	SRM	Structures , Mechanisms, Veh Handling	3.02.01
1629	PSF86	27-MAY-1997:9	B5305	T05 ETA RING CABLE INSTALLATION (TUNNEL(IEA))	7	SRE	Cmd, Ctl & Health Mngmt	3.03.03
1630	PSF86	27-MAY-1997:9	B5308	RIGHT AFT CENTER OFFLOAD	47	SRM	Structures , Mechanisms, Veh Handling	6.01.03
1631	PSF86	27-MAY-1997:17	B5309	T05 SINE BAR/ETA RING INSTALLATION PREPS	37	SRM	Structures , Mechanisms, Veh Handling	3.03.04
1632	PSF86	27-MAY-1997:17	B5309	T12 EXIT CONE BACKFILL INSTL'N/RTV CURE	87	SRM	Structures , Mechanisms, Veh Handling	3.03.05
1633	PSF86	27-MAY-1997:17	B5305	T03 O/I TUNNELCABLE TESTING/ VAL	5	SRE	Cmd, Ctl & Health Mngmt	3.04.01
1634	PSF86	27-MAY-1997:21	B5305	T09 O/I CABLE CONN,ROOSTER TAIL & IEA	4	SRE	Cmd, Ctl & Health Mngmt	3.03.03
1635	PSF86	28-MAY-1997:11	B5309	T16 STIFF RING SPLT PLT INSTALL 90DEG.	5	SRM	Structures , Mechanisms, Veh Handling	3.03.03
1636	PSF86	28-MAY-1997:11	B5309	T19 TUNNEL COVER INSTALLATION	27	SRM	Structures , Mechanisms, Veh Handling	3.03.03
1637	PSF86	28-MAY-1997:11	B5305	T10 NSI CBL CONN/ALLINCO CK	13	SRE	Cmd, Ctl & Health Mngmt	3.04.01
1638	PSF86	29-MAY-1997:11	B5309	T17 PIN RETAINER BAND INSULATION/K5NA CURE	114	SRM	Structures , Mechanisms, Veh Handling	3.03.05
1639	PSF86	29-MAY-1997:11	B5309	T32 NOZZLE CABLE CLOSEOUT	17	SRM	Structures , Mechanisms, Veh Handling	3.03.05
1640	PSF86	29-MAY-1997:17	B5309	T40 PR-855 FOAM INSUL AT STRUTS	45	SRM	Structures , Mechanisms, Veh Handling	3.03.05
1641	PSF86	29-MAY-1997:19	B5309	T06 ETA RING INSTALLATION/K5NA CURE	115	SRM	Structures , Mechanisms, Veh Handling	3.03.03
1642	PSF86	29-MAY-1997:21	B5309	T20 SYS TUN/STIFF RING SEALING/K5NA CURE	74	SRM	Structures , Mechanisms, Veh Handling	3.03.03
1643	PSF86	30-MAY-1997:10	B5309	T11 STIFF RING INSP/INSTALLATION	33	SRM	Structures , Mechanisms, Veh Handling	3.03.03
1644	PSF86	30-MAY-1997:10	B5309	T34 FOAM AFT SKIRT INTRL RING	23	SRM	Structures , Mechanisms, Veh Handling	3.03.05
1645	PSF86	30-MAY-1997:16	B5309	T36 RAIN SHEILD INITIAL INSTALL	7	SRM	Structures , Mechanisms, Veh Handling	3.03.04
1646	PSF86	02-JUN-1997:16	B5309	T21 RAIN/THERMAL CURTAIN INSTALLATION	12	SRM	Structures , Mechanisms, Veh Handling	3.03.05
1647	PSF86	03-JUN-1997:10	B5309	T22 COVER INSTALL AT DIAG STRUT/TEMP INST	38	SRM	Structures , Mechanisms, Veh Handling	3.03.04
1648	PSF86	03-JUN-1997:12	B5309	T31 ETA RING CVR INSTALLATION FROM LOWER STRUT TO 270 DEG.	35	SRM	Structures , Mechanisms, Veh Handling	3.03.03
1649	PSF86	03-JUN-1997:13	B5309	T26 AFT IEA CVR INSTL/RTV & K5NA CURE	33	SRM	Structures , Mechanisms, Veh Handling	3.03.03
1650	PSF86	03-JUN-1997:15	B5309	T35 FOAM AFT SKIRT AFT RING	18	SRM	Structures , Mechanisms, Veh Handling	3.03.05
1651	PSF86	04-JUN-1997:19	B5309	T18 SYS TUN FLR PLATE ASSY/RTV CURE	52	SRM	Structures , Mechanisms, Veh Handling	3.03.05
1652	PSF86	05-JUN-1997:9	B5309	T13 K5NA APPL TO STIFF RING	85	SRM	Structures , Mechanisms, Veh Handling	3.03.05
1653	PSF86	05-JUN-1997:14	B5309	T41 STIFF RING FOAMING OPER	50	SRM	Structures , Mechanisms, Veh Handling	3.03.05
1654	PSF86	06-JUN-1997:9	B5308	T12 AFT SEG CLEVIS CLEAN/INSP	5	SRM	Structures , Mechanisms, Veh Handling	3.01.01
1655	PSF86	06-JUN-1997:11	B5309	T04 INSTALL NOZZLE	30	SRM	Structures , Mechanisms, Veh Handling	3.03.03
1656	PSF86	06-JUN-1997:13	B5309	T42 REMOVE FWD HANDLING RING	5	SRM	Structures , Mechanisms, Veh Handling	3.06.01
1657	PSF86	06-JUN-1997:13	B5308	T12 PROBE INSTALLATION BOND LINE	5	SRM	Structures , Mechanisms, Veh Handling	6.04.02

No.	DATA SET	START TIME DD:MMM-YYYY:H	PROCEDURE NAME	DESCRIPTION	DURATION (Hours)	SYSTEM	DESIGN DISCIPLINE	GENERIC OPS FBS NUMBER (Appx A1)
1658	PSF86	06-JUN-1997:16	B5309	T33 ETA RING FOAMING OPS	21	SRM	Structures , Mechanisms, Veh Handling	3.03.05
1659	PSF86	09-JUN-1997:9	B5309	T24 STRUT PREPS	3	SRM	Structures , Mechanisms, Veh Handling	3.02.01
1660	PSF86	09-JUN-1997:20	B5309	T14 NOZZLE MASS FLOW LEAK CHECK	18	SRM	Structures , Mechanisms, Veh Handling	3.04.01
1661	PSF86	09-JUN-1997:22	B5309	T39 TRANSFER AFT BSTR ASSY FROM B/U TO PALLET	34	SRM	Structures , Mechanisms, Veh Handling	4.01.02
1662	PSF86	10-JUN-1997:11	B5305	T06 ETA RING CABLE INSTALLATION(STRUT)	20	SRE	Cmd, Ctl & Health Mngmt	3.03.03
1663	PSF86	10-JUN-1997:11	B5309	T07 AFT IEA INSTALLATION	11	SRM	Structures , Mechanisms, Veh Handling	3.03.03
1664	PSF86	10-JUN-1997:17	B5305	T02 SYS TUNNEL CABLE INSTALLATION	21	SRE	Cmd, Ctl & Health Mngmt	3.03.03
1665	PSF86	10-JUN-1997:17	B5309	T08 UPPER AND DIAGONAL STRUT INSTALLATION	5	SRM	Structures , Mechanisms, Veh Handling	3.03.03
1666	PSF86	10-JUN-1997:18	B5309	T12 NOZZLE BACKFILL INSTL'N	57	SRM	Structures , Mechanisms, Veh Handling	3.03.03
1667	PSF86	10-JUN-1997:21	B5309	T25 LOWER STRUT INSTALLATION	4	SRM	Structures , Mechanisms, Veh Handling	3.03.03
1668	PSF86	11-JUN-1997:15	B5305	T04 O/I STRUT CBL TEST/VAL	6	SRE	Cmd, Ctl & Health Mngmt	3.04.01
1669	PSF86	11-JUN-1997:19	B5305	T05 ETA RING CABLE INSTALLATION (TUNNEL(IEA))	6	SRE	Cmd, Ctl & Health Mngmt	3.03.03
1670	PSF86	11-JUN-1997:21	B5309	T19 TUNNEL COVER INSTALLATION/RTV CURE	43	SRM	Structures , Mechanisms, Veh Handling	3.03.03
1671	PSF86	11-JUN-1997:21	B5305	T03 O/I TUNNEL CABLE TESTING/VAL	13	SRE	Cmd, Ctl & Health Mngmt	3.04.01
1672	PSF86	12-JUN-1997:15	B5305	T09 O/I CABLE CONN,ROOSTER TAIL & IEA	7	SRE	Cmd, Ctl & Health Mngmt	3.03.03
1673	PSF86	12-JUN-1997:16	B5309	T16 STIFF RING SPLT PLT INSTALL 90DEG.	6	SRM	Structures , Mechanisms, Veh Handling	3.03.03
1674	PSF86	12-JUN-1997:19	B5305	T10 NSI CBL CONN/ALLINCO CK	12	SRE	Cmd, Ctl & Health Mngmt	3.04.01
1675	PSF86	13-JUN-1997:10	B5309	T36 RAIN SHEILD INITIAL INSTALL	4	SRM	Structures , Mechanisms, Veh Handling	3.03.04
1676	PSF86	13-JUN-1997:15	B5309	T32 NOZZLE CABLE CLOSEOUT	17	SRM	Structures , Mechanisms, Veh Handling	3.03.05
1677	PSF86	13-JUN-1997:16	B5309	T40 PR-855 FOAM INSUL AT STRUTS	27	SRM	Structures , Mechanisms, Veh Handling	3.03.05
1678	PSF86	16-JUN-1997:14	B5309	T34 FOAM AFT SKIRT INTRL RING	11	SRM	Structures , Mechanisms, Veh Handling	3.03.05
1679	PSF86	16-JUN-1997:14	B5309	T20 SYS TUN/STIFF RING SEALING/K5NA CURE	34	SRM	Structures , Mechanisms, Veh Handling	3.03.05
1680	PSF86	17-JUN-1997:10	B5309	T21 RAIN/THERMAL CURTAIN INSTALLATION	15	SRM	Structures , Mechanisms, Veh Handling	3.03.05
1681	PSF86	17-JUN-1997:15	B5309	T31 ETA RING CVR INSTALLATION FROM LOWER STRUT TO 270 DEG.	45	SRM	Structures , Mechanisms, Veh Handling	3.03.03
1682	PSF86	17-JUN-1997:15	B5309	T26 AFT IEA CVR INSTL/RTV & K5NA CURE	44	SRM	Structures , Mechanisms, Veh Handling	3.03.03
1683	PSF86	17-JUN-1997:15	B5309	T22 COVER INSTALL AT DIAG STRUT/TEMP INST	54	SRM	Structures , Mechanisms, Veh Handling	3.03.04
1684	PSF86	18-JUN-1997:3	B5309	T41 STIFF RING FOAMING OPER	46	SRM	Structures , Mechanisms, Veh Handling	3.03.05
1685	PSF86	18-JUN-1997:14	B5309	T35 FOAM AFT SKIRT AFT RING	14	SRM	Structures , Mechanisms, Veh Handling	3.03.05
1686	PSF86	18-JUN-1997:17	B5308	T12 AFT SEG CLEVIS CLEAN/INSP	5	SRM	Structures , Mechanisms, Veh Handling	3.01.01
1687	PSF86	18-JUN-1997:21	B5308	T12 PROBE INSTALLATION BOND LINE	4	SRM	Structures , Mechanisms, Veh Handling	6.04.02
1688	PSF86	19-JUN-1997:15	B5309	T39 TRANSFER AFT BSTR ASSY FROM B/U TO PALLET	17	SRM	Structures , Mechanisms, Veh Handling	4.01.02

No.	DATA SET	START TIME DD:MMM-YYYY:H	PROCEDURE NAME	DESCRIPTION	DURATION (Hours)	SYSTEM	DESIGN DISCIPLINE	GENERIC OPS FBS NUMBER (Appx A1)
1689	PSF86	04-NOV-1997:9	B5308	T12 ULTRASONIC INSPECTION	8	SRM	Structures , Mechanisms, Veh Handling	6.03.01
1690	SSV86	11-AUG-1997:9		CTS	0	*	Structures , Mechanisms, Veh Handling	4.01.01
1691	SSV86	11-AUG-1997:9		ORB/ET MATE	0	*	Structures , Mechanisms, Veh Handling	4.01.03
1692	SSV86	11-AUG-1997:9		CONN SLING/PREP TO LIFT	8	*	Structures , Mechanisms, Veh Handling	4.01.03
1693	SSV86	11-AUG-1997:9	S0004	ORB/ET MATE	100	*	Structures , Mechanisms, Veh Handling	4.01.03
1694	SSV86	11-AUG-1997:9		ORB IN TRANSFER AISLE	0	*	Ground Systems & Facilities	5.02.01
1695	SSV86	11-AUG-1997:17		ORB TO VERT/DISC AFT SLING	8	*	Structures , Mechanisms, Veh Handling	4.01.03
1696	SSV86	11-AUG-1997:17		NO LPS	36	*	Ground Systems & Facilities	5.02.01
1697	SSV86	12-AUG-1997:1		ORB/ET SOFTMATE	4	*	Structures , Mechanisms, Veh Handling	4.01.03
1698	SSV86	12-AUG-1997:5		ORB/ET HARDMATE	4	*	Structures , Mechanisms, Veh Handling	4.01.03
1699	SSV86	12-AUG-1997:9		SLING RMVL	4	*	Structures , Mechanisms, Veh Handling	4.01.04
1700	SSV86	12-AUG-1997:9	S0044	SIM RUN	8	*	Cmd, Ctl & Health Mngmt	5.03.04
1701	SSV86	12-AUG-1997:13		TSM CONN	20	MPS	Propulsion	4.02.01
1702	SSV86	12-AUG-1997:21		UMBILICAL MATE	16	TCO	Propulsion	4.02.01
1703	SSV86	13-AUG-1997:9	S0004	MONOBALL CONN/CLOSEOUT	20	TCO	Cmd, Ctl & Health Mngmt	4.01.04
1704	SSV86	13-AUG-1997:9		HAZ GAS LK CKS	8	*	Safety Management & Control	4.02.01
1705	SSV86	13-AUG-1997:9		TSM STATIC MEAS	8	*	Structures , Mechanisms, Veh Handling	4.02.01
1706	SSV86	13-AUG-1997:9	S0008	SHUTTLE I/F TEST	52	*	Cmd, Ctl & Health Mngmt	4.02.02
1707	SSV86	13-AUG-1997:9	V1149	UMB I/F CKS	30	MPS	Propulsion	4.02.02
1708	SSV86	13-AUG-1997:9		PREPS	16	MPS	Propulsion	4.02.02
1709	SSV86	13-AUG-1997:9		PREPS	16	TCO	Cmd, Ctl & Health Mngmt	4.02.02
1710	SSV86	13-AUG-1997:13		ULTRASONICS	4	*	Structures , Mechanisms, Veh Handling	4.02.02
1711	SSV86	13-AUG-1997:17		EXT UMB CAN CLOSEOUT	8	*	Structures , Mechanisms, Veh Handling	4.01.04
1712	SSV86	13-AUG-1997:23		PWR ON	26	*	Power Management	4.05.02
1713	SSV86	14-AUG-1997:1		PURGE CURTAIN INSTL	12	*	Safety Management & Control	4.01.04
1714	SSV86	14-AUG-1997:1		SEQ 6 ORB GHE FILL QD LK CK	2	MPS	Propulsion	4.02.01
1715	SSV86	14-AUG-1997:1		ORB SYS CKS	8	TCO	Cmd, Ctl & Health Mngmt	4.02.02
1716	SSV86	14-AUG-1997:1		CTS	0	MPS	Propulsion	4.02.02
1717	SSV86	14-AUG-1997:1		CTS	0	TCO	Cmd, Ctl & Health Mngmt	4.02.02
1718	SSV86	14-AUG-1997:1		ORB PWR UP	0	TCO	Cmd, Ctl & Health Mngmt	4.02.02
1719	SSV86	14-AUG-1997:3		SEQ 7/8 17 INCH TIMING CKS/200 PSI LK CK	4	MPS	Propulsion	4.02.02

No.	DATA SET	START TIME DD:MMM-YYYY:H	PROCEDURE NAME	DESCRIPTION	DURATION (Hours)	SYSTEM	DESIGN DISCIPLINE	GENERIC OPS FBS NUMBER (Appx A1)
1720	SSV86	14-AUG-1997:5		ET/SRB SYS CKS	16	TCO	Cmd, Ctl & Health Mngmt	4.02.02
1721	SSV86	14-AUG-1997:5		ET/SRB PWR UP	0	TCO	Cmd, Ctl & Health Mngmt	4.02.02
1722	SSV86	14-AUG-1997:7		SEQ 9 LOX F&D QD LK CK	2	MPS	Propulsion	4.02.01
1723	SSV86	14-AUG-1997:9		SEQ 10 LH2 F&D QD LK CK	2	MPS	Propulsion	4.02.01
1724	SSV86	14-AUG-1997:9		SEQ 15/16 SSME CKS	4	MPS	Propulsion	4.02.02
1725	SSV86	14-AUG-1997:11		SEQ 17 MASS SPEC LK CKS	4	MPS	Propulsion	4.02.01
1726	SSV86	14-AUG-1997:13		UMBILICAL FOAMING	24	MPS	Thermal Management	4.01.04
1727	SSV86	14-AUG-1997:21		SRB TVC ACT TEST	4	TCO	Guid, Nav & Ctl	5.03.03
1728	SSV86	15-AUG-1997:1		CONN SRB ACT	8	TCO	Guid, Nav & Ctl	5.06.02
1729	SSV86	15-AUG-1997:9		PWR ON	4	*	Power Management	4.05.02
1730	SSV86	15-AUG-1997:9	A5214	ROLLOUT PREPS	16	TSM	Structures , Mechanisms, Veh Handling	4.05.03
1731	SSV86	15-AUG-1997:9	A5214	ROLLOUT PREPS	16	TSM	Structures , Mechanisms, Veh Handling	4.05.03
1732	SSV86	15-AUG-1997:9		SRB TVC ACT FRT TEST	4	TCO	Guid, Nav & Ctl	5.03.03
1733	SSV86	15-AUG-1997:17	PR-ET-0026	R/H SRB FWD SEG EDDY CURRENT	8	SRB	Structures , Mechanisms, Veh Handling	5.04.01
1734	SSV86	16-AUG-1997:1		PREMIUM DAYS	48		Ground Systems & Facilities	5.02.01
1735	SSV86	17-AUG-1997:17	A5214	ROLLOUT PREPS	7	TSM	Structures , Mechanisms, Veh Handling	4.05.03
1736	SSV86	17-AUG-1997:17	A5214	ROLLOUT PREPS	7	TSM	Structures , Mechanisms, Veh Handling	4.05.03
1737	SSV86	17-AUG-1997:24	A5214	SSV XFER TO PAD	9	*	Structures , Mechanisms, Veh Handling	5.02.01
1738	SSV86	17-AUG-1997:24		CTS	0	*	Ground Systems & Facilities	5.02.01
1739	SSV86	18-AUG-1997:1		PREPS	8	*	Ground Systems & Facilities	5.01.03
1740	SSV86	18-AUG-1997:1	S0009	LAUNCH PAD VAL	40	*	Cmd, Ctl & Health Mngmt	5.03.02
1741	SSV86	18-AUG-1997:3		FIRST MOTION	0	*	Ground Systems & Facilities	5.02.01
1742	SSV86	18-AUG-1997:9		MLP HARDDOWN	6	*	Ground Systems & Facilities	5.02.01
1743	SSV86	18-AUG-1997:9		ET GUCP CONN	12	*	Propulsion	5.02.02
1744	SSV86	18-AUG-1997:9		CTS	0	*	Cmd, Ctl & Health Mngmt	5.02.02
1745	SSV86	18-AUG-1997:9		PWR ON	40	*	Power Management	5.03.02
1746	SSV86	18-AUG-1997:9		LDB CKS	8	*	Cmd, Ctl & Health Mngmt	5.03.02
1747	SSV86	18-AUG-1997:11		RSS EXTEND	0	*	Ground Systems & Facilities	5.02.02
1748	SSV86	18-AUG-1997:17		OPEN PLBDS	4		Ground Systems & Facilities	5.02.01
1749	SSV86	18-AUG-1997:17	S0024	HYPER PREPS	32	TCO	Propulsion	5.02.02
1750	SSV86	18-AUG-1997:17		HYPER PREPS	32	TCO	Ground Systems & Facilities	5.02.02

No.	DATA SET	START TIME DD:MMM-YYYY:H	PROCEDURE NAME	DESCRIPTION	DURATION (Hours)	SYSTEM	DESIGN DISCIPLINE	GENERIC OPS FBS NUMBER (Appx A1)
1751	SSV86	18-AUG-1997:17		SSV PWR UP	0	*	Power Management	5.03.03
1752	SSV86	18-AUG-1997:17	S0600.001	SHUTTLE VERTICAL PL OPS	32	PLO	Payload Accommodations	5.05.02
1753	SSV86	18-AUG-1997:21	IPR-0041	PLBD/GAP HOIST CONTACT	12	PLO	Payload Accommodations	5.04.01
1754	SSV86	19-AUG-1997:5	G2340	ET LOX/LH2 CK-OUT	12	TCO	Propulsion	5.03.03
1755	SSV86	19-AUG-1997:9		OPEN PLBDS	4		Ground Systems & Facilities	5.02.01
1756	SSV86	19-AUG-1997:13		PFR/GAS CAN RMVL	4		Ground Systems & Facilities	5.02.01
1757	SSV86	19-AUG-1997:17		ODS BOLT TORQUE CKS	8		Ground Systems & Facilities	5.02.01
1758	SSV86	19-AUG-1997:17	T1402	GO2 BLKG PLT INSTL	4	LOX	Propulsion	5.03.02
1759	SSV86	20-AUG-1997:5	V1308/	V1046.001/.002/.003 ENG FRT	28	SME	Propulsion	5.03.03
1760	SSV86	20-AUG-1997:5	V9002.06		8	*	Power Management	5.06.02
1761	SSV86	20-AUG-1997:9		ODS BOLT TORQUE CKS	8		Ground Systems & Facilities	5.02.01
1762	SSV86	20-AUG-1997:9	T1401	GH2 BLKG PLT INSTL	4	LH2	Propulsion	5.03.02
1763	SSV86	20-AUG-1997:9		PWR ON	16	*	Power Management	5.03.02
1764	SSV86	20-AUG-1997:9	S0600.001	SHUTTLE VERTICAL PL OPS	8	PLO	Payload Accommodations	5.05.02
1765	SSV86	20-AUG-1997:9		FRCS QD CONN	0	TCO	Propulsion	5.06.02
1766	SSV86	20-AUG-1997:9	B1019	SRB HYD CLOSEOUTS	16	HYD	Power Management	5.07.03
1767	SSV86	20-AUG-1997:13	V1046	SSME FRT	20	SME	Propulsion	5.03.03
1768	SSV86	21-AUG-1997:1		HYPER PREPS	48	TCO	Ground Systems & Facilities	5.02.02
1769	SSV86	21-AUG-1997:1	S0024	HYPER PREPS	48	TCO	Propulsion	5.06.02
1770	SSV86	21-AUG-1997:9	V1202	HE SIG PREPS/TEST	32	MPS	Propulsion	5.03.02
1771	SSV86	21-AUG-1997:9	V1202	HE SIG PREPS	20	MPS	Propulsion	5.03.02
1772	SSV86	21-AUG-1997:9		PWR ON	40		Power Management	5.03.02
1773	SSV86	21-AUG-1997:9		APS QD CONN	0	TCO	Propulsion	5.06.02
1774	SSV86	21-AUG-1997:9	B1019	SRB HYD CLOSEOUTS	16	HYD	Power Management	5.07.03
1775	SSV86	21-AUG-1997:13		CLOSE PLBD'S	4		Ground Systems & Facilities	5.02.01
1776	SSV86	21-AUG-1997:13	S0600.001	SHUTTLE VERTICAL PL OPS	4	PLO	Payload Accommodations	5.05.02
1777	SSV86	22-AUG-1997:5	V1202	HE SIG TEST	12	MPS	Propulsion	5.03.02
1778	SSV86	22-AUG-1997:9		EDDY CURRENT	0		Ground Systems & Facilities	5.02.01
1779	SSV86	22-AUG-1997:9	IPR-0041	PLBD/GAP HOIST CONTACT	4	PLO	Payload Accommodations	5.04.01
1780	SSV86	22-AUG-1997:9	S0600.001	SHUTTLE VERTICAL PL OPS	4	PLO	Payload Accommodations	5.05.02
1781	SSV86	22-AUG-1997:13		BALL SEAL LK CKS	0		Ground Systems & Facilities	5.02.01

No.	DATA SET	START TIME DD:MMM-YYYY:H	PROCEDURE NAME	DESCRIPTION	DURATION (Hours)	SYSTEM	DESIGN DISCIPLINE	GENERIC OPS FBS NUMBER (Appx A1)
1782	SSV86	22-AUG-1997:13	V1308/	V1046.001/.002/.003 ENG FRT	4	SME	Propulsion	5.03.03
1783	SSV86	22-AUG-1997:13	V1046	SSME FRT	4	SME	Propulsion	5.03.03
1784	SSV86	22-AUG-1997:16		APU QD CONN	0	TCO	Power Management	5.02.02
1785	SSV86	22-AUG-1997:21		OMS HI PT BLD QD CONN	0	TCO	Power Management	5.02.02
1786	SSV86	23-AUG-1997:1		PREMIUM DAYS	48		Ground Systems & Facilities	5.02.01
1787	SSV86	25-AUG-1997:1		HYPER PREPS	24		Ground Systems & Facilities	5.02.01
1788	SSV86	25-AUG-1997:1	S0024	HYPER PREPS	24	*	Propulsion	5.06.02
1789	SSV86	25-AUG-1997:9		TPS COM-340 P/L INTERROGATOR RMVL	8		Ground Systems & Facilities	5.02.01
1790	SSV86	25-AUG-1997:9		PWR ON	64	*	Power Management	5.03.02
1791	SSV86	25-AUG-1997:9	T1401	GH2 BLKG PLT RMVL	6	LH2	Propulsion	5.03.02
1792	SSV86	25-AUG-1997:17	T1402	GO2 BLKG PLT RMVL	6	LOX	Propulsion	5.03.02
1793	SSV86	26-AUG-1997:1		CTS	0	TCO	Power Management	5.06.01
1794	SSV86	26-AUG-1997:1		PREPS TO CLEAR PAD TO PERIMETER	4	TCO	Power Management	5.06.01
1795	SSV86	26-AUG-1997:1	S0024	PRELAUNCH PROPELLANT LOAD	64	*	Propulsion	5.06.01
1796	SSV86	26-AUG-1997:5		RJD/RCS DRIVER AND CMD STATUS TEST	4	TCO	Guid, Nav & Ctl	5.03.02
1797	SSV86	26-AUG-1997:9	S0056	SIM LOAD	8	*	Cmd, Ctl & Health Mngmt	5.03.04
1798	SSV86	26-AUG-1997:9	V1045	OMS/RCS OXID LOAD	13	TCO	Propulsion	5.06.01
1799	SSV86	27-AUG-1997:5		FUEL PREPS	4	TCO	Propulsion	5.06.01
1800	SSV86	27-AUG-1997:9	V1045	OMS/RCS FUEL LOAD	14	TCO	Propulsion	5.06.02
1801	SSV86	27-AUG-1997:9	V1329	APU FUEL SERV	10	TCO	Power Management	5.06.02
1802	SSV86	27-AUG-1997:9	B1016	HPU FUEL SERV	8	TCO	Power Management	5.06.02
1803	SSV86	27-AUG-1997:23		CLOSEOUTS	18	TCO	Power Management	5.06.02
1804	SSV86	28-AUG-1997:9		SSV SNAPSHOT	4		Ground Systems & Facilities	5.02.01
1805	SSV86	28-AUG-1997:9		STR-A0270/0271 ORB ET ATTACH BASEPLATE MR REPAIR	8		Ground Systems & Facilities	5.02.01
1806	SSV86	28-AUG-1997:9		PWR ON	4		Power Management	5.03.02
1807	SSV86	28-AUG-1997:17	V1149	SEQ 13 QD CAVITY PURGE VERIF	8	MPS	Propulsion	5.03.03
1808	SSV86	29-AUG-1997:9		STR-A0270/0271 ORB ET ATTACH BASEPLATE MR REPAIR	8		Ground Systems & Facilities	5.02.01
1809	SSV86	29-AUG-1997:9		TPS INS-138 P/L RECORDER RMVL	8		Ground Systems & Facilities	5.02.01
1810	SSV86	30-AUG-1997:1		PREMIUM DAYS	48		Ground Systems & Facilities	5.02.01
1811	SSV86	01-SEP-1997:13		LABOR DAY	0		Ground Systems & Facilities	5.02.01
1812	SSV86	02-SEP-1997:9		SSV SNAPSHOT	4		Ground Systems & Facilities	5.02.01

No.	DATA SET	START TIME DD:MMM-YYYY:H	PROCEDURE NAME	DESCRIPTION	DURATION (Hours)	SYSTEM	DESIGN DISCIPLINE	GENERIC OPS FBS NUMBER (Appx A1)
1813	SSV86	02-SEP-1997:9		STR-A0270/0271 ORB ET ATTACH BASEPLATE MR REPAIR	8		Ground Systems & Facilities	5.02.01
1814	SSV86	02-SEP-1997:9		PWR ON	4		Power Management	5.03.02
1815	SSV86	03-SEP-1997:9		STR-A0270/0271 ORB ET ATTACH BASEPLATE MR REPAIR	8		Ground Systems & Facilities	5.02.01
1816	SSV86	03-SEP-1997:21		RETRACT RSS	0		Ground Systems & Facilities	5.02.01
1817	SSV86	03-SEP-1997:21		CANISTER INSTL - RSS	8	*	Ground Systems & Facilities	5.02.01
1818	SSV86	03-SEP-1997:21	S0600.001	SHUTTLE VERTICAL PL OPS - PCR	24	*	Payload Accommodations	5.05.02
1819	SSV86	04-SEP-1997:5		CANISTER INSTL - RSS	4	*	Ground Systems & Facilities	5.02.01
1820	SSV86	04-SEP-1997:9		PL INSTL - PCR	8	PLO	Payload Accommodations	4.04.04
1821	SSV86	04-SEP-1997:17		LOWER CANISTER	4	*	Ground Systems & Facilities	5.02.01
1822	SSV86	04-SEP-1997:21		EXTEND RSS	0		Ground Systems & Facilities	5.02.01
1823	SSV86	05-SEP-1997:1		OPEN PLBD'S	4	*	Ground Systems & Facilities	5.02.01
1824	SSV86	05-SEP-1997:1		PWR ON	16	*	Power Management	5.03.02
1825	SSV86	05-SEP-1997:1	S0600.001	SHUTTLE VERTICAL PL OPS - PAD	20	*	Payload Accommodations	5.05.02
1826	SSV86	05-SEP-1997:5		PL INSTL - PLB	16	*	Ground Systems & Facilities	5.02.01
1827	SSV86	05-SEP-1997:9		SSV SNAPSHOT	4		Ground Systems & Facilities	5.02.01
1828	SSV86	05-SEP-1997:9		STR-A0270/0271 ORB ET ATTACH BASEPLATE MR REPAIR	8		Ground Systems & Facilities	5.02.01
1829	SSV86	05-SEP-1997:9		OMBUU MATE/LK CKS	8	*	Power Management	5.02.02
1830	SSV86	05-SEP-1997:9	S0009	LAUNCH PAD VAL	8	*	Cmd, Ctl & Health Mngmt	5.03.02
1831	SSV86	06-SEP-1997:1		PREMIUM DAYS	48		Ground Systems & Facilities	5.02.01
1832	SSV86	08-SEP-1997:5		SETUP MVAK	4	*	Ground Systems & Facilities	5.02.01
1833	SSV86	08-SEP-1997:5	S0886	SHUTTLE VERTICAL PL CLOSEOUTS	20	*	Payload Accommodations	5.07.01
1834	SSV86	08-SEP-1997:9		SSV SNAPSHOT	4		Ground Systems & Facilities	5.02.01
1835	SSV86	08-SEP-1997:9		CE-195 H20/ELEC MATES	8	*	Ground Systems & Facilities	5.02.01
1836	SSV86	08-SEP-1997:9		STR-A0270/0271 ORB ET ATTACH BASEPLATE MR REPAIR	8		Ground Systems & Facilities	5.02.01
1837	SSV86	08-SEP-1997:9		PWR ON	4	*	Power Management	5.03.02
1838	SSV86	08-SEP-1997:13		TUNNEL MATES/LK CKS	8	*	Ground Systems & Facilities	5.02.01
1839	SSV86	08-SEP-1997:17		MLI INSTL	8	*	Ground Systems & Facilities	5.02.01
1840	SSV86	08-SEP-1997:21		DECONFIG MVAK	4	*	Ground Systems & Facilities	5.02.01
1841	SSV86	09-SEP-1997:9		STR-A0270/0271 ORB ET ATTACH BASEPLATE MR REPAIR	8		Ground Systems & Facilities	5.02.01
1842	SSV86	09-SEP-1997:9		PWR ON	56	*	Power Management	5.03.02
1843	SSV86	09-SEP-1997:9	S1025	EMERG EGRESS TRNG	8	*	Safety Management & Control	5.03.04

No.	DATA SET	START TIME DD:MMM-YYYY:H	PROCEDURE NAME	DESCRIPTION	DURATION (Hours)	SYSTEM	DESIGN DISCIPLINE	GENERIC OPS FBS NUMBER (Appx A1)
1844	SSV86	09-SEP-1997:9	S0017	TCDT	32	TCO	Cmd, Ctl & Health Mngmt	5.03.04
1845	SSV86	09-SEP-1997:9		CTS	0	*	Cockpit & Crew Cabin	5.03.04
1846	SSV86	10-SEP-1997:9		CTS	0	TCO	Structures , Mechanisms, Veh Handling	5.03.02
1847	SSV86	10-SEP-1997:9	S1287	AFT CLOSEOUT	64	TCO	Structures , Mechanisms, Veh Handling	5.07.02
1848	SSV86	10-SEP-1997:9	S1287	ORB AFT CLOSEOUT	64	*	Structures , Mechanisms, Veh Handling	5.07.02
1849	SSV86	10-SEP-1997:12		TCDT SECURING	5	*	Cockpit & Crew Cabin	5.03.04
1850	SSV86	10-SEP-1997:12		T-0	0	*	Cockpit & Crew Cabin	5.03.04
1851	SSV86	10-SEP-1997:12		TCDT	0	TCO	Cockpit & Crew Cabin	5.03.04
1852	SSV86	10-SEP-1997:13	V1043	IMU CAL	16	GNC	Guid, Nav & Ctl	5.03.03
1853	SSV86	11-SEP-1997:5		SETUP MVAK	4		Ground Systems & Facilities	5.02.01
1854	SSV86	11-SEP-1997:5	S0786	SHUTTLE VERTICAL PL TESTING	16	*	Payload Accommodations	5.05.03
1855	SSV86	11-SEP-1997:9		PL IVT	12	*	Ground Systems & Facilities	5.02.01
1856	SSV86	11-SEP-1997:17		LH SRB IRD R&R	8		Ground Systems & Facilities	5.02.01
1857	SSV86	12-SEP-1997:9		PL IVT	12	*	Ground Systems & Facilities	5.02.01
1858	SSV86	12-SEP-1997:9	V1053.006	PP02 SENSOR ADJ	8	ECL	Environmental Ctl & Life Spt	5.03.01
1859	SSV86	12-SEP-1997:9		PWR ON	12	*	Power Management	5.03.02
1860	SSV86	12-SEP-1997:9	S0786	SHUTTLE VERTICAL PL TESTING	12	*	Payload Accommodations	5.05.03
1861	SSV86	13-SEP-1997:9	TPS-ECL-416	/IPR-0060 CABIN/SH PRESS DECAY CKS/T/S	4	ECL	Environmental Ctl & Life Spt	5.04.01
1862	SSV86	13-SEP-1997:13	IPR-0061/0062	UPPER ODS/C HATCH LK CKS	8	PLO	Payload Accommodations	5.04.01
1863	SSV86	14-SEP-1997:1		PREMIUM DAYS	24		Ground Systems & Facilities	5.02.01
1864	SSV86	14-SEP-1997:9	TPS-ECL-416	/IPR-0060 CABIN/SH PRESS DECAY CKS/T/S	16	ECL	Environmental Ctl & Life Spt	5.04.01
1865	SSV86	15-SEP-1997:9		QD CONN	8	TCO	Power Management	5.02.02
1866	SSV86	15-SEP-1997:9	TPS-ECL-416	/IPR-0060 CABIN/SH PRESS DECAY CKS/T/S	4	ECL	Environmental Ctl & Life Spt	5.04.01
1867	SSV86	15-SEP-1997:9	S0024	HYPER PRESS/CLOSEOUT	8	*	Propulsion	5.06.02
1868	SSV86	15-SEP-1997:9	S1287	ORB AFT CLOSEOUT	40	VOP	Structures , Mechanisms, Veh Handling	5.07.02
1869	SSV86	15-SEP-1997:9	S1287	AFT CLOSEOUT	40	TCO	Structures , Mechanisms, Veh Handling	5.07.02
1870	SSV86	15-SEP-1997:17	IPR-0061/0062	UPPER ODS/C HATCH LK CKS	8	PLO	Payload Accommodations	5.04.01
1871	SSV86	16-SEP-1997:9		SH EARLY STOW	12	*	Ground Systems & Facilities	5.02.01
1872	SSV86	16-SEP-1997:9		RECOVERY/SRSS SYS TEST	6	*	Cmd, Ctl & Health Mngmt	5.03.03
1873	SSV86	16-SEP-1997:9	S5009	ORD INSTL - PART 3	6	TCO	Structures , Mechanisms, Veh Handling	5.06.03
1874	SSV86	16-SEP-1997:9		CTS	0	*	Cmd, Ctl & Health Mngmt	5.06.03

No.	DATA SET	START TIME DD:MMM-YYYY:H	PROCEDURE NAME	DESCRIPTION	DURATION (Hours)	SYSTEM	DESIGN DISCIPLINE	GENERIC OPS FBS NUMBER (Appx A1)
1875	SSV86	16-SEP-1997:9	S0886	SHUTTLE VERTICAL PL CLOSEOUTS	12	*	Payload Accommodations	5.07.01
1876	SSV86	16-SEP-1997:9	V1077	FUEL CELL ULLAGE VERIFICATION	8	FCP	Power Management	5.08.07
1877	SSV86	16-SEP-1997:9		PWR ON	8	*	Cmd, Ctl & Health Mngmt	5.09.02
1878	SSV86	16-SEP-1997:15		LWR APU CARTS	2		Ground Systems & Facilities	5.02.01
1879	SSV86	16-SEP-1997:15		SSV PWR DOWN	0	*	Cmd, Ctl & Health Mngmt	5.06.03
1880	SSV86	16-SEP-1997:17	IPR-0061/0062	UPPER ODS/C HATCH LK CKS	8	PLO	Payload Accommodations	5.04.01
1881	SSV86	17-SEP-1997:1		ORD INSTL/CONN	8	*	Cmd, Ctl & Health Mngmt	5.06.03
1882	SSV86	17-SEP-1997:1	S5009	ORD INSTL - PART 3	24	TCO	Structures , Mechanisms, Veh Handling	5.06.03
1883	SSV86	17-SEP-1997:1	V1184	MMU LOAD	12	FSW	Cmd, Ctl & Health Mngmt	5.09.04
1884	SSV86	17-SEP-1997:9		SSV PWR UP	0	*	Cmd, Ctl & Health Mngmt	5.06.03
1885	SSV86	17-SEP-1997:9		PIC RESIST CKS	2	*	Cmd, Ctl & Health Mngmt	5.06.03
1886	SSV86	17-SEP-1997:9		PWR ON	72	*	Cmd, Ctl & Health Mngmt	5.09.02
1887	SSV86	17-SEP-1997:11		OMS/RCS PRESS TO REG LOCK UP	2	TCO	Propulsion	5.03.03
1888	SSV86	17-SEP-1997:11	S0024	HYPER PRESS/CLOSEOUT	54	*	Propulsion	5.06.02
1889	SSV86	17-SEP-1997:13		STOW ODS BAG	4		Ground Systems & Facilities	5.02.01
1890	SSV86	17-SEP-1997:13	S5009	ORD CLOSEOUT	28	TCO	Structures , Mechanisms, Veh Handling	5.06.03
1891	SSV86	17-SEP-1997:13	S0886	SHUTTLE VERTICAL PL CLOSEOUTS	8	*	Payload Accommodations	5.07.01
1892	SSV86	17-SEP-1997:13	S1287	ORB AFT CLOSEOUT	68	*	Structures , Mechanisms, Veh Handling	5.07.02
1893	SSV86	17-SEP-1997:13	S1287	AFT CLOSEOUT	20	TCO	Structures , Mechanisms, Veh Handling	5.07.02
1894	SSV86	17-SEP-1997:13		OMS/RCS PRESS TO FLT MASS	4	TCO	Propulsion	5.08.04
1895	SSV86	17-SEP-1997:17		DECONFIG MVAK	4	*	Ground Systems & Facilities	5.02.01
1896	SSV86	17-SEP-1997:17		OMS/RCS PRESS STABILIZATION	48	OMS	Propulsion	5.08.03
1897	SSV86	17-SEP-1997:21	V1103.02	EMU FUNCTIONAL/CKOUT	4	TPE	Payload Accommodations	5.05.03
1898	SSV86	18-SEP-1997:5	V1103.02	EMU FUNCTIONAL/CKOUT	12	TPE	Payload Accommodations	5.05.03
1899	SSV86	18-SEP-1997:9		PLB CLOSEOUTS	12	*	Ground Systems & Facilities	5.02.01
1900	SSV86	18-SEP-1997:9	S1005/S1006	LOX/LH2 SYS DEWPOINT AND COND	16	*	Propulsion	5.06.01
1901	SSV86	18-SEP-1997:9	S0886	SHUTTLE VERTICAL PL CLOSEOUTS	16	*	Payload Accommodations	5.07.01
1902	SSV86	18-SEP-1997:9	S1287	AFT INSP	20	TCO	Structures , Mechanisms, Veh Handling	5.09.04
1903	SSV86	18-SEP-1997:17	PR-COM-0238	SIG BATT R&R/RETEST	8	COM	Power Management	5.04.01
1904	SSV86	18-SEP-1997:17	S1287	AFT ACCESS RMVL	12	TCO	Structures , Mechanisms, Veh Handling	5.07.02
1905	SSV86	18-SEP-1997:21		CLOSE PLBD'S FOR FLT	4	*	Ground Systems & Facilities	5.02.01

No.	DATA SET	START TIME DD:MMM-YYYY:H	PROCEDURE NAME	DESCRIPTION	DURATION (Hours)	SYSTEM	DESIGN DISCIPLINE	GENERIC OPS FBS NUMBER (Appx A1)
1906	SSV86	19-SEP-1997:5		AFT CONFIDENCE TEST	12	TCO	Safety Management & Control	5.03.02
1907	SSV86	19-SEP-1997:5		FLT DR INSTL	4	*	Structures , Mechanisms, Veh Handling	5.09.04
1908	SSV86	19-SEP-1997:9	V5097	CREW SYS VERT STOW	16	FCS	Cockpit & Crew Cabin	5.05.02
1909	SSV86	19-SEP-1997:9		QD/DISC/PNL CLOSEOUT	8	TCO	Propulsion	5.07.03
1910	SSV86	19-SEP-1997:17		RMV FLT DOORS	4		Structures , Mechanisms, Veh Handling	5.02.02
1911	SSV86	19-SEP-1997:21		E-1 MDS HE REG A R&R	4		Safety Management & Control	5.04.01
1912	SSV86	20-SEP-1997:1		AFT CONFIDENCE TEST	8	TCO	Safety Management & Control	5.03.02
1913	SSV86	20-SEP-1997:1		FLT DR INSTL	4	*	Structures , Mechanisms, Veh Handling	5.09.04
1914	SSV86	20-SEP-1997:5	V5097	CREW SYS VERT STOW	24	FCS	Cockpit & Crew Cabin	5.05.02
1915	SSV86	21-SEP-1997:1		PREMIUM DAYS	24		Ground Systems & Facilities	5.02.01
1916	SSV86	22-SEP-1997:5	V5097	CREW SYS VERT STOW	4	FCS	Cockpit & Crew Cabin	5.05.02
1917	SSV86	22-SEP-1997:17		PWR ON	80	*	Cmd, Ctl & Health Mngmt	5.09.02
1918	SSV86	22-SEP-1997:19		CTS (T-43 HRS)	0	TCO	Cmd, Ctl & Health Mngmt	5.09.01
1919	SSV86	22-SEP-1997:19	S0007	LAUNCH COUNTDOWN	76	*	Cmd, Ctl & Health Mngmt	5.09.02
1920	SSV86	22-SEP-1997:19		CALL TO STATION/PWR UP	2	TCO	Cmd, Ctl & Health Mngmt	5.09.02
1921	SSV86	22-SEP-1997:19		BFC TEST & MMU/DEU VERIF	8	TCO	Cmd, Ctl & Health Mngmt	5.09.04
1922	SSV86	23-SEP-1997:3		SH LATE STOW	2	*	Ground Systems & Facilities	5.02.01
1923	SSV86	23-SEP-1997:3	S0886	SHUTTLE VERTICAL PL CLOSEOUTS	2	*	Payload Accommodations	5.07.01
1924	SSV86	23-SEP-1997:3		PRSD LOAD PREPS	8	TCO	Power Management	5.08.07
1925	SSV86	23-SEP-1997:11		BIH (T-27 HRS)	4	TCO	Ground Systems & Facilities	5.06.02
1926	SSV86	23-SEP-1997:12		SSV PIC RESISTANCE TEST	2	TCO	Safety Management & Control	5.09.02
1927	SSV86	23-SEP-1997:15		PRSD CRYO LOAD	8	TCO	Power Management	5.06.01
1928	SSV86	23-SEP-1997:23		SH LATE STOW	12	*	Ground Systems & Facilities	5.02.01
1929	SSV86	23-SEP-1997:23		BIH (T-19 HRS)	4	TCO	Power Management	5.06.02
1930	SSV86	23-SEP-1997:23	S0886	SHUTTLE VERTICAL PL CLOSEOUTS	12	*	Payload Accommodations	5.07.01
1931	SSV86	24-SEP-1997:3		MPS & SSME PREPS	8	TCO	Propulsion	5.07.02
1932	SSV86	24-SEP-1997:11		BIH (T-11 HRS)	21	TCO	Ground Systems & Facilities	5.09.04
1933	SSV86	24-SEP-1997:23		COMM ACT	0	TCO	Communications	5.09.04
1934	SSV86	25-SEP-1997:5		SWITCHLIST	2	TCO	Cockpit & Crew Cabin	5.09.04
1935	SSV86	25-SEP-1997:8		TERMINAL COUNTDOWN	15	TCO	Cmd, Ctl & Health Mngmt	5.09.05
1936	SSV86	25-SEP-1997:23		T-0	0	TCO	Cmd, Ctl & Health Mngmt	5.09.05

No.	DATA SET	START TIME DD:MMM-YYYY:H	PROCEDURE NAME	DESCRIPTION	DURATION (Hours)	SYSTEM	DESIGN DISCIPLINE	GENERIC OPS FBS NUMBER (Appx A1)
1937	SSV86	26-SEP-1997:6	S2005	POST LAUNCH SECURING	19	*	Cmd, Ctl & Health Mngmt	5.10.01
1938	SSV86	29-SEP-1997:9	S2005	POST LAUNCH SECURING	16	*	Cmd, Ctl & Health Mngmt	5.10.01
1939	STK86	24-JUN-1997:1	B5303-T31	T-31 LEAK TEST PANEL (MFM) VAL.	24	SRM	Structures , Mechanisms, Veh Handling	4.01.01
1940	STK86	24-JUN-1997:1	B5303	PRE-OPS/PREPS TO STACK	48	SRM	Structures , Mechanisms, Veh Handling	4.01.01
1941	STK86	26-JUN-1997:6	B5303-T33	T-33 HYDRASET PREPS/CONNECT TO HOOK/BEAM	3	SRM	Structures , Mechanisms, Veh Handling	4.01.03
1942	STK86	26-JUN-1997:10	B5303-T22	T-22 BEAM PREPS/CONNECT TO BOOSTER	12	SRM	Structures , Mechanisms, Veh Handling	4.01.03
1943	STK86	26-JUN-1997:21	B5303-T05	T-05 RESHIM (IF REQ'D)	4	SRM	Structures , Mechanisms, Veh Handling	4.01.03
1944	STK86	26-JUN-1997:24	B5303-T05	T-05 STACK AFT BOOSTER ASSY.	3	SRM	Structures , Mechanisms, Veh Handling	4.01.03
1945	STK86	27-JUN-1997:2	B5303-T05	T-05 ROTATION MEAS.	1	SRM	Structures , Mechanisms, Veh Handling	4.01.03
1946	STK86	27-JUN-1997:7	B5303-T3A	T-03A SRB JOINT HTR UMB (GCA) INSTL.	15	SRM	Structures , Mechanisms, Veh Handling	4.01.03
1947	STK86	27-JUN-1997:8	B5303-T22	T-22 BEAM PREPS/CONNECT TO SEG./LIFT	3	SRM	Structures , Mechanisms, Veh Handling	3.01.03
1948	STK86	27-JUN-1997:8	B5303 T-23	T-23 H77-0384-03 4 POINT LIFTING BEAM ADJ. (IF REQ'D)	1	SRM	Structures , Mechanisms, Veh Handling	3.02.01
1949	STK86	27-JUN-1997:8	B5303-T30	T-30 FJAF PREPS	6	SRM	Structures , Mechanisms, Veh Handling	3.02.02
1950	STK86	27-JUN-1997:9	S3001	S3001 PLATFORM OPS.	1	SRM	Ground Systems & Facilities	3.06.01
1951	STK86	27-JUN-1997:9	B5303-T22	T-22 (H77-0384-03)HANDLING BEAM OPS./BEAM TO AISLE	2	SRM	Structures , Mechanisms, Veh Handling	4.01.01
1952	STK86	27-JUN-1997:10	B5303-T14	T-14 HOLDDOWN HARDWARE PRETENSIONING	10	SRM	Structures , Mechanisms, Veh Handling	4.01.01
1953	STK86	27-JUN-1997:11	B5303-T08	T-08 LAC TANG/MATE OPS	13	SRM	Structures , Mechanisms, Veh Handling	3.02.02
1954	STK86	27-JUN-1997:11	B5303-T34	T-34 REMOVE PROTECTIVE COVERS	1	SRM	Structures , Mechanisms, Veh Handling	4.01.03
1955	STK86	27-JUN-1997:11	B5303-T35	T-35 INSTL. OF MODIFIED PROTECTIVE CVRS	1	SRM	Structures , Mechanisms, Veh Handling	4.01.04
1956	STK86	27-JUN-1997:13	B5303-T06	T-06 INSTL SRB HOLDDOWN HDWR	2	SRM	Structures , Mechanisms, Veh Handling	4.01.01
1957	STK86	27-JUN-1997:19	B5303-T07	T-07 CLEVIS PREPS	7	SRM	Structures , Mechanisms, Veh Handling	3.02.02
1958	STK86	27-JUN-1997:20	B5303-T05	T-05 POST OPS	5	SRM	Structures , Mechanisms, Veh Handling	4.01.04
1959	STK86	27-JUN-1997:23	B5303-T3B	T-03B SRB JOINT HTR UMB (GCA) CLOSEOUT	3	SRM	Structures , Mechanisms, Veh Handling	4.01.03
1960	STK86	27-JUN-1997:24	B5303-T25	T-25 ENCLOSURE SKIRT INSTLN.	1	SRM	Structures , Mechanisms, Veh Handling	3.01.04
1961	STK86	28-JUN-1997:1	B5303-T08	T-08 LAC LIFT TO HB/MATE	17	SRM	Structures , Mechanisms, Veh Handling	3.02.02
1962	STK86	28-JUN-1997:11	B5303-T21	T-21 TEMPASONIC INSTRUMENTATION	1	SRM	Structures , Mechanisms, Veh Handling	3.03.03
1963	STK86	28-JUN-1997:18	B5303-T22	T-22 (H77-0384-03)HANDLING BEAM OPS./BEAM TO AISLE	2	SRM	Structures , Mechanisms, Veh Handling	3.06.01
1964	STK86	28-JUN-1997:20	S3001	S3001 PLATFORM OPS.	1	SRM	Ground Systems & Facilities	3.06.01
1965	STK86	30-JUN-1997:1	B5303-T22	T-22 BEAM PREPS/CONNECT TO SEG./LIFT	14	SRM	Structures , Mechanisms, Veh Handling	3.01.03
1966	STK86	30-JUN-1997:1	B5303 T-23	T-23 H77-0384-03 4 POINT LIFTING BEAM ADJ. (IF REQ'D)	6	SRM	Structures , Mechanisms, Veh Handling	3.02.01
1967	STK86	30-JUN-1997:1	B5303-T30	T-30 FJAF PREPS	6	SRM	Structures , Mechanisms, Veh Handling	3.02.02

No.	DATA SET	START TIME DD:MMM-YYYY:H	PROCEDURE NAME	DESCRIPTION	DURATION (Hours)	SYSTEM	DESIGN DISCIPLINE	GENERIC OPS FBS NUMBER (Appx A1)
1968	STK86	30-JUN-1997:1	B5303-T31	T-31 ACCURACY TEST	11	SRM	Structures , Mechanisms, Veh Handling	3.02.03
1969	STK86	30-JUN-1997:5	B5303-T34	T-34 RMV PROT. COVERS	1	SRM	Structures , Mechanisms, Veh Handling	3.02.01
1970	STK86	30-JUN-1997:6	B5303-T35	T-35 INSTL. OF MODIFIED PROTECTIVE CVRS.	1	SRM	Structures , Mechanisms, Veh Handling	3.02.01
1971	STK86	30-JUN-1997:6	B5303-T07	T-07 CLEVIS PREPS	11	SRM	Structures , Mechanisms, Veh Handling	3.02.02
1972	STK86	30-JUN-1997:7	B5303-T08	T-08 LFC TANG MATE/OPS	17	SRM	Structures , Mechanisms, Veh Handling	3.02.02
1973	STK86	30-JUN-1997:12	B5303-T29	T-29 SEGMENT JOINT LEAK CHECK (LA)	17	SRM	Structures , Mechanisms, Veh Handling	3.02.03
1974	STK86	30-JUN-1997:23	B5303-T25	T-25 ENCLOSURE SKIRT INSTLN.	2	SRM	Structures , Mechanisms, Veh Handling	3.01.04
1975	STK86	30-JUN-1997:24	B5303-T36	T-36 RMVL OF MOD PROTECTIVE CVRS.	1	SRM	Structures , Mechanisms, Veh Handling	3.02.01
1976	STK86	01-JUL-1997:1	B5303-T08	T-08 LFC LIFT TO HB/MATE	19	SRM	Structures , Mechanisms, Veh Handling	3.02.02
1977	STK86	01-JUL-1997:8	B5303-T10	LA JOINT CLOSEOUTS T-10,-12,-13	4	SRM	Structures , Mechanisms, Veh Handling	3.03.05
1978	STK86	01-JUL-1997:8	B5303-T12	LA JOINT CLOSEOUTS T-12	109	SRM	Structures , Mechanisms, Veh Handling	3.03.05
1979	STK86	01-JUL-1997:19	B5303-T21	T-21 TEMPASONIC INSTRUMENTATION	2	SRM	Structures , Mechanisms, Veh Handling	3.03.03
1980	STK86	01-JUL-1997:20	B5303-22	T-22 (H77-0384-03)HANDLING BEAM OPS./BEAM TO AISLE	2	SRM	Structures , Mechanisms, Veh Handling	3.06.01
1981	STK86	01-JUL-1997:22	B5303 T-22	T-22 BEAM PREPS/CONNECT SO SEG./LIFT	5	SRM	Structures , Mechanisms, Veh Handling	3.01.03
1982	STK86	01-JUL-1997:22	B5303-T34	T-34 REMOVE PROTECTIVE CVRS.	2	SRM	Structures , Mechanisms, Veh Handling	3.02.01
1983	STK86	01-JUL-1997:22	B5303 T-23	T-23 H77-0384-03 4 POINT LIFTING BEAM ADJ. (IF REQ'D)	5	SRM	Structures , Mechanisms, Veh Handling	3.02.01
1984	STK86	01-JUL-1997:22	B5303-T30	T-30 FJAF PREPS	7	SRM	Structures , Mechanisms, Veh Handling	3.02.02
1985	STK86	01-JUL-1997:23	S3001	S3001 PLATFORM OPS.	2	SRM	Ground Systems & Facilities	3.06.01
1986	STK86	01-JUL-1997:24	B5303-T35	T-35 INSTL. OF MODIFIED PROTECTIVE CVRS.	3	SRM	Structures , Mechanisms, Veh Handling	4.01.03
1987	STK86	01-JUL-1997:47	B5303-T29	LA LEAK CHECK T-29 PORT PLUG INSTL.	14	SRM	Structures , Mechanisms, Veh Handling	3.02.03
1988	STK86	02-JUL-1997:1	B5303-T31	T-31 ACCURACY TEST	6	SRM	Structures , Mechanisms, Veh Handling	3.02.03
1989	STK86	02-JUL-1997:3	B5303-T07	T-07 CLEVIS PREPS	13	SRM	Structures , Mechanisms, Veh Handling	3.02.02
1990	STK86	02-JUL-1997:5	B5303-T08	T-08 LF TANG/MATE OPS	12	SRM	Structures , Mechanisms, Veh Handling	3.02.02
1991	STK86	02-JUL-1997:7	B5303-T29	T-29 SEG. JOINT LEAK CHECK (LC) C77-0497	12	SRM	Structures , Mechanisms, Veh Handling	3.02.03
1992	STK86	02-JUL-1997:15	B5303-T36	T-36 RMVL OF MOD PROTECTIVE CVRS.	3	SRM	Structures , Mechanisms, Veh Handling	4.01.03
1993	STK86	02-JUL-1997:16	B5303-T25	T-25 ENCLOSURE SKIRT INSTLN.	2	SRM	Structures , Mechanisms, Veh Handling	3.01.04
1994	STK86	02-JUL-1997:18	B5303-T08	T-08 LF LIFT TO HB/MATE	14	SRM	Structures , Mechanisms, Veh Handling	3.02.02
1995	STK86	02-JUL-1997:19	B5303-T29	T-29 PORT PLUG INSTL.	18	SRM	Structures , Mechanisms, Veh Handling	3.02.03
1996	STK86	03-JUL-1997:4	B5303-T21	T-21 TEMPASONIC INSTRUMENTATION	2	SRM	Structures , Mechanisms, Veh Handling	3.03.03
1997	STK86	03-JUL-1997:8	B5303-T22	T-22 (H77-0384-03)HANDLING BEAM OPS./BEAM TO AISLE	2	SRM	Structures , Mechanisms, Veh Handling	3.06.01
1998	STK86	03-JUL-1997:9	S3001	S3001 PLATFORM OPS	2	SRM	Ground Systems & Facilities	3.06.01

No.	DATA SET	START TIME DD:MMM-YYYY:H	PROCEDURE NAME	DESCRIPTION	DURATION (Hours)	SYSTEM	DESIGN DISCIPLINE	GENERIC OPS FBS NUMBER (Appx A1)
1999	STK86	03-JUL-1997:16	B5303-T31	T-31 ACCURACY TEST	9	SRM	Structures , Mechanisms, Veh Handling	3.02.03
2000	STK86	03-JUL-1997:16	B5303-T10	LC JOINT CLOSEOUTS T-10,-12,-13	3	SRM	Structures , Mechanisms, Veh Handling	3.03.05
2001	STK86	03-JUL-1997:20	B5303-T12	LC JOINT CLOSEOUTS T-12	81	SRM	Structures , Mechanisms, Veh Handling	3.03.05
2002	STK86	07-JUL-1997:1	B5303-T29	T-29 LEAK CHECK SYS. (LF) C77-0497	13	SRM	Structures , Mechanisms, Veh Handling	3.03.05
2003	STK86	07-JUL-1997:9	B5303-T04	T-04 FWD ASSY RECEIVING/INSP.	4	SRM	Structures , Mechanisms, Veh Handling	4.01.02
2004	STK86	07-JUL-1997:13	B5303-T29	LF LEAK CHECK POST OPS/PORT PLUG INSTL.	11	SRM	Structures , Mechanisms, Veh Handling	3.02.03
2005	STK86	07-JUL-1997:23	B5303-T10	LF JOINT CLOSEOUTS T-10,-12,-13	8	SRM	Structures , Mechanisms, Veh Handling	3.03.05
2006	STK86	08-JUL-1997:1	B5303-T12	LA JOINT CLOSEOUTS T-13	12	SRM	Structures , Mechanisms, Veh Handling	3.03.05
2007	STK86	08-JUL-1997:7	B5303-T12	LF JOINT CLOSEOUTS T-12	96	SRM	Structures , Mechanisms, Veh Handling	3.03.05
2008	STK86	08-JUL-1997:20	B5303-T20	T-20 FWD ASSY HANDLING OPS.(SLING PREPS/CONN./LIFT)	3	SRM	Structures , Mechanisms, Veh Handling	3.01.03
2009	STK86	08-JUL-1997:22	B5303-T11	T-11 LFA MATE PREPS (CLEAN/INSP.)	4	SRM	Structures , Mechanisms, Veh Handling	3.01.04
2010	STK86	09-JUL-1997:1	B5303-T11	T-11 LFA MATE PREPS/MATE	9	SRM	Structures , Mechanisms, Veh Handling	3.02.02
2011	STK86	09-JUL-1997:9	S3001	S3001 PLTFM OPS	2	SRM	Ground Systems & Facilities	3.06.01
2012	STK86	09-JUL-1997:9	B5303-T11	T-11 SLING TO AISLE	2	SRM	Structures , Mechanisms, Veh Handling	3.06.01
2013	STK86	09-JUL-1997:10	B5303-T20	T-20 DISC SLING/SECURE/REM LIFT LUGS	2	SRM	Structures , Mechanisms, Veh Handling	3.06.01
2014	STK86	09-JUL-1997:12	B5303-T16	T-16,LFA JOINT CLOSEOUTS (PIN/CORK BAND INSTL)	64	SRM	Structures , Mechanisms, Veh Handling	3.03.05
2015	STK86	09-JUL-1997:12	B5303-T05	T-33 HYDRASET PREPS/CONNECT TO HOOK/BEAM	21	SRM	Structures , Mechanisms, Veh Handling	4.01.03
2016	STK86	10-JUL-1997:1	B5303-T13	LC JOINT CLOSEOUTS T-13	4	SRM	Structures , Mechanisms, Veh Handling	3.03.05
2017	STK86	10-JUL-1997:1	B5303-T13	LF JOINT CLOSEOUTS T-13	4	SRM	Structures , Mechanisms, Veh Handling	3.03.05
2018	STK86	10-JUL-1997:8	B5303-T17	T-17 LFA JOINT CLOSEOUTS (SPLICE PLATE INSTL)	95	SRM	Structures , Mechanisms, Veh Handling	3.03.05
2019	STK86	10-JUL-1997:8	B5303-T05	T-22 BEAM PREPS/CONNECT TO BOOSTER	9	SRM	Structures , Mechanisms, Veh Handling	4.01.03
2020	STK86	10-JUL-1997:16	B5303-T05	T-05 STACK AFT BOOSTER ASSY.	5	SRM	Structures , Mechanisms, Veh Handling	4.01.03
2021	STK86	10-JUL-1997:20	B5303-T05	T-05 RESHIM (IF REQ'D)	2	SRM	Structures , Mechanisms, Veh Handling	3.02.01
2022	STK86	10-JUL-1997:20	B5303-T3A	T-03A SRB JOINT HTR UMB (GCA) INSTL.	9	SRM	Structures , Mechanisms, Veh Handling	4.03.01
2023	STK86	10-JUL-1997:20	B5303-T06	T-06 INSTL SRB HOLDDOWN HDWR	13	SRM	Structures , Mechanisms, Veh Handling	4.03.01
2024	STK86	10-JUL-1997:21	B5303-T05	T-05 ROTATION MEAS.	2	SRM	Structures , Mechanisms, Veh Handling	4.01.03
2025	STK86	10-JUL-1997:22	B5303-T22	T-22 (H77-0384-03)HANDLING BEAM OPS./BEAM TO AISLE	3	SRM	Structures , Mechanisms, Veh Handling	3.06.01
2026	STK86	10-JUL-1997:22	S3001	S3001 PLATFORM OPS.	2	SRM	Ground Systems & Facilities	3.06.01
2027	STK86	11-JUL-1997:2	B5303 T-22	T-22 BEAM PREPS/CONNECT TO SEG./LIFT	6	SRM	Structures , Mechanisms, Veh Handling	3.01.03
2028	STK86	11-JUL-1997:2	B5303-T34	T-34 REMOVE PROTECTIVE CVRS.	2	SRM	Structures , Mechanisms, Veh Handling	3.02.01
2029	STK86	11-JUL-1997:3	B5303-T35	T-35 INSTL. OF MODIFIED PROTECTIVE CVRS.	2	SRM	Structures , Mechanisms, Veh Handling	3.02.01

No.	DATA SET	START TIME DD:MMM-YYYY:H	PROCEDURE NAME	DESCRIPTION	DURATION (Hours)	SYSTEM	DESIGN DISCIPLINE	GENERIC OPS FBS NUMBER (Appx A1)
2030	STK86	11-JUL-1997:5	B5303-T07	T-07 CLEVIS PREPS	11	SRM	Structures , Mechanisms, Veh Handling	3.02.02
2031	STK86	11-JUL-1997:8	B5303-T08	T-08 RAC TANG/MATE OPS	9	SRM	Structures , Mechanisms, Veh Handling	3.02.02
2032	STK86	11-JUL-1997:8	B5303-T30	T-30 FJAF PREPS	7	SRM	Structures , Mechanisms, Veh Handling	3.02.02
2033	STK86	11-JUL-1997:8	B5303-T14	T-14 HOLDDOWN HARDWARE PRETENSIONING	4	SRM	Structures , Mechanisms, Veh Handling	4.03.01
2034	STK86	11-JUL-1997:9	B5303 T-23	T-23 H77-0384-03 4 POINT LIFTING BEAM ADJ. (IF REQ'D)	2	SRM	Structures , Mechanisms, Veh Handling	3.02.01
2035	STK86	11-JUL-1997:12	B5303-T06	T-05 POST OPS	2	SRM	Structures , Mechanisms, Veh Handling	4.01.04
2036	STK86	11-JUL-1997:15	B5303-T25	T-25 ENCLOSURE SKIRT INSTLN.	2	SRM	Structures , Mechanisms, Veh Handling	3.01.04
2037	STK86	11-JUL-1997:15	B5303-T36	T-36 RMVL OF MOD PROTECTIVE CVRS.	3	SRM	Structures , Mechanisms, Veh Handling	3.02.01
2038	STK86	11-JUL-1997:16	B5303-T08	T-08 LAC LIFT TO HB/MATE	20	SRM	Structures , Mechanisms, Veh Handling	3.02.02
2039	STK86	12-JUL-1997:11	B5303-T21	T-21 TEMPASONIC INSTRUMENTATION	2	SRM	Structures , Mechanisms, Veh Handling	3.03.03
2040	STK86	12-JUL-1997:11	B5303-T22	T-22 (H77-0384-03)HANDLING OPS./BEAM TO AISLE	3	SRM	Structures , Mechanisms, Veh Handling	3.06.01
2041	STK86	12-JUL-1997:13	S3001	S3001 PLATFORM OPS	2	SRM	Ground Systems & Facilities	3.06.01
2042	STK86	14-JUL-1997:1	B5303-T34	T-34 REMOVE PROTECTIVE CVRS.	2	SRM	Structures , Mechanisms, Veh Handling	3.02.01
2043	STK86	14-JUL-1997:1	B5303-T30	T-30 FJAF PREPS	6	SRM	Structures , Mechanisms, Veh Handling	3.02.02
2044	STK86	14-JUL-1997:1	B5303-T31	T-31 ACCURACY TEST (RA)	5	SRM	Structures , Mechanisms, Veh Handling	3.02.03
2045	STK86	14-JUL-1997:3	B5303-T35	T-35 INSTL. OF MODIFIED PROTECTIVE CVRS.	2	SRM	Structures , Mechanisms, Veh Handling	3.02.01
2046	STK86	14-JUL-1997:5	B5303-T07	T-07 CLEVIS PREPS	18	SRM	Structures , Mechanisms, Veh Handling	3.02.02
2047	STK86	14-JUL-1997:6	B5303 T-22	T-22 BEAM PREPS/CONNECT TO SEG./LIFT	6	SRM	Structures , Mechanisms, Veh Handling	3.01.03
2048	STK86	14-JUL-1997:6	B5303-T29	T-29 LEAK CHECK SYS (RA) C77-0497	10	SRM	Structures , Mechanisms, Veh Handling	3.02.03
2049	STK86	14-JUL-1997:11	B5303 T-23	T-23 H77-0384-03 4 POINT LIFTING BEAM ADJ. (IF REQ'D)	2	SRM	Structures , Mechanisms, Veh Handling	3.02.01
2050	STK86	14-JUL-1997:11	B5303-T08	T-08 RFC TANG/MATE OPS	15	SRM	Structures , Mechanisms, Veh Handling	3.02.02
2051	STK86	14-JUL-1997:16	B5303-T29	RA LEAK CHECK/PORT PLUG INSTL.	7	SRM	Structures , Mechanisms, Veh Handling	3.02.03
2052	STK86	14-JUL-1997:22	B5303-T36	T-36 RMVL OF MOD PROTECTIVE CVRS.	5	SRM	Structures , Mechanisms, Veh Handling	3.02.01
2053	STK86	15-JUL-1997:1	B5303-T25	T-25 ENCLOSURE SKIRT INSTLN.	2	SRM	Structures , Mechanisms, Veh Handling	3.01.04
2054	STK86	15-JUL-1997:2	S3001	S3001 PLATFORM OPS.	2	SRM	Ground Systems & Facilities	3.06.01
2055	STK86	15-JUL-1997:3	B5303-T08	T-08 RFC LIST TO HB/MATE	21	SRM	Structures , Mechanisms, Veh Handling	3.02.02
2056	STK86	15-JUL-1997:4	B5303-T10	RA JOINT CLOSEOUTS T-10,-12,-13	8	SRM	Structures , Mechanisms, Veh Handling	3.03.05
2057	STK86	15-JUL-1997:12	B5303-T12	RA JOINT CLOSEOUTS T-12	135	SRM	Structures , Mechanisms, Veh Handling	3.03.05
2058	STK86	15-JUL-1997:22	B5303-T21	T-21 TEMPASONIC INSTRUMENTATION	2	SRM	Structures , Mechanisms, Veh Handling	3.03.03
2059	STK86	15-JUL-1997:24	B5303-T22	T-22 (H77-0384-03)HANDLING BEAM OPS./BEAM TO AISLE	21	SRM	Structures , Mechanisms, Veh Handling	3.06.01
2060	STK86	16-JUL-1997:1	B5303-T22	T-22 BEAM PREPS/CONNECT TO SEG./LIFT	6	SRM	Structures , Mechanisms, Veh Handling	3.01.03

No.	DATA SET	START TIME DD:MMM-YYYY:H	PROCEDURE NAME	DESCRIPTION	DURATION (Hours)	SYSTEM	DESIGN DISCIPLINE	GENERIC OPS FBS NUMBER (Appx A1)
2061	STK86	16-JUL-1997:3	B5303-T34	RFC CLEVIS PREPS T-07	2	SRM	Structures , Mechanisms, Veh Handling	3.02.02
2062	STK86	16-JUL-1997:3	B5303-T31	T-31 ACCURACY TEST	5	SRM	Structures , Mechanisms, Veh Handling	3.02.03
2063	STK86	16-JUL-1997:4	B5303-T35	T-35 INSTL. OF MODIFIED PROTECTIVE CVRS.	2	SRM	Structures , Mechanisms, Veh Handling	3.02.01
2064	STK86	16-JUL-1997:6	B5303-T07	T-07 CLEVIS PREPS	10	SRM	Structures , Mechanisms, Veh Handling	3.02.02
2065	STK86	16-JUL-1997:7	B5303 T-23	T-23 H77-0384-03 4 POINT LIFTING BEAM ADJ. (IF REQ'D)	2	SRM	Structures , Mechanisms, Veh Handling	3.02.01
2066	STK86	16-JUL-1997:8	B5303-T30	T-30 FJAF PREPS	30	SRM	Structures , Mechanisms, Veh Handling	3.02.02
2067	STK86	16-JUL-1997:8	B5303-T08	T-08 RF TANG/MATE OPS	10	SRM	Structures , Mechanisms, Veh Handling	3.02.02
2068	STK86	16-JUL-1997:8	B5303-T29	T-29 LEAK CHECK SYS (RC) C77-0497	19	SRM	Structures , Mechanisms, Veh Handling	3.02.03
2069	STK86	16-JUL-1997:16	B5303-T36	T-36 RMVL OF MOD PROTECTIVE CVRS.	3	SRM	Structures , Mechanisms, Veh Handling	3.02.01
2070	STK86	16-JUL-1997:18	B5303-T25	T-25 ENCLOSURE SKIRT INSTLN.	2	SRM	Structures , Mechanisms, Veh Handling	3.01.04
2071	STK86	16-JUL-1997:18	B5303-T08	T-08 RF LIFT TO HB/MATE	22	SRM	Structures , Mechanisms, Veh Handling	3.02.02
2072	STK86	17-JUL-1997:3	B5303-T29	RC LEAK CHECK/PORT PLUG INSTL.	14	SRM	Structures , Mechanisms, Veh Handling	3.02.03
2073	STK86	17-JUL-1997:15	B5303-T21	T-21 TEMPASONIC INSTRUMENTATION	2	SRM	Structures , Mechanisms, Veh Handling	3.03.03
2074	STK86	17-JUL-1997:15	B5303-T22	T-22 (H77-0384-03)HANDLING BEAM OPS./BEAM TO AISLE	3	SRM	Structures , Mechanisms, Veh Handling	3.06.01
2075	STK86	17-JUL-1997:16	B5303-T20	T-20 FWD ASSY HANDLING OPS.(SLING PREPS/CONN./LIFT)	6	SRM	Structures , Mechanisms, Veh Handling	3.01.03
2076	STK86	17-JUL-1997:16	B5303-T31	T-31 ACCURACY TEST	8	SRM	Structures , Mechanisms, Veh Handling	3.02.03
2077	STK86	17-JUL-1997:16	B5303-T04	T-04 FWD ASSY RECEIVING/INSP.	5	SRM	Structures , Mechanisms, Veh Handling	4.01.03
2078	STK86	17-JUL-1997:17	S3001	S3001 PLATFORM OPS.	2	SRM	Ground Systems & Facilities	3.06.01
2079	STK86	17-JUL-1997:18	B5303-T10	RC JOINT CLOSEOUTS T-10,-12,-13	5	SRM	Structures , Mechanisms, Veh Handling	3.03.05
2080	STK86	17-JUL-1997:21	B5303-T11	T-11 RFA MATE PREPS (CLEAN/INSP.)	3	SRM	Structures , Mechanisms, Veh Handling	3.01.04
2081	STK86	17-JUL-1997:23	B5303-T12	RC JOINT CLOSEOUTS T-12	92	SRM	Structures , Mechanisms, Veh Handling	3.03.05
2082	STK86	17-JUL-1997:24	B5303-T11	T-11 RFA MATE PREPS/MATE	5	SRM	Structures , Mechanisms, Veh Handling	3.02.02
2083	STK86	17-JUL-1997:24	B5303-T29	T-29 LEAK CHECK SYS (RF) C77-0497	11	SRM	Structures , Mechanisms, Veh Handling	3.02.03
2084	STK86	18-JUL-1997:1	B5303-T12	RF JOINT CLOSEOUTS T-12	100	SRM	Structures , Mechanisms, Veh Handling	3.03.05
2085	STK86	18-JUL-1997:4	B5303-T17	T-17 RFA JOINT CLOSEOUTS (SPLICE PLATE INSTL)	83	SRM	Structures , Mechanisms, Veh Handling	3.03.05
2086	STK86	18-JUL-1997:4	B5303-T16	T-16,RFA JOINT CLOSEOUTS (PIN/CORK BAND INSTL)	83	SRM	Structures , Mechanisms, Veh Handling	3.03.05
2087	STK86	18-JUL-1997:4	S3001	S3001 PLTFM OPS	2	SRM	Ground Systems & Facilities	3.06.01
2088	STK86	18-JUL-1997:4	B5303-T11	T-11 SLING TO AISLE	2	SRM	Structures , Mechanisms, Veh Handling	3.06.01
2089	STK86	18-JUL-1997:5	B5303-T20	T-20 DISC SLING/SECURE/REM LIFT LUGS	2	SRM	Structures , Mechanisms, Veh Handling	3.06.01
2090	STK86	18-JUL-1997:11	B5303-T29	RF LEAK CHECK/PORT PLUG INSTL	4	SRM	Structures , Mechanisms, Veh Handling	3.02.03
2091	STK86	18-JUL-1997:16	B5303-T10	RF JOINT CLOSEOUTS T-10,-12,-13	3	SRM	Structures , Mechanisms, Veh Handling	3.03.05

No.	DATA SET	START TIME DD:MMM-YYYY:H	PROCEDURE NAME	DESCRIPTION	DURATION (Hours)	SYSTEM	DESIGN DISCIPLINE	GENERIC OPS FBS NUMBER (Appx A1)
2092	STK86	23-JUL-1997:6	S3001	S3001 PLTFM OPS	4	SRM	Ground Systems & Facilities	3.06.01
2093	STK86	23-JUL-1997:8	B5303-T13	RA JOINT CLOSEOUTS T-13	6	SRM	Structures , Mechanisms, Veh Handling	3.03.05
2094	STK86	23-JUL-1997:8	B5303-T13	RC JOINT CLOSEOUTS T-13	7	SRM	Structures , Mechanisms, Veh Handling	3.03.05
2095	STK86	23-JUL-1997:10	B5303-T18	T-18 ALIGNMENT	5	SRM	Structures , Mechanisms, Veh Handling	3.02.03
2096	STK86	23-JUL-1997:14	S3001	S3001 PLTFM OPS	3	SRM	Ground Systems & Facilities	3.06.01
2097	STK86	23-JUL-1997:21	B5303-T13	RF JOINT CLOSEOUTS T-13	4	SRM	Structures , Mechanisms, Veh Handling	3.03.05
2098	STK86	28-JUL-1997:21	B5303-T36	T-36 RMVL OF MODIFIED PROTECTIVE CVRS	119	SRM	Structures , Mechanisms, Veh Handling	3.02.01
2099	TKC86	15-JUL-1997:9	B5304	B5304 T-3/10/11(LH/RH) T1297,S5010 WORK PREPS	48	SRM	Structures , Mechanisms, Veh Handling	3.06.02
2100	TKC86	16-JUL-1997:1	B5303	B5303 T-26 INSTL FWD SKT ACCESS KIT (LH)	5	SRM	Structures , Mechanisms, Veh Handling	3.01.01
2101	TKC86	16-JUL-1997:1	B5307	B5307 T-10 HEATER CABLE CONNECTS (LH)	8	SRE	Cmd, Ctl & Health Mngmt	3.03.05
2102	TKC86	16-JUL-1997:6	B5307	B5307 T-4 FWD DOME CABLE ROUTING (LH)	7	SRE	Cmd, Ctl & Health Mngmt	3.03.03
2103	TKC86	16-JUL-1997:9	B5307	B5307 T-3 CABLE ROUTING (LH)	8	SRE	Cmd, Ctl & Health Mngmt	3.03.03
2104	TKC86	16-JUL-1997:17	B5304	B5304 T-7 PREPS LSC INSTL (LH)	8	SRM	Structures , Mechanisms, Veh Handling	3.03.02
2105	TKC86	18-JUL-1997:9	B5307	B5307 T-3 (LH)	16	SRE	Cmd, Ctl & Health Mngmt	3.03.03
2106	TKC86	18-JUL-1997:9	B5307	B5307 T-4 (LH)	4	SRE	Cmd, Ctl & Health Mngmt	3.03.03
2107	TKC86	18-JUL-1997:13	B5307	B5307 T-10 (LH)	4	SRE	Cmd, Ctl & Health Mngmt	3.03.05
2108	TKC86	18-JUL-1997:17	B5304	B5304 T-13 HEATER CABLE CLOSEOUTS (LH)	12	SRM	Structures , Mechanisms, Veh Handling	3.03.05
2109	TKC86	21-JUL-1997:1	B5304	B5304 T-1 INSTL FN #9 COVER (LH)	8	SRM	Structures , Mechanisms, Veh Handling	3.03.03
2110	TKC86	21-JUL-1997:1	B5307	B5307 T-6,7,9 CABLE C/O (PR-BI090L-14)	8	SRE	Cmd, Ctl & Health Mngmt	3.03.05
2111	TKC86	21-JUL-1997:9	S0003	S0003 T-20 PREPS FOR ET MATE/T-21 SRB ALIGN VERIFY	44	ETM	Structures , Mechanisms, Veh Handling	4.01.01
2112	TKC86	22-JUL-1997:9	PR-BI090L-14	B5304 T-6,7.9 (PR-BI090L-14 R&R CABLE)	12	SRE	Cmd, Ctl & Health Mngmt	3.05.02
2113	TKC86	22-JUL-1997:21	B5304	B5304 T-1 (LH)	4	SRM	Structures , Mechanisms, Veh Handling	3.03.03
2114	TKC86	22-JUL-1997:21	B5304	B5304 T-2 ETA TRANS FOAMING (LH)	4	SRM	Structures , Mechanisms, Veh Handling	3.03.05
2115	TKC86	22-JUL-1997:21	B5304	B5304 T-30 HDP CABLE INSTL/PREPS FOR ORD INSTL	24	SRM	Structures , Mechanisms, Veh Handling	3.06.02
2116	TKC86	23-JUL-1997:17	B5303	B5303 T-26 INSTL FWD SKT ACCESS KIT (RH)	5	SRM	Structures , Mechanisms, Veh Handling	3.01.01
2117	TKC86	23-JUL-1997:17	B5303	B5303 T-15 HDP SHOE RETAINER INSTL (LH/RH)	16	SRM	Structures , Mechanisms, Veh Handling	3.05.02
2118	TKC86	23-JUL-1997:17	B5304	B5304 T-3 ETA COVER INSTL/APPLY K5 (LH)(PR-16)	16	SRM	Structures , Mechanisms, Veh Handling	3.05.02
2119	TKC86	23-JUL-1997:17	B5150	B5150 INSTL/REM JIB HOIST	8	ESG	Ground Systems & Facilities	4.01.01
2120	TKC86	24-JUL-1997:1	B5307	B5307 T-4 FWD DOME CABLE ROUTING (RH)	5	SRE	Cmd, Ctl & Health Mngmt	3.03.03
2121	TKC86	24-JUL-1997:5	S3001	S3001 PLATFORM OPS	4	SRM	Ground Systems & Facilities	4.01.01
2122	TKC86	24-JUL-1997:9	S0003	S0003 ET/SRB MATE	8	ETM	Structures , Mechanisms, Veh Handling	4.01.03

No.	DATA SET	START TIME DD:MMM-YYYY:H	PROCEDURE NAME	DESCRIPTION	DURATION (Hours)	SYSTEM	DESIGN DISCIPLINE	GENERIC OPS FBS NUMBER (Appx A1)
2123	TKC86	24-JUL-1997:17	S0003	S0003 ET/SRB MATE	8	ETM	Structures , Mechanisms, Veh Handling	4.01.03
2124	TKC86	25-JUL-1997:1	S0003	S0003 ET/SRB MATE	24	ETM	Structures , Mechanisms, Veh Handling	4.01.03
2125	TKC86	25-JUL-1997:1	T1203	T1203 (OPS 20) INSTL FLEX HOSES	4	ETP	Propulsion	4.01.04
2126	TKC86	25-JUL-1997:9	B5307	B5307 T-4 (RH)	7	SRE	Cmd, Ctl & Health Mngmt	3.03.03
2127	TKC86	25-JUL-1997:9	B5304	B5304 T-3 ETA COVER INSTL/APPLY K5 (LH)	1	SRM	Structures , Mechanisms, Veh Handling	3.03.03
2128	TKC86	25-JUL-1997:9	B5307	B5307 T-3 CABLE ROUTING (RH)	20	SRE	Cmd, Ctl & Health Mngmt	3.03.03
2129	TKC86	25-JUL-1997:9	B5304	B5304 T-27 HDP ORD INSTL & CONN/BC INSTL	4	SRM	Structures , Mechanisms, Veh Handling	3.03.04
2130	TKC86	25-JUL-1997:9	B5307	B5307 T-10 HEATER CABLE CONNECTS (RH)	7	SRE	Cmd, Ctl & Health Mngmt	3.03.05
2131	TKC86	25-JUL-1997:9	S5010	S5010 FWD CROSSOVER INSTL (LH)	56	SRM	Structures , Mechanisms, Veh Handling	4.01.03
2132	TKC86	25-JUL-1997:9	S5010	S5010 FWD CROSSOVER INSTL (RH)	56	SRM	Structures , Mechanisms, Veh Handling	4.01.03
2133	TKC86	25-JUL-1997:13	B5304	B5304 T-27 HDP ORD INSTL (LH/RH)	4	SRM	Structures , Mechanisms, Veh Handling	3.03.04
2134	TKC86	25-JUL-1997:13	B5304	B5304 T-34 HDP PLUNGER INSTL (LH/RH)	5	SRM	Structures , Mechanisms, Veh Handling	3.03.04
2135	TKC86	25-JUL-1997:17	S5011	S5011 AFT STRUT ELECT MATE (RH)	28		Power Management	4.01.03
2136	TKC86	25-JUL-1997:17	S5011	S5011 AFT STRUT ELECT MATE (LH)	28		Power Management	4.01.03
2137	TKC86	28-JUL-1997:1	B5304	B5304 T-34 HDP PLUNGER INSTL (LH/RH)	16	SRM	Structures , Mechanisms, Veh Handling	3.03.04
2138	TKC86	28-JUL-1997:1	B5304	B5304 T-26 BLAST SHIELD INSTALLATION (LH/RH)	18	SRM	Structures , Mechanisms, Veh Handling	3.03.04
2139	TKC86	28-JUL-1997:5	B5307	B5307 T-6,7,9 CABLE VERIFY & CONN (RH)	12	SRE	Cmd, Ctl & Health Mngmt	3.03.05
2140	TKC86	28-JUL-1997:9	B5304	B5304 T-6 LOWER STRUT C/O (RH)	33	SRM	Structures , Mechanisms, Veh Handling	3.03.05
2141	TKC86	28-JUL-1997:9	B5304	B5304 T-6 LOWER STRUT C/O (LH)	33	SRM	Structures , Mechanisms, Veh Handling	3.03.05
2142	TKC86	28-JUL-1997:17	B5304	B5304 T-7 PREPS LSC INSTL (RH)	8	SRM	Structures , Mechanisms Veh Handling	3.03.02
2143	TKC86	28-JUL-1997:17	B5304	B5304 T-1 INSTL FN #9 COVER (RH)	8	SRM	Structures , Mechanisms, Veh Handling	3.03.03
2144	TKC86	28-JUL-1997:17	B5304	B5304 T-2 ETA TRANS FOAMING (RH)	4	SRM	Structures , Mechanisms, Veh Handling	3.03.05
2145	TKC86	28-JUL-1997:17	B5304	B5304 T-13 HEATER CABLE CLOSEOUTS (RH)	25	SRM	Structures , Mechanisms, Veh Handling	3.03.05
2146	TKC86	28-JUL-1997:17	B5150	B5150 REM JIB HOIST	4	ESG	Ground Systems & Facilities	4.01.01
2147	TKC86	28-JUL-1997:21	S5011	S5011 (RH) BUS ISO CHECKS	4		Power Management	4.01.03
2148	TKC86	28-JUL-1997:21	S5011	S5011 (LH) BUS ISO CHECKS	4		Power Management	4.01.03
2149	TKC86	29-JUL-1997:1	B5304	B5304 T-3 ETA COVER INSTL/APPLY K5 (RH)	24	SRM	Structures , Mechanisms, Veh Handling	3.03.03
2150	TKC86	29-JUL-1997:1	B5304	B5304 T-5 UPPER FAIRING INSTL (LH)	24	SRM	Structures , Mechanisms, Veh Handling	3.03.05
2151	TKC86	29-JUL-1997:1	B5304	B5304 T-5 UPPER FAIRING INSTL (RH)(PR-20)	60	SRM	Structures , Mechanisms, Veh Handling	3.03.05
2152	TKC86	29-JUL-1997:13	B5304	B5304 T-34 POST 2 PLUNGER DAMAGED	8	SRM	Structures , Mechanisms, Veh Handling	3.05.02
2153	TKC86	30-JUL-1997:1	B5304	B5304 T-16 UPPER & DIAG STRUT FOAM C/O (LH)	37	SRM	Structures , Mechanisms, Veh Handling	3.03.05

No.	DATA SET	START TIME DD:MMM-YYYY:H	PROCEDURE NAME	DESCRIPTION	DURATION (Hours)	SYSTEM	DESIGN DISCIPLINE	GENERIC OPS FBS NUMBER (Appx A1)
2154	TKC86	30-JUL-1997:1	B5304	B5304 T-29C HDP FIRING LINE CHECKS (LH/RH)	5	SRM	Structures , Mechanisms, Veh Handling	3.04.01
2155	TKC86	30-JUL-1997:9	B5304	B5304 T-23 LWR STRUTS FOAM C/O (LH/RH)	52	SRM	Structures , Mechanisms, Veh Handling	3.03.05
2156	TKC86	30-JUL-1997:9	B5307	B5307 T-5 GEI SNAP SHOT/HTR FUNC TEST (LH/RH)	5	SRE	Cmd, Ctl & Health Mngmt	3.04.02
2157	TKC86	31-JUL-1997:13	B5304	B5304 T-16 UPPER & DIAG STRUT FOAM C/O (RH)	41	SRM	Structures , Mechanisms, Veh Handling	3.03.05
2158	TKC86	31-JUL-1997:14	B5304	B5304 T-17 EPDM COVER INSTL (PR-BI090L-21)	14	SRM	Structures , Mechanisms, Veh Handling	3.03.05
2159	TKC86	01-AUG-1997:9	ET-88-FP-0006	PR-ET-88-FP-0006 4" DISCONNECT MEAS	5		Propulsion	4.01.03
2160	TKC86	01-AUG-1997:9	TPS-SB-BI090-001	TPS-SB-BI090-001 FWD SKT CAM CAN EVAL	4		Structures , Mechanisms, Veh Handling	4.03.03
2161	TKC86	01-AUG-1997:13	B5304	B5304 T-32 LWR STRUT EPDM CVR INSTL (LH/RH)	12	SRM	Structures , Mechanisms, Veh Handling	3.03.03
2162	TKC86	01-AUG-1997:17	B5304	B5304 T-17 EPDM UPPER & DIAG COVER INSTL (LH)	10	SRM	Structures , Mechanisms, Veh Handling	3.03.05
2163	TKC86	04-AUG-1997:6	B5304	B5304 T-17 EPDM COVER INSTL (RH)	9	SRM	Structures , Mechanisms, Veh Handling	3.03.05
2164	TKC86	04-AUG-1997:9	B5304	B5304 T-24 PREPS FOR ORD(LH/RH)	8	SRM	Structures , Mechanisms, Veh Handling	3.03.02
2165	TKC86	04-AUG-1997:9	T1297	T1297 AFT FAIRING INSTL(PR-ST-10)	18	ETM	Structures , Mechanisms, Veh Handling	4.01.04
2166	TKC86	04-AUG-1997:17	T1248	T1248 BIPOD INSTL (PR-TS-7/8)	16	ETM	Structures , Mechanisms, Veh Handling	3.06.02
2167	TKC86	05-AUG-1997:1	B5304	B5304 T-18 S & A INSTL (LH/RH)	8	SRM	Structures , Mechanisms, Veh Handling	3.03.03
2168	TKC86	05-AUG-1997:1	B5304	B5304 T-25,28 ORD INSTL & C/O (PR-23R/24R)	4	SRM	Structures , Mechanisms, Veh Handling	3.03.03
2169	TKC86	05-AUG-1997:5	B5304	B5304 T-25,28 ORD INSTL	18	SRM	Structures , Mechanisms, Veh Handling	3.03.03
2170	TKC86	05-AUG-1997:9	B5304	B5304 T-22 BSM COVER CLOSEOUTS (LH/RH)	16	SRM	Structures , Mechanisms, Veh Handling	3.03.05
2171	TKC86	05-AUG-1997:9	S5010	S5010 FWD CROSSOVER INSTL (LH)	2	SRM	Structures , Mechanisms, Veh Handling	4.01.03
2172	TKC86	05-AUG-1997:9	S5010	S5010 FWD CROSSOVER INSTL (RH)	2	SRM	Structures , Mechanisms, Veh Handling	4.01.03
2173	TKC86	05-AUG-1997:17	B5304	B5304 T-31 REMOVE RSS COVERS/APPLY RTV (LH/RH)	8	SRM	Structures , Mechanisms, Veh Handling	3.03.05
2174	TKC86	05-AUG-1997:17	B5304	B5304 T-17 EPDM UPPER & DIAG COVER INSTL (RH)	3	SRM	Structures , Mechanisms, Veh Handling	3.03.05
2175	TKC86	05-AUG-1997:17		B53-4 T-17 K5 (LH)	2	SRM	Structures , Mechanisms, Veh Handling	4.05.03
2176	TKC86	06-AUG-1997:1	B5304	B5304 T29A/T29B FIRING LINE/SEP BOLT (LH/RH)	5	SRM	Structures , Mechanisms, Veh Handling	3.03.04
2177	TKC86	06-AUG-1997:1	S1031	S0008 PREPS/S1031 SRB BATT INSTL	5	SRE	Power Management	4.01.01
2178	TKC86	06-AUG-1997:9	B5304	B5304 T-8/9 LSC INSTL (LH/RH)	4	SRM	Structures , Mechanisms, Veh Handling	3.03.03
2179	TKC86	06-AUG-1997:13	B5304	B5304 T-8,9 LSC INSTL	4	SRM	Structures , Mechanisms, Veh Handling	3.03.03
2180	TKC86	06-AUG-1997:13	B5304	B5304 T-11/12 SYS TUNNEL COVER INSTL(RH)	12	SRM	Structures , Mechanisms, Veh Handling	3.03.03
2181	TKC86	06-AUG-1997:13	B5304	B5304 T-11/12 SYS TUNNEL COVER INSTL (LH)	12	SRM	Structures , Mechanisms, Veh Handling	3.03.03
2182	TKC86	06-AUG-1997:17		S1031 SRB BATT INSTL	4	SRE	Cmd, Ctl & Health Mngmt	3.03.03
2183	TKC86	06-AUG-1997:17	T5249	TPS-ET-88-TS-002 AFT FAIRING PDL C/O(PR-TS-10/11)	24		Thermal Management	3.05.02
2184	TKC86	07-AUG-1997:1	B5304	B5304 T-13A K5NA FN# 8&9 COVERS (LH/RH)	8	SRM	Structures , Mechanisms, Veh Handling	3.03.05

No.	DATA SET	START TIME DD:MMM-YYYY:H	PROCEDURE NAME	DESCRIPTION	DURATION (Hours)	SYSTEM	DESIGN DISCIPLINE	GENERIC OPS FBS NUMBER (Appx A1)
2185	TKC86	07-AUG-1997:1	B5304	B5304 T-14 SYS TUNNEL BOND CHECKS (RH)	5	SRM	Structures , Mechanisms, Veh Handling	3.04.01
2186	TKC86	07-AUG-1997:1	B5304	B5304 T-14 SYS TUNNEL BOND CHECKS (LH)	5	SRM	Structures , Mechanisms, Veh Handling	3.04.01
2187	TKC86	07-AUG-1997:1	B5304	B5304 T-4 S & A LEAK CKS(MFM) (LH)	16	SRM	Structures , Mechanisms, Veh Handling	3.04.01
2188	TKC86	07-AUG-1997:9	B5304	B5304 T-13A ETA RING FOAM C/O (LH/RH)	16	SRM	Structures , Mechanisms, Veh Handling	3.03.05
2189	TKC86	07-AUG-1997:9	S0004	S0004 PREPS FOR ORB/ET MATE T1248 UMB PLATE REM	16	VOP	Structures , Mechanisms, Veh Handling	4.01.01
2190	TKC86	07-AUG-1997:17	B5304	B5304 T-4 S & A LEAK CKS(MFM) (RH)	18	SRM	Structures , Mechanisms, Veh Handling	3.04.01
2191	TKC86	08-AUG-1997:9	B5304	B5304 T-13A ETA RING FOAM C/O (LH/RH)	10	SRM	Structures , Mechanisms, Veh Handling	3.03.05
2192	TKC86	08-AUG-1997:9	B5304	B5304 T-15 K5NA SYS TUNNEL C/O (LH/RH)	20	SRM	Structures , Mechanisms, Veh Handling	3.03.05
2193	TKC86	08-AUG-1997:9	ET-88-TS-002	TPS - AFT FAIRING PDL C/O	16	ETM	Thermal Management	3.05.03
2194	TKC86	08-AUG-1997:16	B5304	B5304 T-28 PR-SB-BI090R-23/24	5	SRM	Structures , Mechanisms, Veh Handling	3.03.05
2195	TKC86	08-AUG-1997:21	B5304	B5304 T-29A FIRING LINE CKS	4	SRM	Structures , Mechanisms, Veh Handling	3.04.01
2196	TKC86	11-AUG-1997:5	S3001	S3002 PLATFORM OPS FOR ORB MATE	4	SRM	Ground Systems & Facilities	3.06.01
2197	TKC86	11-AUG-1997:11	S0004	S0004 ORB MATE PREPS	6	VOP	Structures , Mechanisms, Veh Handling	4.01.03
2198	TKC86	11-AUG-1997:17	S0004	S0004 ORBITER MATE	12	VOP	Structures , Mechanisms, Veh Handling	4.01.03
2199	TKC86	12-AUG-1997:5	S0004	S0004 ORBITER MATE POST OPS	36	VOP	Structures , Mechanisms, Veh Handling	4.01.03
2200	TKC86	12-AUG-1997:9	B5304	B5304 T-15 K5NA SYS TUNNEL (LH/RH)	22	SRM	Structures , Mechanisms, Veh Handling	3.03.05
2201	TKC86	12-AUG-1997:17	T5244	T5244 JACK PAD CLOSEOUT	14	ETM	Thermal Management	3.03.05
2202	TKC86	13-AUG-1997:9	B5304	B5304 T-33 SRB ENG WALKDOWN INSP(PR-26R)	40	SRM	Structures , Mechanisms, Veh Handling	3.04.03
2203	TKC86	13-AUG-1997:9	ET-88-TS-002	TPS-ET-88-TS-002 AFT TPS C/O	21		Thermal Management	4.01.04
2204	TKC86	13-AUG-1997:9	T1203	T1203 (OPS 30) V/V FLEX HOSE CONN TO GUCP	4	ETP	Propulsion	4.01.04
2205	TKC86	13-AUG-1997:17		S0004 POST OPS	44	VOP	Structures , Mechanisms, Veh Handling	4.01.04
2206	TKC86	14-AUG-1997:1	S1030	S0008 S.I.T. T1201,S1030	36	SRE	Power Management	4.02.02
2207	TKC86	14-AUG-1997:17	B5304	B5304 T-31 INSTL IEA END COVERS	4	SRM	Structures , Mechanisms, Veh Handling	3.03.03
2208	TKC86	14-AUG-1997:17	T1203	T1203 (OPS 40) MONITOR/PRESS PANEL REM	2	ETP	Propulsion	4.01.04
2209	TKC86	14-AUG-1997:17	T1203	T1203 (OPS 50/60) TEST PORT LC CK/ F/H REM	2	ETP	Propulsion	4.02.01
2210	TKC86	14-AUG-1997:17	S0008	S0008 PO #2 CABLE DISC/T1302 GUCP INSP.	4		Cmd, Ctl & Health Mngmt	4.02.01
2211	TKC86	14-AUG-1997:17	S6016	S6016 ROLLOUT INSPECTION/B5048 I/T INSP.	31	ETM	Thermal Management	4.05.03
2212	TKC86	15-AUG-1997:17		S3002 RETRACT PLTFM"S FOR ROLLOUT	8	SRM	Ground Systems & Facilities	3.06.01
2213	TKC86	15-AUG-1997:18	PR-SB-BI090R-0026	PR-SB-BI090R-0026 GOUGE IN RF SEGMENT	5		Structures , Mechanisms, Veh Handling	4.01.03
2214	TKC86	17-AUG-1997:20	S3002	S3002 RETRACT PLTFMS FOR ROLLOUT	4	SRM	Ground Systems & Facilities	3.06.01
2215	TKC86	18-AUG-1997:3	A5214	A5214 ROLL OUT	9		Structures , Mechanisms, Veh Handling	5.02.01

No.	DATA SET	START TIME DD:MMM-YYYY:H	PROCEDURE NAME	DESCRIPTION	DURATION (Hours)	SYSTEM	DESIGN DISCIPLINE	GENERIC OPS FBS NUMBER (Appx A1)
2216	TKP86	29-APR-1997:9	T5002	ET OFFLOAD AND SECURE (SEE MINI)	8		Structures , Mechanisms, Veh Handling	6.01.02
2217	TKP86	30-APR-1997:9		ET-88 ON DOCK	8		Structures , Mechanisms, Veh Handling	6.01.02
2218	TKP86	30-APR-1997:9	T5002	PUT ET IN VAB	8		Structures , Mechanisms, Veh Handling	6.02.01
2219	TKP86	30-APR-1997:9	T1002	MONITOR TANK PRESSURE	2	ETP	Propulsion	6.04.02
2220	TKP86	02-MAY-1997:9	T1002	MONITOR TANK PRESSURE	44	ETP	Propulsion	6.04.02
2221	TKP86	19-MAY-1997:9	T5002	PUT IN CELL	8		Structures , Mechanisms, Veh Handling	6.02.01
2222	TKP86	20-MAY-1997:9	T5002	OFFLOAD	8	ETM	Structures , Mechanisms, Veh Handling	6.01.03
2223	TKP86	20-MAY-1997:9	T5144	PLATFORM OPS (NEED RADIOS) (10' CLR)	8	ETM	Ground Systems & Facilities	6.02.01
2224	TKP86	20-MAY-1997:9	T1002	MONITOR TANK PRESSURE	2	ETP	Propulsion	6.04.02
2225	TKP86	21-MAY-1997:9	T6149	RECEIVING INSPECTION (I/T CLR)	8	ETM	Structures , Mechanisms, Veh Handling	6.03.01
2226	TKP86	23-MAY-1997:9	T5048	REM FLT DOOR AND INST/REMV ACCESS (I/T CLR)	16	ETM	Structures , Mechanisms, Veh Handling	3.03.02
2227	TKP86	23-MAY-1997:9	T1102/T1103	PREPS FOR CHECKOUT (10'CLR)	24	ETP	Propulsion	3.04.02
2228	TKP86	29-MAY-1997:9		RECEIVING INSPECTION PR'S	40		Structures , Mechanisms, Veh Handling	6.06.01
2229	TKP86	30-MAY-1997:9	T1147	PREPS	16	ETM	Structures , Mechanisms, Veh Handling	3.03.04
2230	TKP86	30-MAY-1997:9	T5141	AFT HARD POINT C/O (10'CLR)	56		Thermal Management	3.03.05
2231	TKP86	03-JUN-1997:9	T1147	GUCP INSTL (10'CLR)	8	ETM	Structures , Mechanisms, Veh Handling	3.03.04
2232	TKP86	04-JUN-1997:9	T1107	ANCILLARY LK&FLOW CK.(10'CLR)	24	ETP	Propulsion	3.04.01
2233	TKP86	05-JUN-1997:9	ET-88-TS-0001	DELAMINATED SLA	8	ETM	Thermal Management	3.05.02
2234	TKP86	05-JUN-1997:9	ET-88-TS-0002	CHIPPED SLA ON CABLE TRAY SUPT FTGS	8	ETM	Thermal Management	3.05.02
2235	TKP86	10-JUN-1997:9	T5048	REM ACCESS KIT AND PREFORM I/T INSP (I/T CLR)	8	ETM	Structures , Mechanisms, Veh Handling	3.01.03
2236	TKP86	10-JUN-1997:9	T1147	GUCP INSTL (10'CLR)	16	ETM	Structures , Mechanisms, Veh Handling	3.03.04
2237	TKP86	12-JUN-1997:9	T1107	ANCILLARY LK&FLOW CK.(10'CLR)	16	ETP	Propulsion	3.04.01
2238	TKP86	12-JUN-1997:13	ET-88-TS-0002	CHIPPED SLA ON CABLE TRAY SUPT FTGS	20	ETM	Thermal Management	3.05.02
2239	TKP86	13-JUN-1997:9	ET-88-TS-0001	DELAMINATED SLA	16	ETM	Thermal Management	3.05.02
2240	TKP86	16-JUN-1997:9	T1107	LEAK CHECKS (10' CLR)	8	ETP	Propulsion	3.04.01
2241	TKP86	16-JUN-1997:9	ET-88-TS-0003	FOAM DAMAGE ON CORNER OF VERTICAL STRUT	8	ETM	Thermal Management	3.05.02
2242	TKP86	17-JUN-1997:9	T1107	LK CKS CONT (10' CLR)	40	ETP	Propulsion	3.04.01
2243	TKP86	18-JUN-1997:9	T1110	LH2/LO2 17 IN DISC/MEAS/ADJ (10'CLR)	32	ETP	Propulsion	6.03.01
2244	TKP86	24-JUN-1997:9	T1110	LH2/LO2 17 IN DISC/MEAS/ADJ (10'CLR)	28	ETP	Propulsion	6.03.01
2245	TKP86	26-JUN-1997:13	T1110	LH2/LO2 17 IN DISC/MEAS/ADJ (10'CLR)	12	ETP	Propulsion	6.03.01
2246	TKP86	27-JUN-1997:9	T5048	REM ACCESS KIT AND PERFORM I/T INSP (I/T CLR)	8	ETM	Structures , Mechanisms, Veh Handling	3.01.03

No.	DATA SET	START TIME DD:MMM-YYYY:H	PROCEDURE NAME	DESCRIPTION	DURATION (Hours)	SYSTEM	DESIGN DISCIPLINE	GENERIC OPS FBS NUMBER (Appx A1)
2247	TKP86	27-JUN-1997:9	T1107	ANCILLARY LK&FLOW CK.(10'CLR)	8	ETP	Propulsion	3.04.01
2248	TKP86	27-JUN-1997:9	T1160	ALL SYSTEMS TEST	8	ETE	Power Management	3.04.02
2249	TKP86	30-JUN-1997:9	T5051	FLAPPER VALVE ACCESS PORT C/O	8	ETM	Thermal Management	3.03.05
2250	TKP86	30-JUN-1997:9	T1160	ALL SYSTEMS TEST (10'CLR)	8	ETE	Power Management	3.04.02
2251	TKP86	30-JUN-1997:9	T1101	LH2/LO2 VENT DISC (10' CLR)	4	ETP	Propulsion	3.04.02
2252	TKP86	30-JUN-1997:9	T1101	LO2 TANK VENT (10'CLR)	8	ETP	Propulsion	3.04.02
2253	TKP86	01-JUL-1997:9	T1102/T1103	PREPS FOR TRANSFER (10' CLR)	8	ETP	Propulsion	3.01.03
2254	TKP86	01-JUL-1997:9	T5051	FLAPPER VALVE ACCESS PORT C/O	64	ETM	Thermal Management	3.03.05
2255	TKP86	03-JUL-1997:9	T5048	REM ACCESS KIT AND PREFORM I/T INSP (I/T CLR)	4	ETM	Structures , Mechanisms, Veh Handling	3.01.03
2256	TKP86	03-JUL-1997:16	T1002	MONITOR TANK PRESSURE	1	ETP	Propulsion	6.04.02
2257	TKP86	08-JUL-1997:9	ET-88-TS-001	ET ALIGNMENT CONTROL	8	ETM	Ground Systems & Facilities	3.05.02
2258	TKP86	09-JUL-1997:9	T5048	REM ACCESS KIT AND PREFORM I/T INSP (I/T CLR)	8	ETM	Structures , Mechanisms, Veh Handling	3.01.03
2259	TKP86	09-JUL-1997:9	T6248	INSP/PRE-MOVE	72	ETM	Thermal Management	4.01.01
2260	TKP86	10-JUL-1997:9	T5144	PLATFORM OPS (NEED RADIOS) (10'CLR)	64	ETM	Ground Systems & Facilities	3.01.01
2261	TKP86	24-JUL-1997:9	T5002/S0003	ET/SRB MATE	4		Structures , Mechanisms, Veh Handling	4.01.03

APPENDIX E. SYSTEM CODES USED IN SOURCE DATA

SYSTEM	Description
APU	Auxiliary Power Unit (ORB & SRB)
ARM	Pad Swing Arm Systems
CCE	Contamination Control
CCS	Crew Compartment Sampling
CDP	Facility Cranes, Doors & Platforms
CME	SSME Electrical
COM	Orbiter Communications Systems
DPS	Orbiter Data Processing Systems
DRM	SRB Retrieval & Disassembly
ECL	Environmental Control - Orbiter Cooling & Life Support
ECS	Environmental Control - Purge Systems
EFC	Flight Control Systems
ELC	Electrical-Cargo
EPD	Electrical Power Distribution
ESG	ET/SRB Support Systems Calibration & Maintenance
ETE	External Tank Electrical
ETM	External Tank Mechanical & TPS
ETP	External Tank Pneumatics
FC	Fuel Cells
FCP	Fuel Cell Systems
FCS	Flight-Crew Systems
FSW	Flight Software
GNC	Guidance, Navigation and Control Systems
GSE	Ground Support Equipment (Non-Specific)
GSP	Ground Power Systems for Vehicle
HEE	Mobile Equipment
HGD	Hazardous Gas Detection Systems
HGS	Hazardous Gas Detection Systems (CAPSS)
HOP	Horizontal Move Operations
HPU	SRB Auxilliary Power Systems
HVA	HVAC Systems
HWS	Hazard Warning Systems
HYD	Hydraulic Systems (ORB & SRB)
INS	Orbiter Instrumentation Systems
LAA	Vehicle Umbilical Systems
LH2	Liquid Hydrogen Systems
LOX	Liquid Oxygen Systems
LPS	Launch Processing System (Software)
MEQ	Mechanical Systems
MIL	Schedule MILESTONE (CAPSS Code)
MPS	Main Propulsion Systems
NAV	Navigation Support Systems (Ground)
NDE	Non-Destructive Evaluation (Non-Specific)
OEL	Orbiter Electrical
OHE	Orbiter Handling Equipment
OLO	Orbiter landing Operations
OMS	Orbiter OMS/RCS Systems

OPT	Optical Systems
ORH	OMS/RCS HMF Operations
ORO	OMS/RCS OPF Operations
ORP	OMS/RCS Pad Operations
OSO	Orbiter Systems Observer
OTC	Orbiter Test Conductor Operations (Console Ops)
PCR	Payload Changeout Room (CAPSS Code)
PEL	Pad Electrical Systems
PEO	Project Engineering Office
PLE	Payload Electrical
PLM	Payload Mechanical
PLO	Payload Installation/Removal Operations
PLT	Orbiter Mission Kits
PNE	Ground Pneumatic Systems
PRS	Power, Reactants, Storage & Distribution
PVD	Orbiter Purge, Vent and Drain Systems
PYR	Pyrotechnic Systems
QC	Quality Engineering
QEN	Quality Engineering
QEV	Orbiter Receiving/Shakedown Inspection
RME	SSME Mechanical Systems
RMS	Remote Manipulator Systems
SAF	Safety
SME	SSME Engineering
SOF	Software
SPE	Specialized Ground Power Systems
SRE	Range Safety Systems
SRM	SRB Mechanical
STE	Transporters, Elevators and Pad Mechanisms
STR	Orbiter Structures
TCO	Test Conductor Operations (Console Ops)
TCS	Orbiter Thermal Protection - Blankets
TPE	Test Project Engineering
TPS	Orbiter Thermal Protection - Tile
TSM	MLP Orbiter Tail Service Mast Systems
VOP	Vertical Integration Operations
VPL	Vehicle Payload Operations
WEN	Ground Water Systems

APPENDIX F. ABBREVIATIONS, ACRONYMS, AND SYMBOLS

ac	alternating current
APU	auxiliary power unit
ASE	airborne support equipment
ASTWG	Advanced Spaceport Technology Working Group
CG	center of gravity
CIG	Connector Integrity Group
CIL	Critical Items List
CRT	cathode ray tube
dc	direct current
EDU	Extended-Duration Orbiter
ELV	Expendable Launch Vehicle
ESP	engine service platform
ESR&T	Exploration Systems Research and Technology
ET	External Tank
ETO	Earth to orbit
EVA	extravehicular activity
flt	flight
FBS	functional breakdown structure
FMEA	Failure Modes and Effects Analysis
FOD	foreign-object debris
FORP	fuel-oxidizer reaction product
FPPP	Flight Preparation and Process Planning
FRCS	Forward Reaction Control System
FSS	Fixed Service Structure
ft	foot
GEM-FLO	Generic Model for Future Launch Operations
GOX	gaseous oxygen
GPC	general-purpose computer
GSE	ground support equipment
GUCP	ground umbilical carrier panel

GVA	GOX vent arm
HCU	Hydraulic Charger Unit
HGDS	hazardous-gas detection systems
HMF	Hypergolic Maintenance Facility
HPU	hydrolic power unit
HRST	Highly Reusable Space Transportation
ICD	Interface Control Document
IFA	in-flight anomaly
ISS	International Space Station
JSC	Johnson Space Center
km	kilometer
KSC	John F. Kennedy Space Center
lb	pound
LC	Launch Complex
LCC	Launch Control Center
LOX	liquid oxygen
LPS	Launch Processing System
LRU	line replaceable unit
LUT	Launch Umbilical Tower
m	meter
MLP	Mobile Launcher Platform
MMH	monomethylhydrazine
MPS	Main Propulsion System
MR	Material Review
mt	metric ton
MTBF	mean time between failures
NASA	National Aeronautics and Space Administration
NDE	nondestructive evaluation
NGLT	Next-Generation Launch Technologies
NSLD	NASA Shuttle Logistics Depot
OAA	Orbiter Access Arm
OMBUU	Orbiter Midbody Umbilical Unit
OMDP	Orbiter Maintenance Depot Period

OMS	Orbital Maneuvering System
OPF	Orbiter Processing Facility
OWPS	Orbiter Weather Protection System
PCR	Payload Changeout Room
PGHM	Payload Ground Handling Mechanism
PMS	Permanent Measurement System
PRACA	Problem Reporting and Corrective Action
PRSD	Power Reactant Storage and Distribution (System)
psia	pound per square inch (absolute)
PVT	pressure-volume-temperature
RCA	Root Cause Analysis
RCS	Reaction Control System
RF	radio frequency
RMS	Remote Manipulator System
RPSF	Rotation, Processing, and Surge Facility
RSS	Rotating Service Structure
RTV	room-temperature vulcanization
SBS	Systems Breakdown Structure
SCAN	Shuttle Connector Analysis Network
SCAPE	Self-Contained Atmospheric Protection Ensemble
SIP	strain isolation pad
SLI	Space Launch Initiative
SPAS	Shuttle Pallet Satellite
SPC	Shuttle Processing Contract
SPST	Space Propulsion Synergy Team
SRB	Solid Rocket Booster
SR&QA	Safety, Reliability, and Quality Assurance
SSME	Space Shuttle Main Engine
SSV	Space Shuttle Vehicle
STS	Space Transportation System
TDR	time delay reflectometer
TPS	Thermal Protection System
TSM	Tail Service Mast

TVC	thrust vector control
TVS	toxic-vapor suit
USA	United Space Alliance
VAB	Vehicle Assembly Building
VAFB	Vandenberg Air Force Base
WCCS	Window Cavity Conditioning System

APPENDIX G. BIBLIOGRAPHY

Cates, C., "Space Shuttle Processing and Metrics Overview," NASA John F. Kennedy Space Center, July 26, 2003.

Cates, G.R., et al., "Modeling the Space Shuttle," in E. Yucesan et al., (eds.), *Proc. 2002 Winter Simulation Conference.*

Christenson, R.L. and D.R. Komar, "Reusable Rocket Engine Operability Modeling and Analysis," NASA TM/TP-1998-208530, Marshall Space Flight Center, Huntsville, AL, July, 1998.

"Countdown! NASA Space Shuttles and Facilities," IS-2003-09-006-KSC, NASA Information Summaries, September 2003.

"Flight Production Generic Templates," JSC 25187, Appendix A., NASA Johnson Space Center, Houston, TX.

Gillespie,T., "Atlas V Evolved Expendable Launch Vehicle: The Evolution of Responsive Launch;" AIAA-RS2-2004-8001, *2nd Responsive Space Conference,* Los Angeles, CA, April 19-22, 2004.

Guidi, C., et al, *Advanced Spaceport Technology Working Group: Baseline Report,* November 2003.

John F. Kennedy Space Center Research and Technology 2003 Annual Report, NASA-TM-2003-211190, Kennedy Space Center, FL, 2003.

Koelle, H.H., *Saturn System Study,* Redstone Arsenal, AL, November 13, 1959.

McCleskey, C.M., "Strategic Space Launch Concept and Technology Roadmaps to Develop Visionary Spaceports," IAF-99-V-5-06, *50th International Astronautical Congress,* Amsterdam, The Netherlands, October 1999.

McCleskey, C.M., and C. Guidi, "Potential Strategies for Spaceport Systems Towards Airport-Like Operations," AIAA-RS2-2004-6003, *2nd Responsive Space Conference,* Los Angeles, CA, April 19-22, 2004.

McCleskey, C.M., et al., "STS Root Cause Analysis to Derive SLI Operations Requirements," NASA SLI Board Presentation, CR KSC-013, April 2002.

Morgan, D. and A. Jacobs, "NASA's Space Shuttle Program: Space Shuttle Appropriations FY1992-FY2002," Congressional Research Service, Library of Congress, RS21411, February 13, 2003.

Morris, W.D. et al., "Analysis of Shuttle Orbiter Reliability and Maintainability Data for Conceptual Studies;" AIAA 96-4245, *1996 AIAA Space Programs and Technologies Conference,* Huntsville, AL, September 24-6, 1996.

NASA MSFC Engineering Cost Group, Access to Space Study, "STS Baseline Ops Cost Based on FY94 OMB STS Cost Per Flight (with Updates from Project Offices)." Applied Research, Inc., Huntsville, AL, April 13, 1993.

NASA SP-4XX Space Shuttle, ca. 1976.

Raymond, E.T., and C.C. Chenoweth, "Aircraft Flight Control Actuation System Design," Society of Automotive Engineers, Inc., Warrendale, PA, (SAE Publication R-123) 1993.

"Shuttle Operations Zero-Base Cost Study," presentation to Dr. Lenoir, NASA Headquarters, Office of Space Flight, Washington, D.C., July 2, 1991.

"Space Shuttle," NASA SP-407, Lyndon B. Johnson Space Center, NASA, Washington, D.C., 1976.

"Space Shuttle Program Description and Requirements Baseline," NSTS 07700.

"Space Shuttle Program Plan," Updated November 13, 2002.

Space Shuttle Program's "Requirements and Procedures for Certification of Flight Readiness," NSTS 08117 Rev. M.

Spaceport Synergy Team, "Catalog of Spaceport Elements with Functional Definition," NASA *Highly Reusable Space Transportation (HRST) Study,* October 1997.

www.ingramcontent.com/pod-product-compliance
Ingram Content Group UK Ltd.
Pitfield, Milton Keynes, MK11 3LW, UK
UKHW051040091026
15624UKWH00011B/602

9 781782 660569